Lecture Notes in Computer Science 794

Edited by G. Goos and J. Hartmanis

Advisory Board: W. Brauer   D. Gries   J. Stoer

Günter Haring Gabriele Kotsis (Eds.)

Computer Performance Evaluation

Modelling Techniques and Tools

7th International Conference
Vienna, Austria, May 3–6, 1994
Proceedings

Springer-Verlag
Berlin Heidelberg New York
London Paris Tokyo
Hong Kong Barcelona
Budapest

Günter Haring   Gabriele Kotsis (Eds.)

Computer Performance Evaluation

Modelling Techniques and Tools

7th International Conference
Vienna, Austria, May 3-6, 1994
Proceedings

Springer-Verlag

Berlin Heidelberg New York
London Paris Tokyo
Hong Kong Barcelona
Budapest

Series Editors

Gerhard Goos
Universität Karlsruhe
Postfach 69 80
Vincenz-Priessnitz-Straße 1
D-76131 Karlsruhe, Germany

Juris Hartmanis
Cornell University
Department of Computer Science
4130 Upson Hall
Ithaca, NY 14853, USA

Volume Editors

Günter Haring
Gabriele Kotsis
Institute for Applied Computer Science and Information Systems
University of Vienna
Lenaugasse 2/8, A-1080 Vienna, Austria

CR Subject Classification (1991): C.4

ISBN 3-540-58021-2 Springer-Verlag Berlin Heidelberg New York
ISBN 0-387-58021-2 Springer-Verlag New York Berlin Heidelberg

CIP data applied for

Typesetting: Camera-ready by author
SPIN: 10132061   45/3140-543210 - Printed on acid-free paper

Preface

Performance evaluation, reliability and performability are key factors in the development and improvement of computer systems and computer networks. The 7th International Conference on Modelling Techniques and Tools for Computer Performance Evaluation, held in Vienna in May 1994, was organized in cooperation with both international working groups and national special interest groups on performance evaluation. It was an ideal forum for the presentation of recent results in this area, based on the growing interest in the applicability of new techniques and the further development of existing techniques for performance and reliability analysis. A special focus was put on tools to support these aims in all kinds of applications and environments.

1984 Paris	1989 Palma
1985 Sophia Antipolis	1991 Torino
1987 Paris	1992 Edinburgh

In keeping with the tradition of this series of conferences (see the above list of previous conferences) the international performance evaluation community was invited to submit actual research contributions on modelling techniques and tools for performance and reliability analysis of computer systems. Sixty-six papers were submitted in the response to the call for papers, which was distributed worldwide. The following table gives an overview of the geographical distribution of the submissions, where the assignment to the countries follows the affiliation of the authors.

country	submitted	accepted	invited
Australia	1	0	
Austria	4 1/2	2 1/2	
Belgium	1	0	
France	4	1	
Germany	9 2/3	2 2/3	1
Greece	2	0	
Hungary	1 1/2	0	
Italy	7 2/3	4	
Japan	1	0	
Korea	2	0	
Kuwait	1	0	
Rep. of Trinidad and Tobago	1	0	
Spain	1	1	1
Taiwan	1 1/2	0	
The Netherlands	5	3	
UK	3 2/3	1	
USA	18 1/2	5 5/6	2
total	66	21	4

It was not an easy job for the scientific program committee to select 21 papers for presentation at the conference. These papers, together with the papers of the four invited speakers, are included in this volume.

The papers address important problems with special emphasis on both centralized and distributed/parallel computer systems. The techniques presented for performance analysis of parallel systems reflect the growing need for reliable methods for performance analysis of such systems. The tools presented for the analysis of communication networks and distibuted/parallel systems give an excellent overview of the methods and techniques implemented for performance and reliability investigation of state-of-the-art computer systems. Case studies report on the applicability of and experience with these techniques and tools.

We would like to thank all those persons and institutions who and which have contributed to this conference, first of all the authors of the submitted and invited papers, who have considered this conference to be an appropriate opportunity for the presentation of their scientific work. Furthermore, we thank the members of the scientific programme committee and the external referees for their immense effort in reviewing and selecting the papers under serious time constraints. Last but not least, we would like to thank the sponsoring and cooperating institutions for their support of this event and the staff of the Institute of Applied Computer Science and Information Systems at the University of Vienna and those of the Austrian Computer Society for their devoted work, which was a critical factor for the success of this conference.

Vienna, May 1994

Günter Haring
Gabriele Kotsis

This conference was organized by the

Institute of Applied Computer Science and Information Systems
University of Vienna, Austria

in cooperation with

AICA Working Group on Performance Evaluation, Italy

Asociacion de Tecnicos de Informatica (ATI), Spain

CEPIS (The Council of European Professional Informatics Societies)

Dutch User Group
of Queueing/Performance Analysis Software (QPASS)

GI/ITG-Fachgruppe für
Messung, Modellierung und Bewertung von Rechnersystemen, Germany

IFIP Task Group "Performance of Communication Systems"

IFIP Working Group 7.3
(Computer System Modelling)

IFIP Working Group 10.3
(Concurrent Systems)

IFIP Working Group 10.4
(Dependable Computing and Fault Tolerance)

OCG (Austrian Computer Society)

The British Computer Society
Performance Engineering Specialist Group (BCS/PESG)

Program Committee

Referees

Contents

Invited Papers

Full Papers

Techniques and Tools for Reliability and Performance Evaluation: Problems and Perspectives

Kishor S. Trivedi[1]* and Boudewijn R. Haverkort[2] and Andy Rindos[3] and Varsha Mainkar[4]

[1] Dept. of Electrical Engineering, Duke University, Durham, USA
[2] Department of Computer Science, University of Twente, The Netherlands
[3] Networking Systems Hardware, IBM Corp., RTP, USA
[4] Department of Computer Science, Duke University, Durham, USA

Abstract. Modelling techniques and tools of the future must meet the challenges presented by today's highly demanding and schedule-oriented developing environment. With the emergence of high performance and reliability systems the problem of how to analyze such systems has become increasingly more difficult. Traditional assumptions of independent events, exponential distributions and other such "convenient" assumptions no longer model systems realistically. Nevertheless, the demand for answering performance and reliability related questions during the design process has increased. In this paper we discuss some of the issues involved in integrating modeling and design during a product development process. We present a broad range of existing techniques of systems analysis. We also describe a variety of tools that have been developed to make the analysis process simpler.

1 Introduction

One of the greatest challenges facing today's performance, reliability and performability tool designers is to meet the rapidly changing needs of computer and communications product designers and capacity planners. These designers and planners must develop those systems that will satisfy ever more demanding customer expectations of reliability and performance. These challenges should invigorate the area of systems analysis research by introducing a vast array of real problems that stress existing theoretical techniques.

By examining these real problems commonly encountered by such designers, we should gain an insight into the capabilities an ideal analysis tool should provide. Likewise, by examining the pressures experienced by designers as they develop a product, as well as designer-analyst interactions, we should gain insight into additional features (e.g., ease of use, reusability of submodels, etc.) that a successful tool should offer. As a result of these examinations, we should be able to suggest successful ways in which such tools, as well as the systems analysis

* This research was sponsored by IBM under the IBM/Duke University Research Agreement # RAL-R93010-00.

experts that develop them, can be effectively integrated into the design process. This should lead to an increased viability of such tools, as well as a mutually beneficial relationship between tool developers and tool users.

1.1 How should realistic systems be analyzed?

A large number of real system characteristics cannot be easily modeled using present analytic techniques. Most realistic systems cannot be separated into small independent subsystems. Their detailed state representation then may be excessively large. Further, events in real systems are not "memoryless" and event times may differ by several orders of magnitude. All this implies that simple analytic models such as product form queueing networks may not be powerful enough, and detailed analytic models (such as Markov chains) may be prohibitively large and stiff. This unfortunately has led to an over-reliance by designers on discrete event simulation, which is often improperly used, or applied to situations where a simpler analytic model would suffice. When extremely accurate predictions of final product performance/reliability (especially distributions) are needed, discrete event simulation is probably the best method. However, detailed simulation models require a great deal of time to construct, parameterize, validate and solve. As an example, a recently developed detailed simulation model of an actual ATM adapter required 12 hours of machine time on a very powerful server (receiving 100% of the CPU cycles) to achieve acceptable 90% confidence intervals for the desired response time measures. The bottleneck resource in the model was utilized at only 40% of its capacity for the run. Runs with higher utilizations would have required unacceptable amounts of machine time, since run time for the same level of confidence increases exponentially with utilization.

However, when we examine the evolution of the design process of an actual product, it becomes apparent that simulation is often unnecessary. At the earliest stages of this process, a very simple, high-level analytic model of the product design is probably sufficient. Simulation would not be recommended at this stage, since actual values for many system parameters, as well as many of the actual design details, would not be available. However, comparisons of different high-level design alternatives could be made with analytic models so that the best high-level design could be chosen at the outset. As the design matures and more details of the technology, architecture, etc. are defined, more accurate analytic techniques, hierarchical approximation, and finally, detailed discrete-event simulation, might be sequentially used to analyze the system.

Using such an evolutionary modeling approach, a nearly optimal design can be developed from the beginning, avoiding costly 'improvements' made when the design is less flexible (especially when the implementation has begun). Redesign work late in the development process may also lead to delayed release of the product. This points to the value of modeling the system early on and throughout development. Such models typically alert the designer to potentially detrimental design flaws, hopefully, at a time when they can be easily fixed.

1.2 The needs of today's product developers

Today's product designers need to get out the best possible product in the shortest period of time using the least amount of resources. This driving force pulls the performance and reliability of the resultant product in two opposite directions. On the one hand, in today's very competitive market, the developer who is first to market may capture the largest share of the market. Therefore, schedules for product release are very aggressive and sacrosanct. This means that product developers cannot afford a lot of time to develop and analyze models of their product's performance and reliability. On the other hand, customers are requiring far greater reliability and performance which would suggest a need for more careful (and, possibly, more time consuming) design and analysis.

Tools that require very little time to learn to use and allow models to be constructed very quickly will therefore be highly sought after. This suggests that easy-to-use graphical user interfaces and clear, self-explanatory menus should be developed for all future tools that target an industrial user-base. Many performance/reliability tools on the market already provide some of these features. Furthermore, the ability to define reusable submodels of system and network components that can be combined to produce large models and that provide the capability of multiple instances in a given model are an absolute must. A number of tools are avoided in industry because of their inability to provide these capabilities.

Another important aspect in model specification is the ability to specify models in a hierarchical way since this allows one to stay close to the generally hierarchically designed systems. When supported in a flexible way, this also allows for the building of a database (a model-base) of predefined model building blocks. Preferably, the hierarchical model structure should be exploited in the model solution phase, however, this is not always easily done and it almost always introduces approximation error.

Tools that provide alternative analytic methods that require less time and memory to generate and solve the models, in conjunction with simulation, are extremely desirable. Such tools with multiple analysis capabilities would fit very well with the evolution of the design process of an actual product as described before. The conjunction of several techniques in a single tool might allow the user to input the appropriate details of his product only once, and then analyze it using different methods, depending upon his needs for accuracy vs. speed of execution. It would also facilitate the use of hybrid techniques and hierarchical model development.

Ideally a tool should also have an integrated method of calibration for its models. That is, it should be closely coupled with databases maintained by manufactures about their products. These databases could provide information for model calibration such as failure rates, device speeds, repair times and so on. Using the measures contained within these databases, it should not only be easy to parametrize a single run, but also a series of experiments with changing model parameters allowing for parametric studies. Tools with such multiple capabilities could therefore satisfy the pull by today's market forces to develop models quickly and in a timely manner so that products are brought to market in the shortest period of time.

1.3 Relationship Between Designers and Performance Experts

In the past, especially in large corporations, one or more performance departments have existed independently of those departments responsible for the actual design and implementation of products. The justification for such an arrangement has typically been that such separate departments serve as centers of competency, whereby the members can consult with one another when difficult problems arise in their analyses. Such an arrangement also fosters departmental experts who can provide focus specifically on techniques that will make the work of theoretically analyzing products more accurate and efficient.

However, some inherent problems exist in this arrangement. Because analysis and design departments are separate, conflict often arises when the analyst exposes serious performance/reliability flaws in the product design. Although sometimes the feedback is appreciated, it more often leads to a sudden lack of cooperation between the two groups. The design team may suddenly make it very difficult for the performance team to obtain additional design details. This points to another drawback of the arrangement of separate departments. The analyst typically requires time from the designer's often busy schedule to verify his understanding of the design. As an outsider, it is a very difficult task for the analyst to obtain an accurate picture of the details of the design from reams of poorly indexed design documents that are constantly changing as the product design evolves. It is especially difficult to keep up with such changes in both the hardware and the software (some knowledge of both are needed for such models).

The availability of time from a product developer's schedule has become increasingly more difficult to obtain in today's schedule-driven, "more efficient" (a euphemism for over-worked) working environment. Given the cost to a designer of taking the time to cooperate, one might be sympathetic to their reluctance to assist the analyst. If the analyst simply verifies the correctness of the design (that is, it meets performance specifications, etc.), his work is often viewed as superfluous; if he exposes problems, his work is viewed with hostility. Therefore, an analyst should also provide insight into or solutions to problems encountered during the analyses, as well as improvements to the existing design, if his work is to be properly appreciated. Furthermore, those improvements should be available in a timely manner during the design cycle so that they can be implemented.

This has led to several strategies to reduce the problems inherent in a separate department arrangement. For those organizations that were large enough to allow a separate high-level design department for a product, this department was often combined with the analysis department. However, with a movement toward a smaller, more efficient workforce, what is more typically encountered is a small single group with a very broad range of responsibilities for the design and implementation of a single product. In an extreme reaction to needed cuts in work force, some of the groups have no one officially responsible for analyzing the performance/reliability of their product. They therefore operate at the risk of encountering major design problems surfacing late in the implementation or prototype phase.

Because of such potential exposure, these tighter design groups often allocate one person to be responsible either part-time or full-time to these performance-related problems. Since the person is a part of the design and implementation team, he is able to track the design from the very beginning and is not likely to be 'frozen-out' when performance-related problems in the design are encountered, since he is expected to provide some of the solutions. However, one major problem is the availability of enough qualified performance analysts that can be distributed among various product design groups. This inherent variability in talent can be smoothed out either by means of improved training or by means of better tools.

1.4   The focus of this paper

Recently, there have appeared a number of papers with surveys similar to this one. Most notably, we mention Meyer [54] who overviews the history of the concept of performability after he introduced it in the early eighties [51, 52]. De Souza e Silva and Gail [74] discuss the specific technique known as uniformization, which is now known to be the method of choice for transient analysis of continuous-time Markov chains (CTMCs). Trivedi *et al.* [79] present mathematical evaluation techniques for performability. Haverkort and Trivedi [32] overview specification techniques for Markov reward models. Mulazzanni and Trivedi [59] overview tools for dependability, as do Geist and Trivedi [24] and Johnson and Malek [38]. A description of dependability analysis of real-time systems can be found in [78].

This paper differs from the above ones in that it addresses reliability, performance and performability evaluation tools and techniques from the user's perspective. Former papers often were restricted to either performability, or to just tools or to just techniques. In this paper you will find a mix of it all.

The organization of the paper is as follows. In Section 2 we try to answer questions of the form "What are we interested in?" and "Why do we actually model?". Answers to these questions give us directives to approaches towards reliability and performance evaluation, which are discussed in Section 3; this section still focuses on techniques. Tools supporting the various categories of techniques are then discussed in Section 4. Problem areas and future perspectives are discussed in Section 5. The paper is concluded in Section 6.

2   What are we interested in?

In this section we try to answer the question "Why are we doing all this?" For simplicity, we split the question in a number of subquestions. In Section 2.1 we discuss system aspects we are interested in, for various classes of systems. Then, in Section 2.2, we discuss measures that quantify the interest expressed earlier. Various approaches towards how to derive these measures are discussed in Section 2.3.

2.1 System aspects of interest

A first and important distinction in answering this question, is whether we take the viewpoint of a system user or of a system provider. This distinction is most clear when thinking of the system as a public data network as provided by many telephone companies. System users want a good *quality of service* (QoS), whatever that may be, whereas the provider wants a high profit. The latter implies that the system should do what the user wants it to do, but at the lowest possible cost.

The QoS asked for by a user is often subjective, e.g., a user wants a good video channel quality, if he is intending to use the network for video transmission. How this good quality is expressed in terms of bit rates, bit error rates or switch blocking probabilities is not easy determined. These latter measures express what is known as the *objective quality of service*. For example, a user of a parallel processing system will most likely be interested in the throughput and turnaround time of his jobs, not so much in the degree of parallelism achieved by the system. The latter is of interest to the system designer or for the cooperation "selling" parallel processing capacity.

The above distinction is important and should be kept in mind when doing practical evaluation studies; the type of viewpoint has implications on the required results.

2.2 Measures of interest

The distinction made in the previous section directly has its implications for the measures we want to evaluate. We distinguish between *task-oriented* measures and *system-oriented* measures. Task-oriented measures typically say something about the end-to-end performance as perceived by system users. Examples are the end-to-end throughput or delay, or the expected performance level over some time interval of system usage. System-oriented measures say something about how, internally, the system performs its tasks. Examples are the average queue length, the number of operational components at some time, or the utilization of a server. As such, these measures are not so much of interest to the system user, although they are intimately related with the task-oriented measures.

With pure performance evaluation, both task- and system-oriented measures can be obtained. Examples of the former are job response or waiting times, examples of the latter are average number of occupied buffers or utilizations. These measures suffice if the assumption that the system never fails is acceptable.

With pure dependability evaluation, the emphasis is on deriving system-oriented measures, although safety measures are user-oriented as well. However, we think that system users are not so much interested in high availability as in a high probability that the task they want to be performed are actually performed. For that reason, performability measures seem more suitable.

Performability evaluation mainly aims at providing user-oriented measures. It has been claimed by other authors as well, that the evaluation of performability comes closest to the evaluation of objective QoS.

Another distinction that one often encounters is whether the measures are derived over a time interval or for some particular time-instance. Interval-of-time measures include the steady-state measures, when the interval is taken to be infinitely long. Also, cumulative measures over finite time horizons (see Section 3) belong to this category. Instant-of-time measures are also called transient measures; they express the performability (or performance, or reliability) of a system at some time point t.

2.3 Methods of evaluation

We distinguish three classes of methods for system performance, dependability or performability evaluation: measurement-based, model-based and hybrid methods.

Measurement-based evaluation (also called empirical evaluation) requires one to have at one's disposal a measurable system. Apart from the fact that this is often not the case in the design phase of the system, performing measurements is often expensive since it requires special purpose hardware and software. Also, for dependability modeling purposes, measurement is difficult. Dependability events, i.e., system failures, do not occur that often in highly-reliable systems, requiring extremely long measurement sessions.

Measurement studies are often done to determine system parameters that are later to be used in a modeling study, or to validate a model.

As an alternative to measurement-based evaluation, a model-based evaluation can be used. A system model can be very simple, e.g., some mathematical formula relating the system performance to the system parameters, or very complex, e.g., a large system of differential equations or a complex simulation program.

Once a model has been constructed, it needs to be solved. This can be done using discrete-event simulation or using analytical techniques. Analytical techniques can be fully symbolic, semi-symbolic or numerical. Fully symbolic analytical techniques provide simple functional relations between system parameters and the measure of interest ($E[N] = \rho/(1 - \rho)$ in the M/M/1 queue). Semi-symbolic analytical techniques provide mathematical relations between system parameters and measures of interest, however, some parameters in the relation are to be determined by numerical technique (ACE [50]). Lastly, some analytical techniques require numerical solution such as linear-system solution or the solution of a differential equation.

Another distinction that sometimes is made is whether the solution of the model requires the whole model state space to be explicitly generated or not. The most well-known example of the former is the solution of a large but finite Markov model. An example of the latter is the use of an MVA solution procedure [65] in a product-form queueing network or the use of fault-trees for reliability analysis [23].

Most of the useful evaluations in practice use a judicious combination of different modeling approaches with measurements. For example, fault-injection simulation (or actual measurements on a prototype) can provide coverage-like parameters in an analytic reliability model. A performability model of a multi-processor system may have a Markov reliability submodel and a product-form queueing network as a performance submodel.

3 Approaches to performance and reliability evaluation

In this section we discuss four approaches to performance, reliability and performability analysis. In Section 3.1 we discuss so-called *non-state space* methods, thereby meaning that explicit knowledge and enumeration of the state space of the model is not needed for evaluation purposes. In Section 3.2 we discuss Markov chain based methods, and in Section 3.3 stochastic Petri net based models. In Section 3.4 we discuss hierarchical and approximate modeling approaches.

3.1 Non-state space

With this class of models, our aim is to compute required performance and reliability measures without explicitly generating overall state space. This is a very nice property as state-space sizes tend to increase exponentially with the problem size. Three of the most well-known non-state space methods are *product form queueing networks* (PFQNs), *fault trees* (FTs) and *matrix geometric methods* (MGMs).

With PFQNs one needs to specify a number of resources (the queues and servers) as well as the way in which customers make use of these resources. The queues form the active elements that can serve customers in an order governed by one of the scheduling discipline: FCFS, LCFSPR, PS, or IS. The customers travel through the QN according to routing chains. Customers may be grouped in classes. At every queue, customers belonging to a specific class request a general differential service time distribution (at FCFS stations only exponential service time distributions are allowed). After service completion, the customer proceeds to the next queue along its routing chain. The state of a PFQN model is a vector consisting of the number of customers of each class residing at each queue. The completion of a service at a particular queue or the arrival of new jobs causes a state change. Instead of solving such a model at the state space level, one can employ special techniques that exploit the specific model structure and that are much less computation and memory intensive. The convolution approach introduced by Buzen [9] and the mean-value analysis (MVA) introduced by Reiser and Lavenberg [65] and their derivatives constitute the common techniques. With the former an efficient recursive scheme is used to calculate normalizing constants which can be used in straightforward calculations for derived performance measures such as average queue length, utilizations and throughputs. In the MVA approach, a recursive scheme in terms of the average performance measures is developed.

Fault-trees(FT) are a commonly used non-state space (also called combinatorial) method for reliability (availability, safety) analysis. With FTs the conditions under which a system fails, are expressed as a tree structure containing logic gates. Component failure events form the leaves of the tree. Subsystems and components must have stochastically independent failure behavior.

The measures of interest are normally computed using combinatorial methods, that is, the system failure event is expressed as a logical function of the failure events of subsystems and components. Dependability measures (such as reliability or safety at time t or mean time to system failure) of interest are then computed numerically or symbolically directly from the tree. Algorithms for fault-tree analysis can be found in [55].

Specification techniques that are very closely related to FTs are reliability block diagrams and reliability graphs. For further details, the reader is referred to [55] and [66].

With MGMs, the repetitive structure of the underlying Markov chain in many queueing models is exploited. When observing the (embedded) generator matrix $\mathbf{Q}$ of many queueing models, it appears that apart from a number of so-called boundary columns, from some point onwards, all columns are the same, except for the fact that they "shift down". This appears most notably in the M/M/1 queue, however, also in more complex queueing systems this structure can be observed. In the latter case, the columns are often columns of matrices rather than scalars. Due to this special structure, the steady-state probabilities can be grouped in so-called *levels* and the steady state probability vector for level i, i.e., $\underline{z}_i$ can be expressed as

$$\underline{z}_i = \underline{z}_0 \mathbf{R}^i, \quad i = 0, 1, \cdots \tag{1}$$

i.e., the steady state probability vectors per level exhibit a geometric solution in terms of the matrix $\mathbf{R}$. The basis of the recursive solution is obtained by solving a system of linear equations corresponding to the repeating portion of the global balance equations and the normalization equation. The matrix $\mathbf{R}$ follows from a quadratic equation that can easily be solved iteratively. The size of $\mathbf{R}$ equals the number of states per level, typically small and finite. By contrast, the original Markov model solution would have required an infinite system of linear equations.

3.2 Markov reward models

In this section, we present a unified framework for performance, reliability and performability models in terms of Markov reward models. A comprehensive account of Markov reward models for performability analysis appears in [79]. Several references on solution methods for the measures defined below can be found in [15, 29, 64, 74, 79].

Definitions

Let $\{\Theta(t), t \geq 0\}$ be a continuous-time finite-state homogeneous Markov chain (CTMC) with state space Ψ. A constant reward rate r_i is associated with each state i of the Markov chain. With the reward rate specifications, the CTMC can be termed as *Markov reward model* (MRM). If the MRM spends τ_i time units in state i, then $r_i\tau_i$ is the reward accumulated. It is also possible to associate reward rates with the transitions of the CTMC. For more basic information on MRMs, refer to [37].

Let $\mathbf{Q}$ be the generator matrix and $\underline{P}(t)$ be the state probability vector of the MRM. Here $P_i(t)$ denotes the probability of the MRM being in state i at time t. The transient behavior of this MRM is given by the Kolmogorov differential equation:

$$\frac{d\underline{P}(t)}{dt} = \underline{P}(t)\mathbf{Q} \; , \tag{2}$$

given the initial state probability vector $\underline{P}(0)$. The steady-state probability vector $\underline{\pi}$, assuming that it exists and is unique, is obtained by setting the l.h.s. in Equation 2 to zero:

$$\underline{\pi}\mathbf{Q} = 0 \; , \tag{3}$$

subject to the condition $\sum_{i \in \Psi} \pi_i = 1$. Here π_i is the steady-state probability of the MRM being in state i. Let us now define a cumulative state vector of the MRM as $\underline{L}(t) = \int_0^t \underline{P}(x)dx$. $L_i(t)$ denotes the expected total time spent by the MRM in state i during the interval $[0, t)$. Integrating Equation 2, we obtain:

$$\frac{d\underline{L}(t)}{dt} = \underline{L}(t)\mathbf{Q} + \underline{P}(0) \; . \tag{4}$$

For MRMs with absorbing states, the state space Ψ can be partitioned into two subsets: Ψ_A (absorbing states) and Ψ_T (transient states). Corresponding to the non-absorbing states, the submatrix $\mathbf{Q}_T$ of $\mathbf{Q}$ can be defined. The mean time spent by the MRM in state i is given by $\tau_i = \int_0^\infty P_i(x)dx$, which can be computed by integrating Equation 2 from 0 to ∞:

$$\underline{\tau}\mathbf{Q}_T + \underline{P}_T(0) = 0 \; . \tag{5}$$

The mean time to absorption in such a Markov chain is given by:

$$MTTA = \sum_{i \in \Psi_T} \tau_i \; . \tag{6}$$

Performability Measures

Let $\Upsilon(t) = r_{\Theta(t)}$ be the instantaneous reward rate of the MRM. The accumulated reward over a period of time $[0, t)$ is given by:

$$\Phi(t) = \int_0^t \Upsilon(x)dx = \int_0^t r_{\Theta(x)}dx \; . \tag{7}$$

The expected instantaneous reward rate at time t of the MRM is:

$$E[\Upsilon(t)] = \sum_{i \in \Psi} r_i P_i(t) \; . \tag{8}$$

The expected reward rate in steady-state of the MRM is:

$$E[\Upsilon_{ss}] = \sum_{i \in \Psi} r_i \pi_i \ . \tag{9}$$

The expected accumulated reward in the interval $[0, t)$ of the MRM is:

$$E[\Phi(t)] = \sum_{i \in \Psi} r_i L_i(t) \ . \tag{10}$$

The expected time-averaged reward in the interval $[0, t)$ is given by $\sum_i r_i L_i(t)/t$. For an MRM with absorbing states, expected accumulated reward until absorption is:

$$E[\Phi(\infty)] = \sum_{i \in \Psi_T} r_i \tau_i \ . \tag{11}$$

The distribution of $\Upsilon(t)$ is computed as:

$$P[\Upsilon(t) \leq x] = \sum_{r_i \leq x, i \in \Psi} P_i(t) \ . \tag{12}$$

The distribution of accumulated reward until absorption and distribution of accumulated reward over a finite period of time can also be computed.

Let the time to accumulate a given reward r be denoted by $\Gamma(r)$. Then the distribution of $\Gamma(r)$ is known once the distribution of accumulated reward is known [40]:

$$P[\Gamma(r) \leq t] = 1 - P[\Phi(t) < r] \ . \tag{13}$$

For example, the distribution of time to complete a job that requires r units of processing time on a system which is modeled by an MRM can be computed in this manner.

Dependability Measures

In a dependability model, a reward rate of 1 is assigned to all the system operational states and reward rate 0 is assigned to all the system failure states. The instantaneous availability of the system is then $E[\Upsilon(t)]$ and steady-state availability is $E[\Upsilon_{ss}]$. The cumulative operational time of the system in time interval $[0, t)$ is $E[\Phi(t)]$. Interval availability is the proportion of time a system is operational in a given interval of time and it is given by $E[\Phi(t)]/t$. Measures related to time to first system failure are also of interest. To compute these measures, all the failure states are made absorbing (outgoing arcs from these states are removed). Reliability is then given by $E[\Upsilon(t)]$. The lifetime (analogous to cumulative operational time) [74] of the system in interval $[0, t)$ is $E[\Phi(t)]$ and mean time to system failure (MTTF) is $E[\Phi(\infty)]$. The repairability of the system is computed by making all the operational states absorbing and reversing the reward rates (i.e., making reward rate 1 to 0 and vice-versa) and computing $E[\Upsilon(t)]$.

Performance Measures

In a performance model, queue length at a resource may be the reward assignment to a state. Then $E[\Upsilon_{ss}]$ and $E[\Upsilon(t)]$ will yield the average steady state and average transient queue length, respectively. In a like manner, throughput, buffer overflow probability etc. can be obtained as reward measures.

In a performability model, reward assignment is typically computed from a performance model (throughput, probability of violating a response time deadline) which is evaluated for different states of a failure/repair model. Throughput and response time deadline violation probability can then be computed including the effects of failure/repair.

3.3 Stochastic Petri net models

Stochastic Petri nets (SPNs) have been developed as extensions to the non-timed Petri nets by Molloy [56], Ajmone Marsan *et al.* [2] and Meyer *et al.* [53]. Although at first primarily used for the performance analysis of computer systems, SPNs are increasingly being used in other application areas such as performability and dependability evaluation.

When using an SPN specification technique, one has to define a set of places P, a set of transitions T, and a set A of arcs between transitions and places or *vice versa*: $A \subseteq (P \times T) \cup (T \times P)$. Each place can contain zero or more tokens. Graphically, places are depicted as circles, transitions as bars, tokens as dots (or integers) inside circles, and arcs as arrows.

The distribution of tokens over the places is called a marking and corresponds to the notion of state in a Markov chain. All places from which arcs go to a particular transition are called the input places of that transition. All places to which arcs go from a particular transition are called the output places of the transition. A transition is said to be enabled when all of its input places contain at least one token. If a transition is enabled it may fire. Upon firing, a transition removes one token from all of its input places and puts one token in all of its output places, possibly causing a change of marking, i.e., a change of state.

The firing of transitions is assumed to take an exponentially distributed time. Given the initial marking of an SPN, all the markings as well as the transition rates can be derived, under the condition that the number of tokens in every place is bounded. Thus a finite Markov chain is obtained.

The reward rates are described as a function of the markings, i.e., at the SPN level. The reward rates and the Markov chain together yield a MRM [16].

Various extensions have been made to the basic SPN model described above [2, 16, 53]. These include arcs with multiplicity, a shorthand notation for multiple arcs between a place-transition pair, immediate transitions that take no time at all to fire (depicted as thin bars), and inhibitor arcs from places to transitions that prevent the transition to fire as long as there are tokens in the place (depicted as lines with a small circle as head). Also, more flexible firing rules have been proposed, most notably the introduction of *gates* in stochastic activity networks (SANs) [53, 68] and guards or enabling functions in Stochastic Reward Nets (SRNs) [16].

Normally, SPN (look-alike) models are solved via an underlying MRM which can automatically be derived, thereby using the wide variety of available techniques as indicated in the previous section. For very large models when state-space generation is prohibitive, simulation can be used as well. Especially in the field of dependability and performability evaluation there might be a need for the incorporation of fast-simulation techniques such as importance sampling [26] or injection simulation [57]. For a restricted class of SPN models product form solutions are available, see e.g., [36]. Given such a structure, MVA [19] and convolution [18] schemes have recently been devised.

3.4 Hierarchical and Approximate Models

With hybrid approaches, two or more techniques are combined in the construction and solution of a single model. Very often this takes the form of hierarchical modeling. Submodels are specified in one formalism and the result of the submodel analysis are embedded in a higher-level model. Hierarchical modeling, however, is not always hybrid modeling. The decomposition result in PFQN is a form of non-hybrid, hierarchical modeling. Beginnings of a theory of hierarchical models of SPN type can be found in [17]. There is less general theory for hybrid modeling. We therefore mention some published approaches.

For pure performance studies, Balbo *et al.* combined queueing networks and GSPNs for the analysis of system models with non-product form characteristics [3]. The non-product form parts of the model are solved using GSPNs, the results of which are used in load-dependent queueing stations that fall in the category of PFQNs.

With the software tool SHARPE (see Section 4) many model types can be analyzed, using a variety of techniques. The result of one analysis can be embedded in other models. This can be done in a cyclic way as well; in that case fixed-point iteration techniques are needed to solve the overall model.

In the dynamic queueing network concept proposed by Haverkort *et al.* [28, 29] queueing networks are used for describing performance aspects, and GSPNs are used to describe dependability aspects of fault-tolerant computer systems. An overall model is not explicitly constructed, instead, an approximate solution based on behavioral decomposition as is common in performability evaluation is utilized.

For non-product-form networks (NPFQN), a number of approximate techniques exist. A major concern with these techniques is in the characterization of their error under a wide variety of realistic network situations. A number of methods exist for closed non-product-form networks, including Marie's algorithm [49]. A wide variety of methods also exist for general open networks. A popular method, on which a number of tools are based (e.g., QNA [81]) is a two-moment decomposition method that was developed by Whitt to handle networks with general independent interarrival and service time distributions.

Mean waiting times are predicted using a two-moments of service and interarrival times. Mean interarrival times are computed by solving the standard traffic equations. The coefficient of variation of the interdeparture time is computed by means of Marshal's formula, using an approximation for the mean waiting time. As the use of Marshal's formula implies, all interarrival processes to resources within the network are assumed to be renewal processes. Under this assumption, using heuristic methods developed by Albin and Whitt [81] that are based on large amount of empirical evidence, a linear system of simultaneous equations is derived to solve for the coefficient of variation of interarrival times to all resources.

4 Tools for performance, reliability and performability evaluation

For the basic approaches distinguished in Section 3, we discuss a number of software tools that support them. Due to space limitations we can not go into much detail, however, we provide references that can be tracked down for further study.

4.1 Non-state space

In the reliability domain, the tools SHARPE [66] and HARP [22] can both solve fault-tree models. SHARPE also solves reliability block diagrams and reliability graphs. SHARPE can provide semi-symbolic expressions (in terms of t) for the reliability function.

Performance modeling packages QNAP [63] , RESQ [47], and HIT [5] support a wide variety of PFQN analyses, such as MVA and convolution algorithms. The tool HIT also supports nice facilities for hierarchical modeling, both exact (Norton's theorem, and approximate). SHARPE also solves multiclass PFQN models using the MVA algorithm and series-parallel task precedence graphs.

Two tools that make use of MGMs are MAGIC developed by Squillante [76] and Xmgmtool developed by Haverkort [34] respectively. MAGIC allows one users to input the regular (repeating) block-structure of the Markov generator matrix. It subsequently calculates the matrix $\mathbf{R}$ and the initial vector of the recursion $\underline{z}_0$. Xmgmtool provides similar facilities, however, it also provides facilities to specify queueing systems in terms of their interarrival and service time distributions (both of phase-type). The underlying regular matrix structure is subsequently generated and solved. The output is also given in terms of the queueing systems originally specified.

4.2 Markov reward models

With SHARPE [67], MRMs can be input at the state level by a simple enumeration of the state-change rates and the reward rates per state. SHARPE then solves Markov and semi-Markov reward models for their steady-state, transient and cumulative behavior. Specification is textual, but abilities include solving the model for many different parameters using "loop" specifications.

The textual tool MARCA has been developed by Stewart [77] at North Carolina State University. Although not really an SPN tool, its modeling constructs, i.e., buckets, balls, and transitions, can easily be interpreted in an SPN context as places, tokens and transitions. Emphasis in the the tool is on advanced steady state numerical solvers.

There are many tools available that solve models via the underlying MRM. The largest class of such tools exists in the context of SPNs; that is why we discuss them separately. However, some other tools based on other modeling paradigms are discussed below. In particular we address queueing network (QN) based tools and tools based on production rule systems (PRS).

The tool QNAP2, developed at INRIA by Potier et al. [63, 80], is a general QN-based performance analysis tool which supports simulation, (approximate) product-form solutions as well as a numerical solution based on an underlying MRM. In fact, given a textual representation of a QN, the QNAP2 model is transformed to an intermediate model similar to the MARCA model . Only steady-state measures are computed.

The performance analysis tool NUMAS, developed at the University of Dortmund by Müller-Clostermann [60], is a textual tool for Markovian queueing network analysis. As an extension, NUMAS allows the modeling of queues with server breakdowns and repairs. NUMAS thus allows for steady state performability analysis.

The graphical performance analysis tool MACOM, developed by Sczittnick et al. at the University of Dortmund [73], is mainly used for the steady state analysis of blocking phenomena in communication networks. MACOM emphasizes advanced techniques for the steady state analysis of large MRMs.

Based on PRSs [32] are the tools METFAC, ASSIST and USENUM. The textual tool METFAC, developed by Carrasco and Figueras at the University of Catalunya [10, 11], supports the use of a PRS specification technique and has been used for performance, dependability and performability modeling of computer systems. Steady state, transient, as well as cumulative measures can be computed.

The tool ASSIST has been developed by Johnson and Butler at NASA [39] as a front-end to the SURE package [8] for reliability analysis of (computer) systems. This textual tool allows for the flexible specification of PRS. By the use of arrays of state variables and loops in the production rules, compact specifications can be written. Also facilities for truncating state spaces are available. The ASSIST program translates the PRS to input for the SURE package. This input is an MRM. The SURE package deals with absorbing semi-Markov models. Therefore, only transient measures are computed.

The textual tool USENUM, developed by Sczittnick et al. at the University of Dortmund [7, 72], allows users to define Markovian models by means of a finite state machine. USENUM can be used stand-alone, or within the QN tool MACOM.

4.3 Stochastic Petri net models

A wide variety of tools for stochastic Petri nets have been developed. We briefly discuss the most well-known tools that are based on MRMs.

GreatSPN, developed by Chiola *et al.* at the University of Torino [12], is a graphical Petri net tool which is primarily used for the performance analysis of computer and communication systems. Analysis techniques are mainly for steady state measures.

ESP, developed by Bobbio and Cumani [6, 21], is a textual SPN tool. In this tool, special emphasis is put on the use of phase-type distributions instead of only exponential distributions, on transient measures and on the aggregation of stiff MRMs.

METASAN, developed by Sanders and Meyer [68, 69] at the University of Michigan, is based on SANs. The tool includes steady state, transient and cumulative analysis methods.

The tool UltraSAN, developed by Sanders *et al.* [20] at the University of Arizona, is also based on the SAN concept. With UltraSAN, the input of the models is totally graphical. UltraSAN allows for a structured form of hierarchical modeling which results in lumped underlying MRMs that are substantially smaller (so-called reduced base-models [70]) than their "flat" counterparts. Steady state as well as transient simulation are also available as solution methods.

SPNP, developed by Ciardo *et al.* [14], is a C-based SPN tool which allows for a flexible definition of a class of SPN models known as stochastic reward nets. Steady state, transient and cumulative measures are supported. By the flexible use of C, it is possible to construct models hierarchically, that is, results of one model can be used in the analysis of another model, even in a fixed point iterative manner [17].

TOMSPIN is a general SPN tool developed at SIEMENS AG [41], for performance and dependability analysis. Steady state and transient measures are supported. An approximate solution for hierarchically structured SPN models based on an aggregation algorithm is also included.

PENPET is a performability modeling tool developed by Lepold [42, 43] at SIEMENS AG. It is a high-level tool built on top of TOMSPIN in which one SPN is used for the specification of system dependability aspects, and another for the system performance aspects.

The graphical tool DSPNexpress has been developed by Lindemann at the Technical University of Berlin [46]. Interesting aspect of this tool is that it allows for DSPNs, i.e., SPNs in which transitions may have deterministic timing. Under certain conditions, an embedded Markov chain can be constructed that allows one to solve for the steady state probabilities.

4.4 Hierarchical and Approximate Models

The performability modeling tools DyQNtool$^+$ [35] has been developed by Haverkort *et al.* at the University of Twente. The tool operates along the lines of the dynamic queueing network concept and is an extension of its predecessor DyQNtool

[28, 29, 30]. DyQNtool+ has been developed as a shell of programs around the packages SHARPE and SPNP. The SHARPE package is used for the solution of series of PFQNs of which the results are used as reward rates in the SRNs specified using SPNP. Thus, in a very flexible way, performability models can be evaluated.

The tool SHARPE has been developed by Sahner and Trivedi [66, 67]. SHARPE allows users to specify SPN, QN and FT like models as well as MRMs directly. Also hierarchical modeling is possible, that is, the results of a model analysis can be used in higher-level model evaluations, possibly using a different modeling approach.

5 Problems and perspectives

In this section, we discuss a number of recurring problems in performance, dependability and performability evaluation. We discuss the issue of largeness in Section 5.1 and the issue of stiffness in Section 5.2. The modeling of non-exponential behavior is addressed in Section 5.3.

5.1 Largeness

Models of practical systems are often very large. We use special specification, generation, storage and solution techniques to deal with the large models (largeness tolerance) or avoid largeness altogether.

First consider the techniques related to largeness tolerance. With this we mean the techniques that aim at being able to handle models as large as possible without affecting the model size itself.

For non-state space models, largeness tolerance techniques would encompass better implementations of MVA and convolution like algorithms, using extra significant digits to keep the normalizing constants accurate. Sparse storage techniques should be used whenever matrices are involved, such as in MGM. Also, whenever possible, special properties should be exploited. As an example of this, the matrix $\mathbf{R}$ that has to be calculated when using MGM has the property that zero entries come in rows, i.e., whenever the first element of a row equals zero, the rest is zero as well. This can be exploited in devising the storage and computational schemes, as has been done in Xmgmtool [34].

For MRM-based modeling techniques, largeness tolerance techniques presuppose the use of a concise specification method (e.g. SPN), automated MRM generation, sparse storage, sparsity preserving numerical techniques, e.g., using SOR instead of Gaussian elimination for the solution of the steady-state behavior of large MRMs, and orthogonal uniformization for the transient solution of acyclic Markov chains [27].

With largeness avoidance, we try to circumvent the generation of very large models. Although the capacity of modern day workstations is enormous, there will always remain systems that yield models that become too large to be handled directly.

As mentioned before, an important largeness avoidance technique widely applicable is hierarchical modeling which is based on the "divide and conquer" principle. The basic idea is to split a large model in smaller ones that can be analyzed in isolation. The results of the submodels are integrated in a single overall model that is small enough to be analyzed.

In the field of PFQN, it has been shown that such a decomposition approach can be performed in an exact way ("Norton's theorem", due to Chandy, Herzog and Woo [71]). For non-PFQNs, the decomposition is approximate. The theory developed by Courtois and which is used for the analysis of the degradable QN in NUMAS establishes bounds on the error made in the steady-state probabilities thus obtained. In fact, the performability evaluation approach based on MRM is motivated by these decomposition properties. An approximate way of solving large SPN models by decomposition is discussed by Ciardo and Trivedi [17]. They use a fixed-point iteration scheme. In this context, the work on automatic lumping as performed in UltraSAN by Sanders et al. is also of interest [20, 70].

The truncation of "the least important states", i.e., those states that have a small probability mass is another way to avoid large models. Work in this direction has been done by Haverkort [33], Li and Silvester [44], Muppala et al. [61], Bavuso et al. [4] and de Souza e Silva and Ochoa [75]. It is especially appropriate in case MRM or SPN models are used.

5.2 Stiffness

Informally speaking, stiffness is a property of a model to take very long to be solved. Often this is caused by the fact that in the model parameters of widely varying order of magnitude play an important role; this clearly is the case in dependability-related models where failure rates are very small and repair rates are orders of magnitudes larger.

Given a particular model type, one can specialize the above informal definition of stiffness. For MRMs, stiffness is often defined as ratio between the largest and smallest rate in the transition rate matrix; the higher this ratio, the more stiff the model. Even more refined are the definitions related to a solution technique for a specific model. In using uniformization for instant-of-time measures of MRMs, the value of qt is often called the stiffness index, where t is the time epoch of interest and q is the maximum over the absolute values of the diagonal entries of the rate matrix, i.e., the uniformization rate. The recently developed extensions of the uniformization technique such as steady-state detection [62] and adaptive uniformization [58] decrease the impact of stiffness.

For very stiff MRMs, implicit integration techniques (such as Runge-Kutta) seem to be the most efficient [48]. The use of these techniques in combination with uniformization also seems fruitful [48].

When discrete-event simulation is the used as a solution method, stiffness can often be circumvented by using importance sampling [26] as implemented in SAVE [25] and UltraSAN [20] or by using injection simulation [57].

Model decomposition, where the fast and slow rates are separated from each other, is another way of avoiding stiff models. This is implicitly done in many hierarchical solution techniques [23].

5.3 Non-exponential behavior

In many model solution techniques, the only allowed time-distributions are exponential distributions. This is the case for the FCFS stations in PFQNs and for the timed transitions in SPNs. To include more variability in a model is generally not much of a problem since a hyper-exponential distribution with two phases can be used to create very large coefficients of variation. More of a problem is the inclusion of (quasi-)deterministic timing.

When simulation is used as a solution technique, non-exponential timing is not a problem. Moreover, it often reduces simulation time as less variance is put in.

One common way to "Markovize" a non-exponential distribution is to use phase-type distributions. In combination with MGMs this is a very attractive approach, although it increases the size of the models to be dealt with. In the context of SPN models or MRMs, this method of phases is often less easy to apply: the state space suffers tremendously.

Recent developments in the field of DSPNs (introduced by Ajmone Marsan and Chiola [1] and further developed by Lindemann [45]) and Markov-regenerative SPNs, introduced by Choi *et al.* [13], alleviate the exponential assumption in SPN models significantly. When one deals with product-form SPNs [36], insensitivity properties known from stochastic-processes theory establish that in many circumstances it does not really matter what the form of the distributions is; only their means matter.

6   Future Work and Concluding Remarks

The challenge in the future lies in developing modeling tools that will operate in a highly constrainted and schedule-oriented environment. The availability of such tools will provide a means of effectively integrating performance experts into the product design team.

In the future, tools must be made sophisticated enough so that an expert system might even eliminate the need for the human expert. Then the designer himself could perform the analyses without any special expertise in the theories of reliability, queues, Markov chains and stochastic Petri nets. At the very least, it would help to smooth variations in the competence of the analyst from one group to the next. A performance/reliability 'center of competency' department, that provides upgrades and maintenance to and advice on, the tools itself might then arise; but the actual analysis would shift to the designers themselves.

A great deal of progress has been made in the last decade in techniques for the generation and solution of large performance, reliability and performability models. Correspondingly, software tools have also been built and distributed. Due to the increased capacity of modern-day workstations, mathematical evaluation techniques for performability and dependability evaluation have become much more feasible. Because modern-day workstations have large internal memories (up to a few hundred MB) and very fast processors (at least 50 MHz clock frequency) the numerical evaluation of large to very large models has become possible. Also, simulation experiments that were unthinkable are now within reach by using only moderately priced workstations.

However, the need to construct and evaluate even larger models continues. Modern computer-communication systems have reached such a complexity that evaluation of their performability and dependability during the design process is an absolute necessity in order to build high-performance systems that provide the requested service for a reasonable price.

Despite the above need, there is still a long way to go to a really integrated design-evaluation path. Still, a lot of progress has been made over the last two decades. In this paper we overviewed that progress and indicated some issues we think will be of importance in the coming years.

References

1. M. Ajmone-Marsan and G. Chiola. On Petri nets with deterministic and exponentially distributed firing times. In *Lecture Notes in Computer Science*, volume 266, pages 132–145. Springer-Verlag, 1987.
2. M. Ajmone Marsan, G. Conte, G. Balbo, "A Class of Generalized Stochastic Petri Nets for the Performance Evaluation of Multiprocessor Systems", *ACM Transactions on Computer Systems* 2(2), pp.93–122, 1984.
3. G. Balbo, S.C. Bruell, S. Ghanta, "Combining Queueing Networks and Stochastic Petri Nets for the Solution of Complex Models of System Behaviour", *IEEE Transactions on Computers* 37(10), pp.1251–1268, 1988.
4. S.J. Bavuso, J. Bechta Dugan, K.S. Trivedi, E.M. Rothmann, W.E. Smith, "Analysis of Typical Fault-Tolerant Architectures using HARP", *IEEE Transactions on Reliability* 36(2), pp.176–185, 1987.
5. H. Beilner, J. Mäter, N. Weissenberg, "Towards a Performance Modelling Environment: News on HIT", in: *Modelling Techniques and Tools for Computer Performance Evaluation*, Editors: D. Potier, R. Puigjaner, Plenum Press, pp.57–75, 1989.
6. A. Bobbio, "Petri Nets Generating Markov Reward Models for Performance/Reliability Analysis of Degradable Systems", in: *Modelling Techniques and Tools for Computer Performance Evaluation*, Editors: D. Potier, R. Puigjaner, Plenum Press, pp.353–365, 1989.
7. P. Buchholz, *Die strukturierte Analyse Markoffscher Modelle, Informatik Fachberichte* 282, Springer Verlag, 1991.
8. R.W. Butler, "The SURE Reliability Analysis Program", *NASA Technical Memorandum* 87593, 1986.
9. J. P. Buzen. Computational algorithms for closed queueing networks with exponential servers. *Commun. ACM.*, 16(9):527–531, Sept. 1973.
10. J.A. Carrasco, *Modelacion y Evaluacion de la Tolerancia a Fallos de Sistemas Distribuidos con Capacidad de Reconfiguracion*, PhD thesis, University of Catalunya, Spain, 1986.
11. J.A. Carrasco, J. Figueras, "Metfac: Design and Implementation of a Software Tool for Modeling and Evaluation of Complex Fault-Tolerant Computing Systems", *Proceedings FTCS 16*, IEEE Computer Society Press, pp.424–429, 1986.

12. G. Chiola, "A Graphical Petri Net Tool for Performance Analysis", in: *Modelling Techniques and Performance Evaluation,* Editors: S. Fdida, G. Pujolle, North-Holland, pp.323–333, 1987.

13. H. Choi, V. G. Kulkarni, and K. S. Trivedi. Markov Regenerative Stochastic Petri Nets. In *16th IFIP W.G. 7.3 Int'l Sym. on Computer Performance Modelling, Measurement and Evaluation (Performance'93)*, Rome, Italy, Sep. 1993.

14. G. Ciardo, J. Muppala, K.S. Trivedi, "SPNP: Stochastic Perti Net Package", *Proceedings of the Third International Workshop on Petri Nets and Performance Models*, IEEE Computer Society Press, pp.142–151, 1989.

15. G. Ciardo, J. Muppala, and K. Trivedi, "Analyzing Concurrent and Fault-Tolerant Software using Stochastic Reward Nets," *Journal of Parallel and Distributed Computing*, Vol. 15, pp. 255-269, 1992.

16. G. Ciardo, A. Blakemore, P.F.J. Chimento, J.K. Muppala, K.S. Trivedi, "Automated Generation and Analysis of Markov Reward Models using Stochastic Reward Nets", in: *Linear Algebra, Markov Chains, and Queueing Models*, Editors: C. Meyer and R. J. Plemmons, Vol.48 of *IMA Volumes in Mathematics and its Applications*, Springer-Verlag, 1992.

17. G. Ciardo, and K. S. Trivedi, " Decomposition Approach for Stochastic Reward Net Models," *Performance Evaluation*, Vol. 18, No. 1, pp. 37-59, July 1993.

18. J. L. Coleman, W. Henderson, P. G. Taylor, *Product Form Equilibrium Distributions and a Convolution Algorithm for Stochastic Petri Nets* Research Report, University of Adelaide, 1992

19. A. J. Coyle, W. Henderson, P. G. Taylor, "Reduced Load Approximations for Loss Networks", to appear in Telecommunications Systems.

20. J. A. Couvillion, R. Freire, R. Johnson, W.D. Obal II, A. Qureshi, M. Rai, W.H. Sanders, J.E. Tvedt, "Performability Modelling with UltraSAN", *IEEE Software*, pp.69–80, September 1991.

21. A. Cumani, "ESP—A Package for the Evaluation of Stochastic Petri Nets with Phase-Type Distributed Transition Times", *Proceedings of the International Workshop on Timed Petri Nets*, IEEE Computer Society Press, pp.144–151, 1985.

22. J. Bechta Dugan, R. Geist and M. Smotherman, "The Hybrid Automated Reliability Predictor", *AIAA Journal on Guidance, Control and Dynamics*, Vol. 9, No. 3, May-June 1986, pp. 319-331.

23. R. Geist, M. Smotherman, K. S. Trivedi, J. Bechta Dugan, "Reliability Analysis of Life-Critical Systems", *Acta Informatica*, Vol. 23, No. 6, Nov. 1986.

24. R. Geist, K.S. Trivedi, "Reliability Estimation of Fault-Tolerant Systems: Tools and Techniques", *IEEE Computer* 23(7), pp.52–61, 1990.

25. A. Goyal, W.C. Carter, E. de Souza e Silva, S.S. Lavenberg, K.S. Trivedi, "The System Availability Estimator", *Proceedings FTCS 16*, IEEE Computer Society Press, pp.84–89, 1986.

26. A. Goyal, P. Heidelberger, and P. Shahabuddin. Measure specific dynamic importance sampling for availability simulations. In A. Thesen, H. Grant, and W. D. Kelton, editors, *Proc. of the 1987 Winter Simulation Conference*, 1987.

27. D. Gross, D.R. Miller, "The Randomization Technique as a Modelling Tool and Solution Procedure for Transient Markov Processes", *Operations Research* 32(2), pp.343–361, 1984.

28. B.R. Haverkort, I.G. Niemegeers, "Using Dynamic Queueing Networks as a Tool for Specifying Performability Models", *ACM Performance Evaluation Review* 17(1), p.225, 1989.

29. B.R. Haverkort, *Performability Modelling Tools, Evaluation Techniques, and Applications*, Ph.D. thesis, University of Twente, 1990.
30. B.R. Haverkort, I.G. Niemegeers, P. Veldhuyzen van Zanten, "DyQNtool—A Performability Modelling Tool Based on the Dynamic Queueing Network Concept", in: *Computer Performance Evaluation: Modelling Techniques and Tools*, Editors: G. Balbo, G. Serazzi, North-Holland, pp.181–195, 1992.
31. B.R. Haverkort, "Approximate Performability Modelling using Generalized Stochastic Petri Nets", *Proceedings of the 1991 International Workshop on Petri Nets and Performance Models*, IEEE Computer Society Press, 1991, pp.300-309.
32. B. Haverkort and K. Trivedi. Specification and generation of Markov reward models. *Discrete-Event Dynamic Systems: Theory and Applications 3* , pp.219–247, 1993.
33. B.R. Haverkort, "Approximate Performability and Dependability Modelling using Generalized Stochastic Petri Nets", *Performance Evaluation* 18(1), pp.61–78, 1993.
34. B. R. Haverkort, A.P.A. van Moorsel, and D-J Speelman. Xmgm: Performance modeling using matrix geometric techniques. In *Proceedings of the 2nd Int'l workshop on modeling, analysis and simulation of computer and telecommunication systems*, 1994.
35. B.R. Haverkort, "Performability Evaluation using DyQNtool$^+$", submitted for publication, 1994.
36. W. Henderson and P. G. Taylor, "Aggregation Methods in exact performance analysis of stochastic Petri nets", *Proceedings of the 3rd Int'l Workshop on Petri Nets and Performance Models*, pp.12–18, 1989.
37. R.A. Howard, *Dynamic Probabilistic Systems, Vol. II: Semi-Markov and Decision Processes*, New York, Wiley, 1971.
38. A.M. Johnson Jr., M. Malek, "Survey of Software Tools for Evaluating Reliability, Availability, and Serviceability", *ACM Computing Surveys* 20(4), pp.227–269, 1988.
39. S.C. Johnson, R.W. Butler, "Automated Generation of Reliability Models", *Proceedings of the 1988 Annual Reliability and Maintainability Symposium*, pp.17–22, 1988.
40. V. Kulkarni, V. F. Nicola, R. M. Smith, and K. S. Trivedi. "Numerical evaluation of performability measures and job completion time in repairable fault-tolerant systems" In *Proc. 16th Intl. Symp. on Fault Tolerant Computing*, Vienna, Austria, July 1986. IEEE.
41. R. Lepold, "Tomspin: Benutzerhandbuch", internal report Siemens AG, 1991.
42. R. Lepold, "PENPETt: A New Approach to Performability Modelling using Stochastic Petri Nets", *Proceedings of the First International Workshop on Performability Modelling of Computer and Communication Systems*, Editors: B.R. Haverkort, I.G. Niemegeers, N.M. van Dijk, University of Twente, pp.3–17, 1991.
43. R. Lepold, "Performability Evaluation of a Fault-Tolerant Computer Systems using Stochastic Petri Nets", *Proceedings of the Fifth International Conference on Fault-Tolerant Computing Systems*, Springer Verlag, Nürnberg, 1991.
44. V.O.K. Li, J.A. Silvester, "Performance Analysis of Networks with Unreliable Components", *IEEE Transactions on Communications* 32(10), pp.1105–1110, 1984.
45. C. Lindemann, "An improved numerical algorithm for calculating steady-state solutions of deterministic and stochastic Petri net models", *Proceedings of the 4th Int'l Workshop on Petri Nets and Performance Models*, 1991.
46. C. Lindemann, R. German, "DSPNexpress: A Software Package for Efficiently Solving Deterministic and Stochastic Petri Nets", in: *Performance Tools 1992*, Editors: R. Pooley, J. Hillston, Edinburgh University Press Ltd., forthcoming, 1992.

47. E. A. MacNair and C. H. Sauer. *Elements of practical performance modeling*. Prentice Hall, Englewood Cliffs, New Jersey, USA, 1985.

48. Manish Malhotra *Specification and Solution of Dependability Models of Fault-Tolerant Systems*, Ph.D. Thesis, Dept. of Comp. Sc., Duke University, April 1993.

49. R. A. Marie. An approximate analytical method for general queueing networks. *IEEE Trans. Software Engg.*, SE-5:530–538, 1979.

50. R. A. Marie, A. L. Reibman, K. S. Trivedi "Transient Analysis of Acyclic Markov Chains", *Performance Evaluation* 7, 1987.

51. J.F. Meyer, "On Evaluating the Performability of Degradable Computer Systems", *IEEE Transactions on Computers* 29(8), pp.720–731, 1980.

52. J.F. Meyer, "Closed-Form Solutions of Performability", *IEEE Transactions on Computers* 31(7), pp.648–657, 1982.

53. J.F. Meyer, A. Movaghar, W.H. Sanders, "Stochastic Activity Networks: Structure, Behavior, and Application", *Proceedings of the International Workshop on Timed Petri Nets*, IEEE Computer Society Press, pp.106–115, 1985.

54. J.F. Meyer, "Performability: A Retrospective and Some Pointers to the Future", *Performance Evaluation*, 14(3&4), pp.139–156, 1992.

55. K. B . Misra (Ed.), *New Trends in System Reliability Evaluation*, Elsevier Science Publishers, 1993.

56. M.K. Molloy, "Performance Analysis using Stochastic Petri Nets", *IEEE transactions on Computers* 31(9), pp.913–917, 1982.

57. A.P.A. van Moorsel, *Performability Evaluation Concepts and Techniques*, Ph.D. thesis, University of Twente, Department of Computer Science, 1993.

58. A.P.A. van Moorsel, W.H. Sanders, "Adaptive Uniformization", forthcoming in *Stochastic Models*, 1994.

59. M. Mulazzani, K.S. Trivedi, "Dependability Prediction: Comparison of Tools and Techniques", *Proceedings IFAC SAFECOMP*, pp.171–178, 1986.

60. B. Müller-Clostermann, "NUMAS—A Tool for the Numerical Analysis of Computer Systems", in: *Modelling Techniques and Tools for Computer Performance Analysis*, Editor: D. Potier, North-Holland, pp.141–154, 1985.

61. J. K. Muppala, A. Sathaye, R. Howe, K. S. Trivedi, "Dependability Modeling of a Heterogeneous VAXcluster System Using Stochastic Reward Nets", in: *Hardware and Software Fault Tolerance in Parallel Computing Systems*, Editor: D. Averesky, Ellis Horwood Ltd., 1992, forthcoming.

62. J. Muppala and K. S. Trivedi, "Numerical Transient Solution of Finite Markovian Queueing Systems", in: *Queueing and Related Models*, U. N. Bhat and I. V. Basawa (ed.), pp. 262-284, Oxford University Press, 1992.

63. D. Potier, M. Veran, "The Markovian Solver of QNAP2 and Examples", in: *Computer Networking and Performance Evaluation*, Editors: T. Hasegawa, H. Takagi, Y. Takahashi, pp.259–279, 1986.

64. A.L. Reibman, K.S. Trivedi, "Transient Analysis of Cumulative Measures of Markov Model Behavior", *Stochastic Models* 5(4), pp.683–710, 1989.

65. M. Reiser and S. S. Lavenberg. Mean value analysis of closed multichain queueing networks. *J. ACM.*, 27(2):313–322, Apr. 1980.

66. R.A. Sahner, K.S. Trivedi, "Reliability Modelling using SHARPE", *IEEE Transactions on Reliability* 36(2), pp.186– 193, 1987.

67. R.A. Sahner, K.S. Trivedi, "A Software Tool for Learning About Stochastic Models", *IEEE Transactions on Education* 36(1), 1993.

68. W.H. Sanders, J.F. Meyer, "Performability Evaluation of Distributed Systems using Stochastic Activity Networks", *Proceedings of the 1987 International Workshop*

on Petri Nets and Performance Models, IEEE Computer Society Press, pp.111–120, 1987.

69. W.H. Sanders, *Construction and Solution of Performability Models Based on Stochastic Activity Networks*, Ph.D. dissertation, University of Michigan, USA, 1988.

70. W.H. Sanders, J.F. Meyer, "Reduced Base Model Construction for Stochastic Activity Networks", *IEEE Journal on Selected Areas in Communications* **9**(1), pp.25–36, 1991.

71. C. H. Sauer and K. M. Chandy. *Computer Systems Performance Modeling*. Prentice-Hall, 1981.

72. M. Sczittnick, *Techniken zur funktionalen und quantitativen Analyse von Markoffschen Rechensystemmodellen*, M.Sc. thesis, University of Dortmund, August 1987.

73. M. Sczittnick, B. Müller-Clostermann, "MACOM—A Tool for the Markovian Analysis of Communication Systems", in: *Proceedings of the Fourth International Conference on Data Communication Systems and Their Performance*, Editor: R. Puigjaner, pp.456–470, 1990.

74. E. de Souza e Silva, H.R. Gail, "Performability Analysis of Computer Systems: from Model Specification to Solution", *Performance Evaluation*, **14**(3&4), pp.157–196, 1992.

75. E. de Souza e Silva, P.M. Ochoa, "State Space Exploration in Markov Models", *ACM Performance Evaluation Review* **20**(1), pp.152–166, 1992.

76. M.F. Squillante, "MAGIC: A Computer Performance Modelling Tool based on Matrix-Geometric Techniques", in: *Computer Performance Evaluation: Modelling Techniques and Tools*, Eds.: G. Balbo, G. Serazzi, North-Holland, pp.411–425, 1992.

77. W. J. Stewart, "MARCA: Markov Chain Analyzer", in: *Numerical Solution of Markov Chains*, Editor: W.J. Stewart, Marcel Dekker, 1991.

78. L. Tomek, V. Mainkar, R. Geist, and K. Trivedi. Reliability analysis of life-critical real-time systems. *Proceedings of the IEEE*, January 1994.

79. K.S. Trivedi, J.K. Muppala, S.P. Woolet, B.R. Haverkort, "Composite Performance and Dependability Analysis", *Performance Evaluation*, **14**(3&4), pp.197–215, 1992.

80. M. Veran, D. Potier, "QNAP2: A Portable Environment for Queueing System Modelling", in: *Modelling Techniques and Tools for Computer Performance Evaluation*, Editor: D. Potier, North-Holland, pp.25–63, 1985.

81. W. Whitt. The queueing network analyzer. *The Bell System Technical Journal*, 62(9), Nov. 1983.

Experimental Analysis of Parallel Systems: Techniques and Open Problems

Daniel A. Reed*

Department of Computer Science
University of Illinois
Urbana, Illinois 61801
USA

Abstract. Massively parallel systems pose daunting performance instrumentation and data analysis problems. Balancing instrumentation detail, application perturbation, data reduction costs, and presentation complexity requires a mix of science, engineering, and art. This paper surveys current techniques for performance instrumentation and data presentation, illustrates one approach to tool extensibility, and discusses the implications of massive parallelism for performance analysis environments.

1 Introduction

The most constant difficulty in contriving the engine has arisen from the desire to reduce the time in which the calculations were executed to the shortest which is possible. Charles Babbage

In the past one hundred and fifty years, little has changed since Babbage's remark. Performance optimization remains a difficult and elusive goal. And as we move from vector supercomputers to parallel systems that scale from tens to thousands of processors, many of the performance instrumentation, data analysis, and presentation techniques we have used successfully on single processor systems must be re-examined, modified, and extended. Despite a solid technical base, building successful (i.e., widely used) performance tools remains part art, part science, and part engineering.

The goal of this paper is to survey current techniques for performance instrumentation and data presentation, illustrate one approach to tool extensibility, and discuss the implications of massive parallelism for performance analysis environments. In §2–§3, we describe the range of performance instrumentation

* Supported in part by the Advanced Research Projects Agency under ARPA contract numbers DAVT63-91-C-0029 and DABT63-93-C-0040, by the National Science Foundation under grants NSF IRI 92-12976 and NSF CDA87-22836, by the National Aeronautics and Space Administration under NASA contract numbers NAG-1-613 and USRA 5555-22, and by a collaborative research agreement with the Intel Supercomputer Systems Division.

approaches and data analysis techniques currently used on parallel systems. We illustrate the use of these techniques in §4 via the Pablo performance analysis environment. In §5, we turn to an analysis of the current state of the art, the problems inherent in capturing and analyzing data from hundreds or thousands of high-speed processors, and a series of research suggestions that might alleviate current problems. Finally, §6 summarizes the current software state and the suggested research avenues.

2 Instrumentation Techniques

A [VAX 11/]780 MIPS is the electron volt of computing. Bill Joy

The ideal performance instrumentation technique would be unobtrusive, accurate, and easily activated and deactivated. Because both theoretical and practical constraints prevent realization of the ideal, a plethora of hardware and software instrumentation techniques have been proposed. Most are variants of four simple techniques: profiling, counting, interval timing, and event tracing. Each has different strengths and weaknesses, each strikes a different balance among dynamic detail, instrumentation overhead, and implementation complexity, and each can be implemented in a variety of ways.

Profiling [5] is by far the most common instrumentation and data reduction technique. In its standard implementation, the program counter is sampled at fixed intervals, and a histogram of program counter values is constructed. Because the interval between each sample is fixed, the total number of samples in a histogram bin is an approximation to the time spent in that code region. When coupled with ancillary symbol table data, post-processing can convert the histogram into estimates of procedure occupancy — the profiles produced by standard Unix tools.

The disadvantages of profiling are its dependence on an external sampling task, typically the operating system, the coarse sampling frequency, usually every 10–20 milliseconds, and the correlation of profiles from multiple processors. Although it is possible to profile operating system code, the implementation can be awkward, as it must avoid using the facilities being profiled. More tellingly, profiling's coarse granularity can underestimate the time spent in small, frequently invoked routines. For massively parallel systems, profile data from multiple processors must be integrated to show the spectrum of behavior.

Despite its limitations, profiling is the most appropriate first step in performance tuning. Not only is its implementation overhead modest, it highlights the most troublesome regions of a code without overwhelming the user with voluminous and unwanted performance data.

Event counting eliminates the potential statistical errors of profiling, but usually at the expense of more invasive instrumentation. Although it is possible to passively count events using hardware counters, counting software events typically requires either user modification of source code or compiler-generated counting code.[2] In addition to the intellectual cost of manually inserted counters, counting code can perturb program execution. On a single processor system, this perturbation is manifest as an increase in program execution time, but on a parallel system it can be more pernicious — the partial order of events may change, changing the total event count on each processor.

Interval timing is the measured analog of the sampling using to compute profiles. By inserting calls to a system clock in code, one can measure the amount of time spent in particular code fragments. By accumulating interval sums and counts, a measured, rather than sampled, profile results.

Finally, event tracing [21] is potentially the most invasive, and the most detailed, of the four performance measurement techniques. Like counting and interval timing, the target code must be modified to include software instrumentation. However, rather than counting or timing event occurrences, event tracing generates a timestamped record each time an event occurs. Because events can occur on a millisecond or even microsecond time scale, even single processor systems can quickly generate megabytes or gigabytes of event data. Hence, event tracing must be used with care, and is often the instrumentation of last resort — it can provide detail but at high cost.

The implementation constraints for profiling, counting, interval timing, and event tracing all differ, yet they have certain common implementation needs. Of these, the most important is a high-resolution, low-overhead global clock. Without a high-resolution global clock, it is not possible to correlate events that occur on disparate processors. The result is causality violations [17] — progenitor events on one processor appear to have occurred after their progeny on another processor. Similarly, if the clock access overhead is high, the cost of instrumentation will be high, and it will be difficult to accurately measure events separated by small time intervals.

3 Analysis and Presentation Techniques

The purpose of computing is insight, not numbers. Richard Hamming

The range of proposed and implemented performance data analysis and presentation techniques is as broad as statistics, computer graphics and scientific visualization, and data sonification. Except for sonification, which is still evolving, all have been widely used to reduce and present performance data.

In the past, sonification has drawn on the experience of the computer music community, just as scientific visualization initially drew on the experience of the graphics community before developing it's own language and representations. However, like visualization and graphics, sonification and music differ in important ways, and sonification idioms are beginning to emerge [16].

Regardless of the data presentation idiom, massive parallelism imposes constraints not present on the analysis and presentation of performance data from single processor systems. The computational cost of data reductions must not

---

[2] A count of basic blocks is the best known example.

be excessive, and graphical displays of performance data, if used, must scale to thousands of processors. As an example, critical path calculations [9], which identify the path through an event trace that most limits the computation, do not scale well. The computation cost becomes prohibitive with large event traces and large numbers of processors. Similarly, many types of graphical performance data displays that show the network interconnections or interprocessor communication events cannot scale to more than a few hundred processors — standard workstation displays lack the requisite pixels.[3]

Finally, the performance tool developer violates Hamming's dictum at his or her peril. Performance tools must be simple to use, they must provide insight, they must be efficient, and they must not fixate on detailed behavior. Just as thermodynamics and statistical mechanics explain the movement of energy without tracking individual gas molecules, we need higher-level models of performance that first describe and present system and application behavior in the aggregate.

4 Pablo Performance Analysis Environment

Massive parallelism exacerbates the already difficult problems of bottleneck identification and performance tuning. The primary reason for the increased difficulty is not the number of processors, but rather the greater complexity and frequency of component interactions, as well as the number of potential optimization gradients. Typically, the space of potential performance optimizations is poorly understood, rarely convex, and of very high dimension.

Like an archeologist who must infer physical structure, habitat, and activity from a few physical artifacts, many performance tools often force users and system software designers to infer system or application behavior from data that has only indirect relation to the application or system software model. Given the complexity of component interactions in a massively parallel system, the inference process is tedious and error prone.

An ideal performance analysis environment should support interactive insertion of instrumentation points, as well as subsequent data analysis, reduction, and display. Moreover, because users embrace performance tools only as a last resort, maximizing ease of use and minimizing learning time are critical to a performance environment's success. These constraints dictate performance environments that are *portable* across a range of parallel architectures, minimizing learning time, *scalable* with the size of the system being studied, allowing a single tool to be used, and *extensible*, allowing users to add environment functionality as needed.

Below, we describe the Pablo Performance Analysis Environment[4] [20, 19, 18], a software system designed to provide portable, yet extensible, capture, re-

---

[3] As an example, the communication matrix of Figure 7 requires $O(N^2)$ pixels to show the communication pattern among $O(N)$ processors. With N approaches 1000, a complete workstation screen is required for this single display.

[4] Pablo is a trademark of the Board of Trustees of the University of Illinois.

duction, and display of dynamic performance data. The environment consists of three primary components, an extensible data capture library and associated graphical instrumentation tools, a data meta-format [2] that describes the structure of performance data records without constraining their contents, and a graphical data analysis toolkit that allows users to quickly prototype performance data reductions.

4.1 Software Instrumentation

As Figure 1 suggests, the Pablo instrumentation software captures dynamic performance data via instrumented source code that is linked with a data capture library. During program execution, the instrumentation code generates performance data that can either be recorded by the data capture library or extracted as it is generated [19]. Source code instrumentation points can be specified by manually modifying the source code to call the data capture library, or interactively by using the graphical instrumentation interface and the instrumenting parser.

The instrumenting parser identifies instrumentable source code constructs and passes this information to the graphical instrumentation interface. The graphical interface, based on X and Motif, interprets this information and allows the user to interactively specify source code instrumentation points. By design, the graphical instrumentation interface permits users to instrument only a subset of all source code locations; more general instrumentation is possible by manually inserting calls to the data capture library. By limiting instrumentation to procedure calls and outermost loops, the graphical interface lessens the likelihood that users will inadvertently perturb the computation by generating invasive instrumentation.

Figure 2 shows a snapshot of the Pablo graphical instrumentation interface. The pulldown menus at the top of the interface allow users to load files and apply instrumentation directives to source code. Items in the Global menu can be used to trace or count all procedure or function calls in the code. The Routine menu is used in conjunction with the scrolled list just below the menu bar. This list contains all the procedures and functions in the source code. By selecting a routine, one can selectively trace or count all calls *to* or *in* a particular routine. Finally, via the scrolled source code window, one can selectively enable or disable instrumentation of a single source code line. The highlighted items in the source code window are the instrumentable constructs. Clicking the mouse on a line to select a construct, then pressing one of the push buttons at the bottom of Figure 2, suffices to enable a new instrumentation point or disable an existing one that had earlier been specified using the pulldown menus. In the source code scroll box of Figure 2, there are two instrumentable constructs, the two loops; one has been instrumented to generate event traces, and the other has been instrumented to count event occurrences.

In addition to counting and event tracing, the Pablo data capture library also supports interval timing, adaptive instrumentation control, and several extension interfaces. Pablo's implementation of adaptive instrumentation control asso-

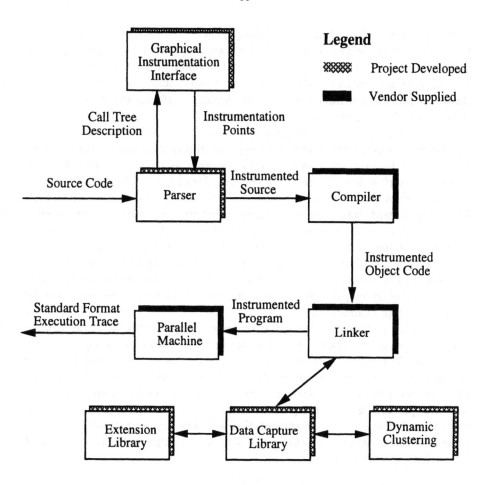

Figure 1 Pablo Instrumentation Software

ciates a user-specified maximum instrumentation level with each event. Higher levels allow greater volumes of event data to be recorded in the performance data file; lower levels bound the maximum volume of event data to lower levels. In addition to instrumentation control via user-specified instrumentation levels, the Pablo instrumentation software will dynamically adjust event levels within the user-specified range (i.e., the instrumentation software may reduce the level below the user-specified point, but it will not increase it beyond that point). The Pablo data capture library also monitors the aggregate event rate using a similar algorithm. Together, the event thresholds and the global threshold allow the Pablo instrumentation software to balance event data volume against application perturbation, maximizing the amount of useful trace data.

Of the Pablo data capture library extension interfaces, the two most notable are those for dynamic statistical clustering and semantic data compression. As §5.3 discusses in greater length and Table 1 illustrates, detailed instrumentation on massively parallel systems can quickly generate tremendous data volumes. However, with the widely-used SPMD programming model, most processors follow nearly identical execution paths and have similar performance metrics. The goal of dynamic statistical clustering is to automatically identify equivalence classes of processor behavior and record detailed performance data only from equivalence class representatives and outliers.

Extension interfaces for semantic data compression allow performance tool developers to extend the Pablo data capture library and reduce the recorded data volume by processing raw performance data as it is generated. After an extension interface is registered with the base data capture library, all events of a specified class are passed to the extension software. The extension software can selectively discard data, compute summary statistics, or synthesize new events. For example, one extension uses procedure entry/exit trace records to compute procedure lifetimes. Another computes histograms of input/output operation latencies. The simplicity of the extension interface mechanism belies its power. By balancing post-mortem data analysis against the overhead of on-line data reduction, tool developers can optimize instrumentation and data reduction for their computing environment.

The Pablo instrumentation software design is necessarily a compromise to maximize portability. Compiler support for instrumentation is clearly preferable (e.g., to avoid precluding certain optimizations), as is hardware and operating system support for data capture (e.g., to reduce the instrumentation perturbation). However, these approaches require access to vendor hardware and system software, precluding portability. Our design intent is that the three independent, though cooperating, instrumentation software components can be quickly ported to a new parallel system or the individual components can be replaced by machine-specific versions.

4.2 Data Meta-format

Many scientific communities have well-established data formats and international databases for storing and distributing information critical to future research. A consistent data format simplifies the dissemination of research results and promotes the development and sharing of tools to manipulate the data. In addition, the adoption of a standard "data language" reduces the time required to exchange and analyze raw data, allowing researchers to devote more time to research.

Unfortunately, there are no standard data formats, either physical or logical, for processing performance data from sequential or parallel computer systems. Not only does this force each group to develop new data management tools, it also limits data sharing. The difficulty in defining a standard medium of data exchange is exacerbated by the diversity of interesting performance data and the plethora of extant and proposed parallel hardware and software architectures.

Given the problem of defining a general purpose performance data format that could represent data from diverse sources, we considered adopting a fixed record format that could represent just those events generated by the current

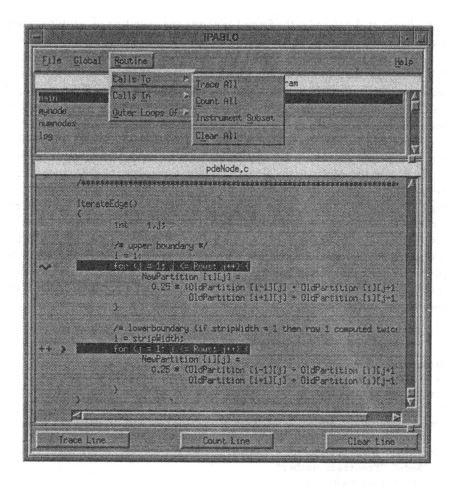

Figure 2 Pablo Graphical Instrumentation Interface

implementation of the Pablo instrumentation software. However, a fixed format was not consistent with Pablo's extensibility design goal. It would have made it impossible to add new event types without major software redesign, and it would have limited Pablo's potential applicability to future generations of massively parallel systems. Recognizing the paramount importance of portability and extensibility, we opted to remove all notion of data semantics from the data format, and instead, embed only a description of the structure of the data records in the file.

The Pablo Self-Describing Data Format (SDDF) is a performance data description language that specifies both the structure of data records and data record instances. The format can describe a wide variety of data records, not just a fixed set. Intuitively, the format supports the definition of records containing scalars and arrays of the base types found in most programming languages (i.e., byte/character, integer, and single and double precision floating point). Self-describing data files include a group of record definitions and a subsequent sequence of tagged data records. The tag identifies the type of the record, allowing the data record byte stream to be interpreted using a particular record definition.

Figure 3 shows an example of an SDDF record descriptor for a procedure exit trace event. This event was synthesized by a procedure extension to the base Pablo data capture library, based on a source instrumentation using the Pablo graphical instrumentation interface. Notice that the record descriptor contains a tag (105), used to identify portions of the following byte stream as procedure exit records. In addition, it specifies the time of procedure exit, processor number, procedure, source code location, and inclusive and exclusive procedure lifetimes (i.e., with and without descendent calls). The procedure extension computes procedure lifetimes by maintaining a stack of active procedure calls and matching procedure and exit events. On detection of an exit event, procedure lifetimes can then be computed.

```
#105:
// "description" "Procedure Exit Trace Record"
"Procedure Exit Trace" {
// "Seconds" "Timestamp"
double "Seconds";
// "ID" "Event ID"
int "Event Identifier";
// "Node" "Processor number"
int "Processor Number";
// "Procedure" "Procedure Index"
int "Procedure Index";
// "Byte" "Source Byte Offset"
int "Source Byte";
// "Line" "Source Line Number"
int "Source Line";
// "Inclusive Seconds" "Inclusive Duration"
double "Inclusive Seconds";
// "Exclusive Seconds" "Exclusive Duration"
double "Exclusive Seconds";
};;
```

Figure 3 Pablo Example SDDF Record Descriptor

The SDDF format supports both ASCII and binary versions, with routines for conversion between versions. In addition, Intel is now using SDDF as the native data format for ParAide,[5] their commercial performance analysis tools for the Paragon XP/S.

4.3 Data Analysis

To date, most performance analysis environments have been designed to process performance data that contains only a fixed set of data types captured on a particular parallel system. Even within such restricted domains, most environments are limited in scope and provide only a small set of simple data reductions. In consequence, the initial excitement of users often gives way to frustration — interpreting the performance data rapidly leads to questions that cannot be answered within the scope of the environment's capabilities.

Given the rapid evolution of massively parallel systems, it is increasingly unrealistic to believe that robust performance analysis tools can be cost-effectively developed for each new software and hardware environment. Moreover, the diffusion of intellectual effort across multiple, one-of-a-kind performance tools perpetuates the problem, namely the lack of standard, portable, and extensible performance analysis software.

To accommodate diversity and extensibility, the Pablo data analysis environment exploits a toolkit of data transformation modules capable of processing the self-describing data format and a graphical programming model that allows users to interactively connect and configure a data analysis graph. Performance data flows through the configured graph and is transformed to yield the desired performance metrics. The details of file input/output, storage allocation, and module execution scheduling are isolated in the environment infrastructure.

The user specifies the desired data transformations and presentations by interactively connecting analysis, data display, and sonification [15, 16] modules to form a directed acyclic graph, then selecting the SDDF data records to be processed by each data analysis module. All data passed among modules in the configured data analysis graph is encapsulated in the self-describing data format. The environment infrastructure uses the SDDF record and field names to prompt the user as individual modules are configured. When the configured data analysis graph begins execution, the Pablo analysis environment infrastructure extracts the specified fields from the data records and passes those values to the modules. By exploiting the flexible SDDF format and isolating the data semantics in the user's analysis graph configuration, a small set of analysis modules can be used for a plethora of data reductions.

Figure 4 shows the first steps in constructing a simple Pablo data analysis graph. One selects modules from the dialogue at the right of Figure 4, instantiating instances on the graph canvas on the left. After all the desired modules have been instantiated, one interconnects the modules to define the data flow through the graph. In this example, an SDDF file input module has been connected to a module that computes profiles based on event durations. A more complex

---

[5] ParAide is a trademark of the Intel Supercomputer Systems Division.

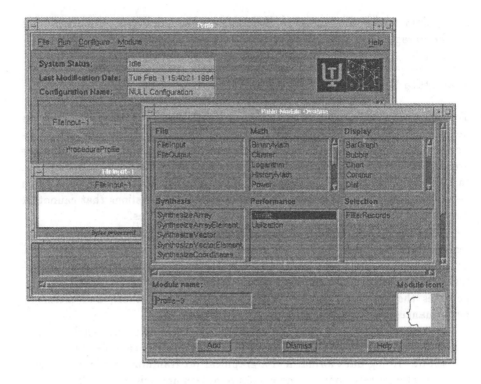

Figure 4 Pablo Data Analysis Interface (Graph Construction)

graph might have multiple file input modules and tens of data transformation, graphical display, and sonification modules.

The second step in analysis graph configuration is specifying what subset of the performance data in an SDDF file should be processed by each graph module. After specifying an SDDF file to be read by the file input module, Figure 5, shows the user selecting SDDF records and record fields to be processed by the profile module. In the figure, the **Records** scroll box shows the names of all the SDDF record descriptors in the input file; the record and field names are from descriptors similar to that in Figure 3. Here, the user has selected a procedure exit record; the fields displayed in the **Fields** scroll box are from that record. By clicking on record fields and the inputs to the profile module, the user interactively specifies that the Pablo infrastructure should extract the specified fields from each instance of a procedure exit record and pass those as input to the profiling module.

The ability to interactively extract data from SDDF records is the key to the Pablo analysis environment's flexibility and extensibility. Because the details of record formats are specified by the SDDF descriptors and the analysis modules are oblivious to this detail, interactive graph configuration exploits a small set

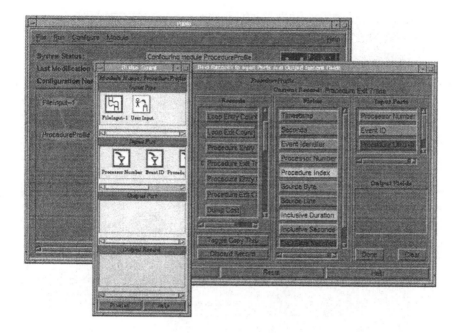

Figure 5 Pablo Data Analysis Interface (Graph Configuration)

of architecture-neutral data analysis modules to construct a wide variety of data reductions and displays.

Finally, Figure 6 shows the execution of the configured data analysis graph. The profile module computes the distribution of event occurrences and intervals for each event and displays that data in both tabular and graphical form. At any time during graph execution, the user can add or delete analysis modules, tap the data stream between modules and write to a file in SDDF format, or reconfigure individual modules to change their function. Once an analysis graph has been constructed, it can be saved and later used to process other SDDF files. The only requirement is that they have the same kinds of SDDF record descriptors.

4.4 Current Research

The Pablo performance analysis environment infrastructure provides a springboard for research and a flexible platform for continued software development. The four foci of current research are

- integration of the Pablo infrastructure with Rice University's data parallel Fortran D compiler,
- exploration of virtual reality techniques for performance data immersion and real-time adaptive control,

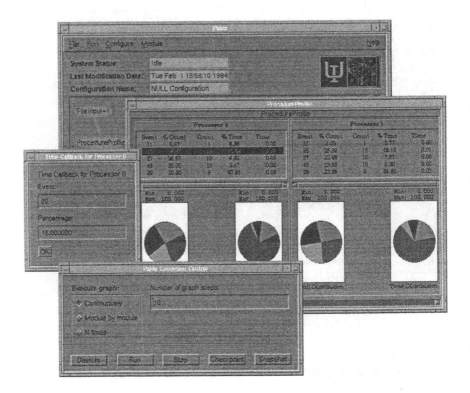

Figure 6 Pablo Data Analysis Interface (Graph Execution)

- dynamic statistical clustering techniques to reduce performance data volume and identify behavioral equivalence classes, and
- instrumentation and data analysis environment extensions to study parallel input/output patterns.

Using the flexibility of the Pablo SDDF data format, we are working with Rice University to export portions of the compiler program analysis data base in the SDDF format. Coupling this information with dynamic performance data, obtained by compiler-synthesized calls to the Pablo data capture library, will permit correlation of performance data from the generated code with the high-level data parallel code created by the application developers.

To understand the input/output patterns of parallel scientific codes, and as a precursor to developing more effective file system software, we are building a set of data capture library extensions to compute input/output access statistics. In addition to input/output event data, these statistics include histograms of file request sizes, interreference times, and request durations.

5 Research Directions

Although the theory of performance instrumentation and data reduction is well established, the emergence of massively parallel systems as the replacement for traditional vector supercomputers has raised new software engineering and performance instrumentation problems. Below, we discuss five issues that we believe will shape the future of performance instrumentation and data analysis for parallel systems.

5.1 User-Friendly Tools

Although most of us would like to believe that application and system software developers use performance tools frequently and look forward to their next "performance safari" with great anticipation, it is not true. Not only do most software developers not enjoy using performance tools, they actively avoid the prospect. Repeated experience has shown that users will eschew sophisticated tools with complex user interfaces and steep learning curves in favor of simple, intuitive tools, even if the simpler tools are technically inferior. Simply put, software developers embrace performance tuning tools only as a last resort, when no other viable alternative exists.

Lest these observations seem overly pessimistic, consider the standard workstation software development style; performance-directed design is rare. Instead, we and others generally develop and debug software without regard to performance, except as guided by our intuition about potential performance bottlenecks. Only if execution times are unacceptable do we deign to analyze the code. Even then, simple, static performance profiles usually suffice to identify those procedures that dominate execution time. We do not study dynamic call graphs, analyze code execution phases, or study procedure call event traces unless there is no other expedient alternative.

Developers of scientific codes on high-performance, parallel systems follow a similar development path. Their objective is scientific results, not software development. Hence, they will embrace the path of minimum human effort, using the simplest tools possible for as long as possible. Their code's performance need not be optimal, nor even near-optimal. It simply must be fast enough to not impede scientific research and to execute within its specified resource constraints.

At tool developers, we must learn two lessons. First, the goal of computing is solutions, not tool use. Second, the most common debugging mechanism is print or write statements, and the most common performance tuning mechanism often is the manual insertion of calls to system timing routines. The implications of these lessons are that tools must be easy to use, they must proceed from simple to complex, and they must satisfy the common case. Ensuring ease of use requires close collaboration with human factors experts, field testing of software prototypes with friendly users, and multiple iterations of software re-design.

Proceeding from simplicity dictates presenting the simplest, most obvious performance data (e.g., profiles) first, then allowing users to explore more detailed data only if they desire. Forcing users to examine unneeded or unwanted performance data is the source of many users' displeasure with performance tools.

Satisfying the common case is only possible if one understands user needs. In an academic environment, this is best accomplished by forming a collaborative, multidisciplinary software development team that creates both application and tool software. For academics, it is critical that the application software developers not be computer scientists; we are too prone to build tools for ourselves, overlooking the fact that for the overwhelming majority of users, computing is a means, not an end. Similarly, vendor tool developers must poll their user base, field test prototypes, and interview trial users. Only by observing what features are used, which are not, which are deprecated, and why, can one accommodate user needs.

From a research perspective, codifying and assembling these experiences, together with descriptions of the application software and hardware environment, would provide a resource base to guide future tool development. Moreover, given a rich enough experience base, we can begin to generalize and identify those features that are critical to user acceptance.

5.2 Performance Data Correlation

Performance tools can capture, present and correlate performance data at many levels, including hardware, system software, library, and application. Often, however, data is not presented in ways that match the tool user's intellectual model. For example, showing detailed, dynamic data on cache utilization is only meaningful if the user both understands the implications of this data *and* has some mechanism to affect change (e.g., by modifying data structure format or reference patterns). Similarly, operating system performance data [21, 14], though invaluable to an operating system designer or system tuner, may be of little help to an application developer. Even application performance data must be related to the application code and data structures in ways that the application developer can understand and exploit to improve performance. Simply put, performance data must be expressed in meaningful ways and related to objects that are part of the software or hardware developer's computation model.

For programs on distributed memory parallel systems that contain explicit calls to message passing routines, the mapping between performance data and source code often is obvious, and many performance tools depend on the software developer to relate message passing data to source code. For example, Figure 7 shows a Pablo display where the number of bytes sent between each pair of possible source and destination nodes is represented by a matrix — a color code or gray scale indicates the data volume in bytes.[6]

Unless the code is large and complex, with many types of messages generated by multiple computation phases, most users have a mental model of the expected interprocessor communication (e.g., ring or nearest neighbor on a grid)

---

[6] ParaGraph [6], Intel's ParAide [12, 22] and many other tools contain similar displays.

and can associate observed patterns with source code locations. However, if there are many types of messages or the same message type is generated from many source code locations, the mapping of performance data to code is more complex, and a simple communication matrix may not provide enough detail. Ideally, by coupling a graphic display like that of Figure 7 with a source code display, the user would be able to selectively display and correlate messages by type, size, length, destination, and source code location. However, such displays must satisfy the dicta of §5.1, notably providing other, simpler performance summaries as a prelude, or they will not be used.

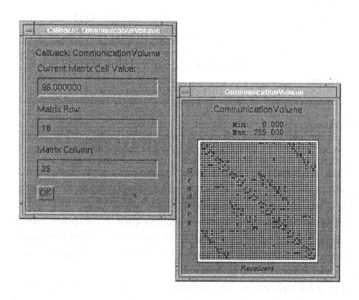

Figure 7 Pablo Communication Matrix (Iterative PDE Solver)

Although correlating dynamic performance data and application source code may be burdensome for some developers of message passing codes, it is extraordinarily difficult when using a data parallel programming model such as that provided by Fortran D [7], Vienna Fortran [3], or High-Performance Fortran (HPF) [11]. These and other data parallel languages raise the programming level from explicit interprocessor communication to operations on arrays and array sections, creating a large semantic gap between the application programming model and the machine execution model.

In most data parallel languages, array distribution constructs allow users to specify data placement across the distributed memories of the parallel machine. Given the data distribution directives, compilers translate array operations into codes that contain explicit message passing. As an example, Figure 8 shows a

Fortran D code fragment [8] with the arrays uu and uud distributed by blocks
in the first dimension.[7]

```
        parameter (n$proc = 8)
        common /vars/ uu(64,64,64)
        decomposition d(64,64,64)
        align uu, uud with d
        distribute d(block,:,:)

     do 41 j=1,64
     do 41 i=1,64
        uud(i,j,1) = c79dz*(uu(i,j,2) - uu(i,j,64))
   1                + c136dz*(uu(i,j,3) - uu(i,j,64-1))
        uud(i,j,2) = c79dz*(uu(i,j,3) - uu(i,j,1))
   1                + c136dz*(uu(i,j,4) - uu(i,j,64))
  41    continue
```

Figure 8 Fortran D Code Fragment

Because only a portion of the array resides on each processor, references
to those portions of the array that are on other processors are via a remote
access protocol realized by compiler-synthesized message passing. Although these
protocols can range from simple to complex, the owner-computes rule [7] is one of
the best known. Under owner-computes, the processor whose memory contains
an array element, the "owner," is responsible for computing any modifications
to that array element.

During program execution, the volume and frequency of interprocessor communication depend on both the application program's array access patterns and
on the sophistication of the compiler optimizations. If the distributed array access pattern is highly data dependent (e.g., a sparse matrix calculation where
the access pattern depends on matrix sparsity and structure), the compiler may
have to synthesize scatter-gather interprocessor communication that is executed
each loop iteration.

Even if the array access patterns are regular and repeated, as in Figure
8, the high communication software latency on most distributed memory parallel systems mandates that compilers balance communication latency against
data transfer bandwidth. By extracting communication from loops, combining
requests for multiple array elements into a single message, and overlapping communication and computation, the compiler can reduce data sharing costs, albeit
by creating code that may contain a multiplicity of special cases and machine-
specific optimizations. As an example, Figure 9 shows a portion of the Fortran

---

[7] The original Fortran 77 code from which this Fortran D was created was provided
by Thomas Eidson of ICASE at the NASA Langley Research Center.

77 and message passing code, generated by the Rice Fortran D compiler, for the Fortran D code fragment of Figure 8. Notice that

- the "boundary processors" zero and seven are special cases because, in general, the number of array elements on each processor need not be the same,
- the buf3D$r routines package array sections for transmission to those other processors that require access,
- only a portion of the computation is performed on each processor and that portion is a function of the array distribution,
- for the regular access pattern of Figure 8, the number and sizes of all messages are known at compilation time.

The complexity of the code in Figure 9, compared to that in Figure 8, as well as large semantic gap separating the two, makes it impossible for application developers to relate dynamic performance data to source code without deep knowledge of both compiler transformations and the execution model. Even then, a careful perusal of the compiler-generated source code is required. Because easing the programming burden is one of the primary motivations for high-level, data parallel languages, requiring software developers to delve into the details of low-level code is self-defeating.

Figure 10 shows a snapshot of dynamic processor utilization and communication traffic when executing the code of Figure 9. Not only would an application developer who wrote the code in Figure 8 find the notion of communication traffic in Figure 10 baffling, he or she would have no notion of its significance (i.e., if this amount of communication indicative of good performance or a severe performance bottleneck). Likewise, the regular code of Figure 8 suggests no reason why there should be such disparity in utilizations across processors.

To redress the growing semantic gap between high-level data parallel languages and current, low-level dynamic performance analysis tools, we need performance analysis software that can integrate the compiler's program analysis data base with dynamic performance data and provide performance guidance in the context of the original source code. In a data parallel programming model, application program developers can affect performance by changing array distributions to increase locality, modifying loop nests to increase parallelism, and exploiting array computation functions to exploit parallel libraries. Performance tools must identify performance bottlenecks in the data parallel source code and suggest source code changes in this context.

Integrating performance tools and compilers not only benefits the application developer, it permits more efficient implementation of performance instrumentation and analysis tools. Only the compiler data base contains a description of the source code transformations and the relation between compiler-synthesized messages and array element references. Because the loop bounds and array references are known at compilation time for some codes, interprocessor communication counts and data volume may be calculable without run-time instrumentation. If not, exploiting compile-time information can reduce the volume of dynamic performance data that must be captured (e.g., if message sizes and destinations are known, this information need not be recorded during execution).

```
C       --<< Send uu(1:64, 1:64, 7:8) >>--
        if (my$p .eq. 7) then
          call buf3D$r(uu, 1, 64, 1, 64, -1, 10, 1, 64,
     *                 1, 1, 64, 1, 7, 8, 1, r$buf6(1))
          id$7 = isend(115, r$buf6, 8192 * 4, 0, my$pid)
        endif
C       --<< Send uu(1:64, 1:64, 1:2) >>--
        if (my$p .eq. 0) then
          call buf3D$r(uu, 1, 64, 1, 64, -1, 10, 1, 64,
     *                 1, 1, 64, 1, 1, 2, 1, r$buf5(1))
          id$5 = isend(114, r$buf5, 8192 * 4, 7, my$pid)
        endif
C       --<< Send uu(1:64, 1:64, 7:8) >>--
        if (my$p .lt. 7) then
          call buf3D$r(uu, 1, 64, 1, 64, -1, 10, 1, 64,
     *                 1, 1, 64, 1, 7, 8, 1, r$buf3(1))
          id$3 = isend(113, r$buf3, 8192 * 4,
     *                 my$p + 1, my$pid)
        endif
C       --<< Send uu(1:64, 1:64, 1:2) >>--
        if (my$p .gt. 0) then
          call buf3D$r(uu, 1, 64, 1, 64, -1, 10, 1, 64,
     *                 1, 1, 64, 1, 1, 2, 1, r$buf1(1))

          id$1 = isend(112, r$buf1, 8192 * 4,
     *                 my$p - 1, my$pid)
        endif


        if (my$p .eq. 0) then
C         --<< Wait Recv uu(1:64, 1:64, 63:64) >>--
          call msgwait(id$8)
          do j = 1, 64
            do i = 1, 64
              uud(i, j, 1) = c79dz * (uu(i, j, 2) -
     *            r$buf6(j * 64 + i + 4032)) +
     *            c136dz * (uu(i, j, 3) -
     *            r$buf6(j * 64 + i - 64))

              uud(i, j, 2) = c79dz * (uu(i, j, 3) -
     *            uu(i, j, 1)) + c136dz * (uu(i, j, 4) -
     *            r$buf6(j * 64 + i + 4032))
            enddo
          enddo

          <*** ADDITIONAL CODE EXCISED ***>
```

Figure 9 Fortran 77 Code Fragment Generated from Fortran D

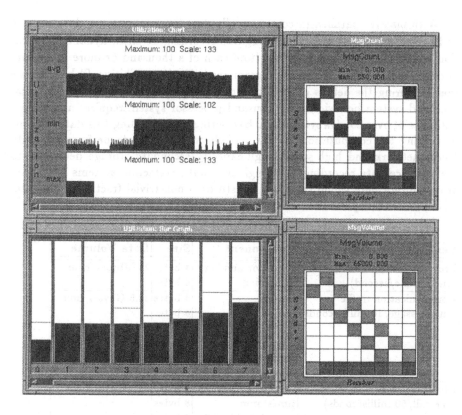

Figure 10 Pablo Performance Data for Fortran D Code

To successfully integrate compilers and performance tools, tool developers must work more closely with compiler developers, and compiler developers must be encouraged to provide access to the compiler data base, either via a set of access functions or via a standard, though extensible, data format. As an example, §4.4 briefly describes our early attempt to integrate the Pablo performance analysis environment with the Rice Fortran D compiler by exploiting the Pablo SDDF data format [2, 19] to describe compiler transformations.

5.3 Scalability

The lure of scalable parallel computer systems is their ability to execute the same code on tens, hundreds, or thousands of processors. Because the performance of an application code is a complex function of computation complexity, interprocessor data sharing, and input/output, and the relative dominance of these factors changes with the number of processors, predicting an application code's performance across a range of parallel system configurations is extraordinarily difficult.

With massive parallelism comes equally daunting performance data capture and presentation problems. Capturing and recording detailed, dynamic application and system performance data from each of a thousand or more processors during a multiple hour parallel program execution is prohibitive. Table 1 shows that capturing the details of execution control flow, interprocessor communication, and system behavior on a thousand processor system requires a sustained data rate of over 25 gigabytes/hour. Extracting and recording this data without perturbing the computation is only possible with a separate performance data access network and high-speed, high-capacity secondary storage dedicated to performance data recording. Although such data extraction systems have been proposed, not only would their cost constitute a non-trivial fraction of the total system, but their potential utility is questionable.

Performance Metric	Component	Size	Data Volume
Processor utilization (one sample/second)	processor number utilization	4 bytes 4 bytes	28 Mbytes/hour
Communication traffic (one message/10 milliseconds)	sender receiver size type timestamp duration	4 bytes 4 bytes 4 bytes 4 bytes 8 bytes 8 bytes	22.5 Gbytes/hour
Procedure calls (one call/50 milliseconds)	processor number timestamp procedure identifier source location inclusive duration exclusive duration	4 bytes 8 bytes 4 bytes 4 bytes 8 bytes 8 bytes	2.5 Gbytes/hour
TOTAL			25.28 Gbytes/hour

Table 1 Detailed Performance Data Volume Estimates (1024 processors)

Even if capturing gigabytes of performance data were not difficult, processing and presenting it is. With data from a distributed memory parallel system, it is not possible to show the communication pattern for every processor. Communication matrices like that in Figure 7 do not scale beyond a few hundred processors — current workstation screens like sufficient pixels. The underlying problem is deeper. Just as thermodynamics and statistical mechanics explain the movement of energy without tracking individual gas molecules, we need higher-level models of performance that first describe processor behavior in the aggregate.

The *prima facia* evidence suggests that capture of detailed, dynamic performance data from all processors is neither desirable nor practical. The dicta that performance tools be easy to use, proceed from simple to complex, and satisfy the common case imply that simpler schemes be exploited to characterize performance on massively parallel systems.

Application profiles can identify the source code associated with the most egregious software performance bottlenecks and are often all that is needed for a software developer to optimize the offending the code. In contrast to the high software overheads and large data volumes of Table 1, profiles are of modest cost, requiring only periodic program counter sampling and histogramming. Even if a profile does not provide sufficient information to suggest a possible source code optimization, the profile data can guide the activation of more detailed, expensive software instrumentation in a subset of the application code.

Not only does a well-designed profiling facility highlight those source code locations that are the dominant contributors to execution time, on a parallel system it delimits the range of processor performance variation (i.e., the slowest, fastest, and "representative" processors). Because the single-program-multiple-data (SPMD) model is the most common on massively parallel systems, one might expect all processors to have similar profiles. However, data-dependent control flow, together with application and operating system interactions, can induce large performance variations across processors. For example, software traps for denormalized arithmetic on processors that implement the IEEE floating point standard can slow normal execution by two or three orders of magnitude. Hence, identifying processor outliers often is a key to performance optimization. Given the identity of processor outliers, one can capture more detailed data from just the subset and representatives of nominal behavior. Finally, if the computation has multiple phases, each with a different bottleneck, multiple profiles per processor, each for a different phase, can be used to enable and disable instrumentation.

Dynamic statistical clustering [19] is one possible mechanism for automatically identifying behavioral equivalence classes. In Figure 11, each processor's position in space is defined by the current values of two performance metrics. Because the dimensionality of the metric space grows with the number of metrics, visually identifying processor clusters becomes problematic. Moreover, as the computation evolves, the metric values and processor locations change. Finally, because most computations are iterative, the processor positions often trace approximately closed curves in the metric space.

By using a partitional clustering scheme [13] to periodically identify processor behavioral outliers (i.e., singleton clusters) and representatives from larger clusters, one can dramatically reduce the volume of performance data that must be captured — as the number of processors increases, the number of clusters grows quite slowly. Simply put, the SPMD programming model, common to both explicit message passing and compiler-synthesized codes, encourages behavioral regularity; anomalies (e.g., denormalized arithmetic) usually arise due to either programming error or unexpected data variability.

Preliminary experiments [19] suggest that dynamic clustering has great promise, but several research questions must first be resolved:

- the balance between clustering computation costs and data extraction overhead,
- clustering interval size and its relation to computation phase changes,

- cluster representative selection and cluster weighting,
- similarity measures for cluster identification, and
- accommodating measurement jitter (i.e., small differences in successive measurements).

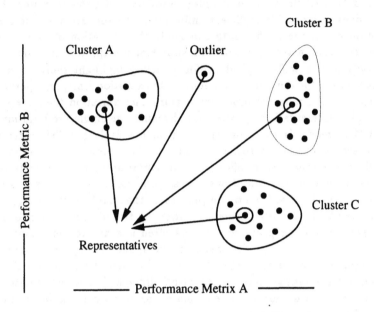

Figure 11 Behavioral Equivalence Classes via Statistical Clustering

A second, potentially effective approach to scalable performance instrumentation and presentation is to abandon the common notion of first capturing data without regard to its potential use, then sifting the data to identify and display anomalies. Instead, if one views the execution of a parallel program as a temporal data base [23], performance tuning becomes a set of temporal queries, with performance data capture driven by the needs of the queries [10]. Not only does this reduce the volume of captured performance data by limiting the code locations and time intervals when data is needed, it also focuses the performance analysis software design on design of effective, high-level performance queries. Moreover, because the queries are specific (e.g., what fraction of time is spent in this procedure, or what is the communication locality of this loop nest), the instrumentation and data capture system can exploit this specificity to minimize the instrumentation overhead.

5.4 Data Immersion and Performance Steering

Because massively parallel systems with thousands of processors can quickly generate enormous amounts of performance data, intelligent, adaptive data capture via, for example, statistical clustering or query-driven instrumentation, will become *de rigueur* for minimizing data volume and instrumentation perturbation.

Given the existence of a flexible performance instrumentation infrastructure that can dynamically enable or disable instrumentation points based on observed behavior, it becomes feasible to "close the loop" by exploiting performance data to guide application performance.

Just as application steering allows scientists to examine intermediate data and guide the code toward more interesting phenomena, performance steering would allow either application developers, performance analysts, or software performance agents to optimize application performance by changing application actions or system resource management policies. For example, changing the granularity of work distribution from an application's centralized task queue might reduce contention or improve load balance. Similarly, changing an array data distribution might increase data locality in a code region.

As a feasibility test of real-time adaptive control, we have constructed a prototype system [19] that extracts performance data from a parallel genome sequencing code [1] and transmits that data to an immersive virtual world that is a three-dimensional generalization of a two-dimensional scatterplot matrix [4]. Though a head-mounted display that shows stereo projections of the dynamic data and three-dimensional audio cues that augment graphics with sound, the user can explore the dynamic data and, using a mouse with a three-dimensional tracker, directly control the granularity of the sequencing code's input/output operations. Because the balance of computation and input/output in the genome sequencing code strongly depends on the sequence being matched against the genome data base, performance is difficult to predict *a priori*, and our preliminary experience suggests that interactive performance tuning can substantially increase application performance.

From interactive control, it is but a small step to adaptive software control. Intelligent performance agents could accept and process dynamic performance data as a part of the ongoing computation and adjust application and system parameters to maximize performance. Given the nascent state of system software on many parallel systems and the large number of potential optimization axes, application flexibility is critical to maximizing performance.

5.5 Operating System Resource Management

Arguably, the current state of system software for scalable parallel systems is analogous to that for sequential systems in the early 1960's — we know little about the dynamics of parallel operating system component interactions, and optimal resource management policies for scalable parallel systems are unknown. Moreover, even many basic questions about traditional resource management remain unanswered in the parallel systems context (e.g., the effect of node scheduling algorithms on aggregate performance, or the interactions of paging with intertask communication).

As an extended example of current problems, consider secondary storage access, the Achilles heel of most parallel systems. Input/output systems for massively parallel machines typically consist of a large number of disk arrays (usually RAIDs), connected to input/output servers that are in turn connected to the processor interconnection network.[8] In this context, the interactions of application parallelism, the number of logically disjoint computation activities, application data access patterns, and paging activity are largely unexplored, as are the potential benefits of physical data contiguity (i.e., preallocating contiguous space for the portion of a file that resides on each disk) and the control parameters for data caching and prefetching algorithms.

Many of the resource management issues that were carefully explored in the 1960's and 1970's for single processor systems must be revisited in the parallel context. Experience may show that many of the answers are the same; however, only measurement and experimentation can confirm or deny such conjectures. Just as detailed, dynamic performance data can illuminate application performance bottlenecks and suggest possible code optimizations, performance instrumentation and analysis can guide resource management policy decisions. Moreover, the potential user community, operating system developers, is much smaller, more receptive to performance tools, and able to exploit detailed data.

Operating system instrumentation does, however pose more difficult problems than those faced when instrumenting application codes. First and most importantly, one must not use the system services being measured (e.g., input/output measurements should not exploit the input/output system to extract performance data). More tellingly, in a parallel operating system, even subtle changes in timing behavior can potentially change the service order for resource requests, reorder the requests across processors, and change the scheduling of application tasks. For example, if tasks are statically assigned to processors, as is common on distributed memory parallel systems, operating system instrumentation is unlikely to substantially perturb the system and application behavior [21]. In contrast, on a cache-coherent, shared-memory parallel system with dynamic task scheduling, perturbing the task schedule can have profound effects. The likelihood of major perturbations, and the expected effects on the partial order of processor events, depend on the intrusiveness of the instrumentation, the types of system services that have been instrumented, and the features of the operating system.

6 Summary

Although the theory of performance instrumentation and data reduction is well established, the emergence of massively parallel systems as the replacement for traditional vector supercomputers, together with an increasing emphasis on software usability, raises new software engineering and performance instrumentation problems. We must build performance tools that are efficient, that can correlate dynamic data with application source code, and that can scale to thousands of processors. In addition, the complexity of parallel systems and their sensitivity to small variations in application behavior suggest that we must study application resource demands and system responses and explore real-time adaptive control

---

[8] Both the Thinking Machines CM-5 and the Intel Paragon XP/S follow this design.

mechanisms. The Pablo performance analysis environment has been designed as a flexible, extensible software infrastructure for examining just these issues.

Acknowledgments

I am indebted to Tom Birkett, Dave Jensen, Bobby Nazief, Ted Nelson, Bob Olson, and Brian Totty for their insights and software contributions to the early development of Pablo. Likewise, Wayne Smith and Judy Guist of the Intel Supercomputer Systems Division were a frequent and valuable source of suggestions for design improvements and software extensions. Current Pablo group members, Ruth Aydt, Roger Noe, Tara Madhyastha, Phil Roth, Keith Shields, Will Scullin, and Luis Tavera have made invaluable contributions, and their enthusiasm has made collaboration a joy. Finally, the design for the integration of Pablo with the Rice Fortran D compiler is the joint result of collaborative research with Ken Kennedy, John Mellor-Crummey, Vikram Adve, and Jhy-Chun Wang.

References

1. ARENDT, J. W. Parallel Genome Sequence Comparison Using an iPSC/2 with a Concurrent File System. Master's thesis, University of Illinois at Urbana–Champaign, Department of Computer Science, Jan. 1991.
2. AYDT, R. A. SDDF: The Pablo Self-Describing Data Format. Tech. rep., University of Illinois at Urbana-Champaign, Department of Computer Science, Sept. 1993.
3. CHAPMAN, B., MEHROTRA, P., AND ZIMA, H. Programming in Vienna Fortran. *Scientific Programming 1*, 1 (Fall 1992), 31–50.
4. CLEVELAND, W. S., AND MIGILL, M. E., Eds. *Dynamic Graphics for Statistics*. Wadsworth & Brooks/Cole, 1988.
5. GRAHAM, S., KESSLER, P., AND McKUSICK, M. gprof: A Call Graph Execution Profiler. In *Proceedings of the SIGPLAN '82 Symposium on Compiler Construction* (Boston, MA, June 1982), Association for Computing Machinery, pp. 120–126.
6. HEATH, M. T., AND ETHERIDGE, J. A. Visualizing the Performance of Parallel Programs. *IEEE Software* (Sept. 1991), 29–39.
7. HIRANANDANI, S., KENNEDY, K., AND TSENG, C.-W. Compiler Optimizations for Fortran D on MIMD Distributed-Memory Machines. In *Supercomputing '91* (Nov. 1991), pp. 86–100.
8. HIRANANDANI, S., KENNEDY, K., AND TSENG, C.-W. Preliminary Experiences with the Fortran D Compiler. In *Supercomputing '93* (Nov. 1993), pp. 338–350.
9. HOLLINGSWORTH, J. K., AND MILLER, B. P. Parallel Program Performance Metrics: A Comparison and Validation. In *Supercomputing '92* (Nov. 1992), pp. 4–13.
10. HOLLINGSWORTH, J. K., AND MILLER, B. P. Dynamic Control of Performance Monitoring on Large Scale Parallel Systems. In *7th ACM International Conference on Supercomputing* (July 1993), pp. 185–194.
11. HPFF. High-Performance Fortran Language Specfication, Version 1.0. Tech. rep., High Performance Fortran Forum, May 1993.
12. INTEL. *Application Tool User's Guide*. Intel Supercomputer Systems Division, Beaverton, Oregon, Oct. 1993.

13. JAIN, A. K., AND DUBES, R. C. *Algorithms for Clustering Data*. Prentice Hall, Englewood Cliffs, NJ, 1988.

14. KOHR, D. R., ZHANG, X., RAHMAN, M., AND REED, D. A. A Performance Study of an object-Oriented Parallel Operating System. In *27th Hawaii International Conference on System Sciences* (Jan. 1994).

15. MADHYASTHA, T. M., AND REED, D. A. A Framework for Sonification Design. In *Data Sonification*, G. Kramer, Ed. Addison-Wesley Publishing Company, 1994.

16. MADHYASTHA, T. M., AND REED, D. A. Data Sonification: Do You See What I Hear? *IEEE Software* (*submitted for publication* 1994).

17. REED, D. A. Performance Instrumentation Techniques for Parallel Systems. In *Models and Techniques for Performance Evaluation of Computer and Communications Systems*, L. Donatiello and R. Nelson, Eds. Springer-Verlag Lecture Notes in Computer Science, 1993.

18. REED, D. A., AYDT, R. A., MADHYASTHA, T. M., NOE, R. J., SHIELDS, K. A., AND SCHWARTZ, B. W. An Overview of the Pablo Performance Analysis environment. Tech. rep., University of Illinois at Urbana-Champaign, Department of Computer Science, Sept. 1993.

19. REED, D. A., AYDT, R. A., NOE, R. J., ROTH, P. C., SHIELDS, K. A., SCHWARTZ, B. W., AND TAVERA, L. F. Scalable Performance Analysis: The Pablo Performance Analysis Environment. In *Proceedings of the Scalable Parallel Libraries Conference*, A. Skjellum, Ed. IEEE Computer Society, 1993.

20. REED, D. A., OLSON, R. D., AYDT, R. A., MADHYASTHA, T. M., BIRKETT, T., JENSEN, D. W., NAZIEF, B. A. A., AND TOTTY, B. K. Scalable Performance Environments for Parallel Systems. In *Proceedings of the Sixth Distributed Memory Computing Conference* (1991), IEEE Computer Society Press.

21. REED, D. A., AND RUDOLPH, D. C. Experiences with Hypercube Operating System Instrumentation. *International Journal of High-Speed Computing* (Dec. 1989), 517–542.

22. RIES, B., ANDERSON, R., AULD, W., BREAZEAL, D., CALLAGHAN, K., RICHARDS, E., AND SMITH, W. The Paragon Performance Monitoring Environment. In *Proceedings of Supercomputing '93* (Nov. 1993), pp. 850–859.

23. SNODGRASS, R. A Relational Approach to Monitoring Complex Systems. *ACM Transactions on Computer Systems 6*, 2 (May 1988), 157–196.

Performance Experiences of the Barcelona Olympic Games Computer System

Antoni Carmona[1], Luis Domingo[2], Rafael Macau[3], Ramon Puigjaner[4], Facundo Rojo[5]

Abstract. This paper presents the performance works done before the Olympic Games of Barcelona'92 in order to correctly size the computer system. Also it presents some measurements taken during the Games and the comparison between the estimation and the reality.

1 Introduction

If anybody looks at this paper with the hope to find some new development or some beautiful theory, he or she can stop immediately. This paper presents the performance works done before and during the Olympic Games of Barcelona'92 in order to correctly size and use the computer system It summarises in few pages much dark work done by the authors and many other people involved in any way in the project.

To understand the problems surrounding the organisation of an Olympic Games in their computer aspect, sections 2 and 3 give a survey of the special characteristics of this project.

Section 4 exposes in a chronological perspective the set of works done to diminish the risk of a bad computer sizing. It shows how step by step the uncertainty was reduced but never eliminated.

Section 5 gives some significant results of the computer system performance during the Olympic Games, before to go to section 6 in which some comparisons between estimations and realities are done. Finally section 7 shortly concludes.

2 Technological Environment of the Olympic Games

Many people looking at the great pyramid of Keops are asking themselves: How will the old Egyptians be able to build it without cranes, bulldozers and all the current building machines?. Similarly, most of the participants and visitors of the last Olympic Games are asking how the organisers succeeded when the satellite

1. Antoni Carmona particpated at the project of the Olympic Games through the agreement between the University of the Baleatica Islands and the COOB'92 developing his Final Project to get the License in Informatics. His current position is: Departament d'Informàtica. Parlament de les Illes Balears. Palau Reial 14. 07001 PALMA (Spain). Phone: +34-71-727334. Fax: +34-71-718201.
2. Luis Domingo particpated at the project of the Olympic Games assigned as member of the IBM staff. His current position is: Consultant of the IBM Consulting Group. IBM, SAE. Via Augusta 212. 08021 BARCELONA (Spain). Phone: +34-3-4018000. Fax: +34-3-2091116.
3. Rafael Macau particpated at the project of the Olympic Games as member of the COOB'92 staff. His current position is: ERITEL, SA. Via Augusta 200. 08021 BARCELONA (Spain). Phone: +34-3-2008533. Fax: +34-3-2008143.
4. Ramon Puigjaner particpated at the project of the Olympic Games through the agreement between the University of the Baleatica Islands and the COOB'92 as Advisor in sizing and performance topics. His current position is: Universitat de les Illes Balears. Carretera de Valldemossa km. 7.6. 07071 PALMA (Spain). Phone: +34-71-173288. Fax: +34-71-173003. E-mail: dmirpt0@ps.uib.es
5. Facundo Rojo particpated at the project of the Olympic Games as member of the COOB'92 staff. His current position is: SELESTA, SA. París 50. 08029 BARCELONA (Spain). Phone: +34-3-4107200. Fax: +34-3-4395180.

communications, the computers and other high technology equipment were not available.

It is a fact that all the Olympic Games of the old Greece and more the two thirds of the modern ones were successfully organised without the help of the electronic technology of the information. What was so different? What did change after the first half of this century?.

In that time there were less sportive disciplines than today and were fewer participants and countries. As the Games had approximately the same duration than today (from 16 to 18 days) or even more, the number of matches and sportive events was much less and the number of sportive venues was also much less. The sports requiring much computation to prepare the classifications (like gymnastics, ice skating, diving jump, etc.) gave the results several hours after the end of the competition . This time was devoted to the computation and verification of the results; the decision of the Judges was normally announced the day after and nobody was impatient except, perhaps, the eventual winners.

There were few journalists following the Games and most of them were happy if their reports, sent on the telephone or telex, reached their countries for the regular issue of the next day. However each time more information concerning the results was asked and the telecommunication users were transformed in radio reporters (since the Games of Amsterdam in 1928) needing the results information in "real-time" and direct lines to transmit "live" their comments.

However the listeners were still unable to directly verify (in a visual form) the rapidity and accuracy of the listened data, easing the task of reporters and organisers.

However in the early sixties the change started. First of all, more and more International sportive Federations joined the Olympic Games. Also new competitions were added to traditional Olympic sports, like athletics, gymnastics or swimming, and women's competitions were added to several sports. In addition many new countries, mainly from Africa and Asia joined the International Olympic Committee. Both trends contributed to a large increase of the program of the Olympics. More sports, more countries and more participants needed more sportive venues and more events to decide the winners.

Secondly the early sixties saw revolutionary improvements in the TV and telecommunication technologies mainly by the introduction of the satellite communications allowing the image transmission from continent to continent. The TV sets became cheapest and more people were able to buy them. The Olympic Games became a universal and popular event and millions and even million million of people all around the world were switching on their TV sets during two weeks. The TV live coverage needs of the Olympic events influenced their organisation. The structure of the competition program had to be changed in order to get more interesting TV programs. The result process and presentation in true real time became an imperative requirement in order to offer intermediate classifications during the event and to announce the final results and the winners before closing the live TV transmissions. The increasing interest of the public opinion around the Olympic Games attracted more writing journalists. The concurrence to the TV coverage forced to improve the timing and the content or the reports. The journalists became more exigent asking that the results were distributed just few minutes later than the end of the competitions, asking direct telephone lines from the report seats and from the press centres, etc.

Thirdly and finally, the rules and techniques to determine many results (by measuring, timing or counting) were change to be improved. So, the photo-finish (to decide the winners), the electronic timing (to be able to detect hundredths of second instead of tenths), etc. have become indispensable in the competitions. The sports in which the results are computed from the referee's evaluation have increased the number of referees and are using more complex algorithms to ensure the impartiality and to avoid the eventual preferences for some athletes. All these aspects have converted the processes of computing and getting the results and the classifications in more exigent and complex.

The increasing size of the Olympic Games has also made more exigent and more complex the organisational and logistic tasks (reception, accreditation, transportation, housing, ticketing, access control, etc.). It is also necessary to remind that the Olympic Games are organised each four year by a different town and country. This implies that the team in charge of this task, make this work just once in his life without having the possibility to test and repeat complete operations in a real environment. The organisation of the Barcelona Olympic Games required around 40000 people. From them, around 5000 people were related to technological (computer, communications, TV, electronics, etc.) activities.

General management	50
Local management	300
Technical staff	1070
Operators	2600
External services	1500
TOTAL	5520

2.1 Computer system

Concerning the computer system, the Barcelona Olympic Games had a system composed schematically by:
- One central computer IBM 9021-720 with six processors giving 120 Mips and a peripheral equipment composed by five disk subsystems with control units allowing four simultaneous transfers and around sixty disks able to contain mirror copies of all the critical files of the applications.
- One back-up central computer sharing the peripheral equipment with the previous one.
- One mirror back-up computer centre placed in a different location.
- 4600 intelligent terminals (IBM PSs) distributed over more than 100 different locations; most of them were in Barcelona but some of them were quite far from Barcelona (even to more than 300 Km), and half of them were used in a client-server architecture.
- 1000 stand-alone intelligent terminals.
- The basic software of this system was the typical one of the IBM systems: MVS/ESA as operating system, CICS as transaction manager and DB2 (version 2.2) as data base manager.

2.2 Application software

In order to avoid data duplicities (mainly concerning the participant's identification) all the applications shared the same data base, whose size was around 2.5 Gbytes (plus an on line copy of another 2.5 Gbytes). However only a 20% of these data will be of critical importance during the games.

SIFO (Olympic Family Information System). This application was able to be accessed from more than 2000 PSs (some of them used at different periods by SIGO applications) located at all the Olympic installations (sportive venues, hotels, Olympic villages, airports, hospitals, operation management facilities, etc.) to give information to all people (participants, trainers, journalists, IOC, NOCs, etc.) concerned in any way with the Games. It was composed of two main applications:
- Electronic mail
- Access to the information stored in the data base, mainly results, event's calendar and participant biographies.

This application was of free access (according to their accreditation's) to all people concerned by the games. In consequence the estimations said that it would generate most of the computer workload during the games. Later this application was renamed as AMIC (friend, in Catalan).

SIGO (Operations Management Information System). This application was able to be accessed from 500 PS devices (some of them used at different periods by SIFO applications) operated by COOB people and located at all the Olympic installations (sportive venues, hotels, Olympic villages, airports, hospitals, operations management facilities, etc.) to allow organising committee people to do their work. Main aspects concerned by this application were:
- accreditation of all people concerned by the games
- reception and leaving of participants
- housing
- transportation
- etc.

The estimations were that its highest load would be during the days before the games and during the games its load should be low and non coincident in time with peak loads of other applications.

SIR (Results Information System). This application was mainly a data entry of all the results and information produced during the celebration of the sportive events in order to furnish information:
- to update the data base in the central system
- to distribute written information concerning the results to the journalists
- to allow that SICO application gives on line information to journalists.

The information generated at each venue passed to the SICO LAN of the same venue and went to the central system in order to update the central data base, to be broadcast to all other active (according the calendar) SIR LANs and to be printed at IBC (television) and MPC (journalists). It was served by 500 PS devices located at sportive venues and operated by COOB people. The estimations were that its load would not be very high but it would be very critical, specially in response time.

SICO (Journalists Information System). This application was intended to give real time information to journalists about the development of some sportive events. It was served by 1100 PS devices with a very simple keyboard allowing just to change the information displayed among several pre-established tables. It allowed to connect with SIFO application in order to ask about the biographies of participants of the events that were currently being active.

It received information from SIR and sent information to the central system in order to update the central data base and to be broadcast to all other active venues, IBC and MPC.

2.3 Communication system

All the venues were connected to the computer system by means of two different communication lines (either private or leased) of different capacity (from 9600 bps to 2 Mbps) depending on the traffic generated at each location.

The basic communication structure of each sportive venue was composed by a duplicated token-ring LAN connecting al the terminals assigned to each application (SIFO, SIGO, SIR and SICO). All these LANs were connected to a duplicated backbone token-ring LAN in charge to transmit the information to the lines connecting the venue to the computer system (figure 1).

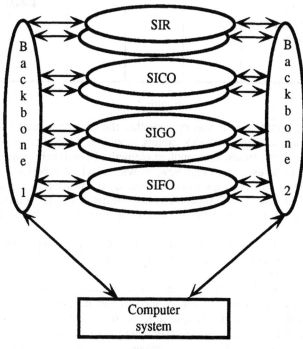

Figure 1

3 Another important problem (still one more)

This complex system was developed from scratch in Barcelona'92. The experiences of the previous Olympic Games were used just as prototypes because the functional requirements of the users and because the technological differences due to the accelerated development of the computer and communication systems during the eighties.

One of the more complex problems of the organisation of the Barcelona Olympic Games was to size the computer system and the communication network to be used by unknown people using non existing applications.
The scheduling of the global computer and communication system project (figure 2) shows clearly the problem.

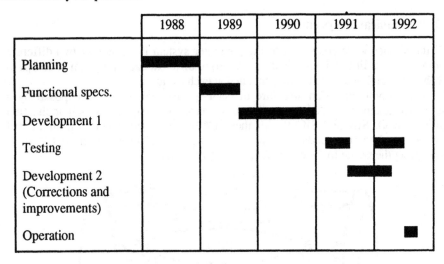

	1988	1989	1990	1991	1992
Planning					
Functional specs.					
Development 1					
Testing					
Development 2 (Corrections and improvements)					
Operation					

Figure 2

This scheduling shows that it was evident that the computer installation and the computer installation and the network cabling should be started before the end of the software development. In addition, another element of complexity was that some of the components to be used were very new, so with a lack of experience in their use. Also, the testing in 1991 was not exactly a testing; it consisted in the simultaneous organisation of 13 World or European championships of several sports. Obviously, the service level should be at the normal high level of this kind of events. In conclusion:
1. The process of configuration and sizing the computer system and the communications network should be started in parallel with the application software development.
2. The installation of the computer system and the communication network should be started before the end of the application software development.
3. This tight time scheduling (impossible to be modified) did not allow any misconception error. There were several no return decision points without the possibility of any error. There was a project of four and a half years with more then 500 technical people involved in which an 1% of deviation meant to start the operation of the system after the end of the Olympic Games.

4. How the problem was tackled?

Two basic strategies and a complementary one were chosen:
- Basic strategies:
    * Overlapping of activities
    * Step by step solution refinement

- Complementary strategy:
  * Oversizing level greater than the normal in a project of this dimension.

In consequence, the sizing of the computer system and of the communication network was started at the same time than the definition of the functional specifications. The installation of those elements also started before the end of the first phase of the application software development. At each step the tools used were appropriate at the information available and the quality of the results was according the reliability of the input data. Also very frequently the reasonable oversizing was decided based on worst case studies.

4.1 1988: Planning

The goals of this period were:
- To budget the computer system and the communication network taking into account that this budget would not be increased (actually it did not increase).
- To establish a first draft of the general architecture of the computer and communication system.
- To set a first general command to the suppliers.

The technique used was purely conceptual and mainly qualitative. A reduced technical staff analysed four different sources of information:
- The functionality offered in the last two summer Olympic Games (Los Angeles'84 and Seoul'88).
- The computer and communications infrastructure used in Los Angeles'84 and Seoul'88 and the one used in the winter Olympic Games of Calgary'88. Remark: The statistics of these Olympic Games were very useful to set a correct approximation.
- The changes in the information technology field during the last years.
- The functionality and the service level to be offered in Barcelona'92.

The general architecture of the computer and communication system for Barcelona'92 issued from these analyses. It was the conceptual frame to be respected by the hardware and software developers.

Among the decisions taken in this phase, some of the more important were:
- a unique and centralised computer system linked to all the venues by an appropriate communication network taking into account the fact that most of the event was located in Barcelona or close to it. The distributed system technology did not seem enough reliable.
- a unique data base for all the applications in order to ease the information entry procedure and to avoid bad replications.

4.2 1989/1990: Early development

The goal of this phase was to transform the general frame in a detailed project allowing to start the installation of the computer and communications infrastructure but without closing completely their volumes.

To attain this goal the development team start to define the functionality of the different applications. With this information it was possible to refine the architecture defined in the previous phase.

However, from the performance view point an important problem was detected: It was quite easy to build a model of the computer system, but what would be the workload of the system?

Both problems were tackled separately.

Performance estimation: Even with a rough estimation of the workload, a performance model of the real time aspects of the computer system was started. Its main hypotheses were:

Hardware configuration:
- A biprocessor of 15 MIPS each one
- Two disk configurations; their difference was the number of disks (2 or 10) devoted to the files with information to be accessed by SIFO.
- Disk access time independent of the load and not considering the connection path (control units, channels, etc.).
- CPU path length proportional per application to the number of logical I/O accesses plus and estimated overhead.

Workload:
- SIFO, SIGO and SICO were represented as conversational workloads defined by their number of terminals, think time distribution, number and distribution of logical I/O accesses, distribution of the disk time accesses, number and distribution of instructions per logical I/O and estimated overhead.
- Taking into account the locality of accesses of the SIFO queries to the data base it was scanned for this application several probabilities to find the desired information in the main memory.
- SIR was represented as throughput workload defined by its arrival rate distribution, think time distribution, number and distribution of logical I/O accesses, distribution of the disk time accesses, number and distribution of instructions per logical I/O and estimated overhead.

Model: A central server simulation model was built using QNAP2 [1].

Conclusions: From the model result analysis the following conclusions were extracted:
- To obtain a correct performance of the system and in particular of the SIFO application, its information ought to be split in, at least, 10 disks.
- Taking into account the amount of information to be stored for the SIFO application it was not reasonable to make this distribution in more than 2 disks. In consequence it ought to be necessary:
  * to have a large main memory in order to increase the probability to find in this support the desired information without the need to access the disk.
  * to design the data base in such a way that the information was able to remain in main memory after its use if there was free space in it or even it could be fixed in main memory.

This model was improved several times during 1990 and the beginning of 1991 to take into account the information coming from the application design mainly from SIFO, the most critical one from the performance view point. The main refinement was to detail the estimated dialogues (in number of interactions and exploring the effect of different think times) between the SIFO user and the computer system

considering also the fact that some interactions would be local (between the user and the PS). The conclusions were roughly the same:
- criticality of SIFO
- need of a large main memory

Moreover, the range of estimated think times had a minor influence on the performance.

These studies were based on many hypotheses (some of them, as we will see later, were far from the final realities); however they had, as consequence, to detect the main critical points of the system and to suggest appropriate policies to obtain the desired performance.

Workload estimation: In the 1990 second quarter the need of verifying the communication network sizing appears. It was not easy to imagine the traffic across the LANs and MANs if the timing of the different applications ought to be taken into account. So the first objective was to know the amount of traffic, expressed in bytes per second, flowing through the communications (LANs and lines between every venue, either sportive or operational), at any time during the Games. A load simulator was built to make these calculations under different assumptions with the characteristic that it should run on a compatible PC in order to allow different components of the COOB staff to use it and obtaining the results in a reasonably short computing time.

Taking into account the number of sports (around 30), venues (around 300), sportive events (close to 2000) and the use characteristics of query terminals, it was decided to develop this simulator in DBASE IV recently appeared at that time. The general structure of the load simulator is shown in figure 3.

Besides the load generation procedure, the analyser should be able to give enough information to compute the expected mean response time in every computer LAN of any venue. For this analysis, the method of Bux [2, 3] was used.

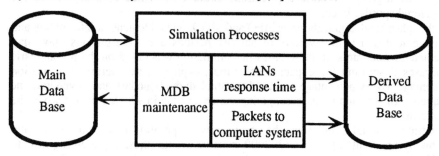

Events calendar	Daily traffic chronograms
Venues	Global traffic chronograms
Sports	Packet extraction intervals
Sport frequencies	Detailed traffic to the
Terminal user expected profile tables	computer system
Computing parameters	LAN response times

Figure 3

Finally, in 1991 this load generator was expanded to generate all the information related to every expected packet sent to the computer system, to validate the load hypotheses used to size it. This process was too heavy for the platform (hardware and software) used and it was no more possible to get the results in real-time but in deferred time (batch processing). Perhaps if this need was known at the beginning another platform would have been chosen because the final version of this simulator exhausted the DBASE IV available memory.

All the information related to the Games, the input data to the analyser, was stored on the Main Data Base, MDB. All input parameters were able to be modified. Any change over the predefined sport events, dates and timetables, workload and user expected profiles or any other parameters were easy to modify normally by means of browse menus in order to simulate a new condition. The main file, the event calendar, was joined with assistant tools to help the user in the task of events editing, minimising the possibility to input incoherent events.

The results of every simulation were consistently stored to allow their later study without the need of repeating the same simulation in the Derived Data Base, DDB.

Because of the significant application behaviour differences, each one of them had a particular computing process to calculate the input and output traffic of each venue. The analyser gave separate chronograms of every application, but there was an option to accumulate the traffic generated by each application to obtain the total traffic over the venue lines (figure 4). The accuracy period could be adjusted between five and 20 minutes in order to shadow peak loads or to reduce the simulation time.

The study of applications related to sportive events as SIR and SICO was driven by the begin and end time of the competition in that place. An iterative process based on the sportive event calendar, decomposed it in predefined subevents, each one with a particular information generation profile and specified frequencies. This hierarchical structure finished at packet level, giving the time in which a packet was sent to the line. Chronograms of these applications could be oriented to a day, or to the considered sportive event.

On the other hand, applications as SIFO and SIGO, related to general information and management and with a variety of activity periods along the day just gave day oriented chronograms. In these cases the simulation was based on the probabilistic processes describing the expected behaviour, expressed as a terminal utilisation percentage. Every percentage was used to establish the boundary and to generate the frequency of the random process, giving the exact time in which a packet would be sent to the line.

In a traffic global simulation, the analyser was operating over 300 venues and following an event calendar with 2000 sportive events (those representing near to 40.000 individual elements) and probabilistic processes with close to 100 different distribution time profiles. Global extraction spent more than 10 hours to compute all the results creating a 40 Mbytes file in the DDB. Obviously, the large amount of output data obliged to a late collapsing process to allow result's readability (figure 5). Moreover, to simplify their study, the user could choose from one minute to several hours of the final report.

The results of all these simulations allowed to improve the configuration of both computer system and communication network. So, the works to install and to arrange the computer centre started in mid 1990. Approximately at the same date the installation of the communication lines belonging to the COOB also started.

Resum de la càrrega d'una línia Venue Host per a totes les Aplicacions

Codi lloc	1				Data	Hora	
Nom Lloc	Estadi Olímpic de Montjuic				Codi Joc --	Interval	5

Data	Hora	CD	»»»	««	X:»»»	-:««
01/08/92	08:00		17	171	--	
01/08/92	08:05		18	187	--	
01/08/92	08:10		20	206	--	
01/08/92	08:15		22	229	---	
01/08/92	08:20		25	257	---	
01/08/92	08:25		29	294	---	
01/08/92	08:30		33	343	----	
01/08/92	08:35		40	411	-----	
01/08/92	08:40		50	514	x-----	
01/08/92	08:45		67	686	x-------	
01/08/92	08:50		100	1028	x-----------	
01/08/92	08:55		167	1740	xx------------------	
01/08/92	09:00		206	2608	xx----------------------------	
01/08/92	09:05		206	2179	xx-----------------------	
01/08/92	09:10		199	2148	xx---------------------	
01/08/92	09:15		199	2148	xx---------------------	
01/08/92	09:20		206	2303	xx------------------------	
01/08/92	09:25		199	2248	xx-----------------------	
01/08/92	09:30		212	1691	xx-----------------	
01/08/92	09:35		164	1540	xx---------------	
01/08/92	09:40		164	1688	xx-----------------	
01/08/92	09:45		161	1637	xx----------------	
01/08/92	09:50		171	1619	xx----------------	
01/08/92	09:55		198	1605	xx----------------	
01/08/92	10:00		262	2661	xxx------------------------	
01/08/92	10:05		221	2062	xxx--------------------	
01/08/92	10:10		255	2027	xxx--------------------	
01/08/92	10:15		214	2003	xx--------------------	
01/08/92	10:20		631	2498	xxxxxx---------------------	
01/08/92	10:25		165	2298	xx-----------------------	
01/08/92	10:30		202	2454	xx------------------------	
01/08/92	10:35		154	1920	xx--------------------	
01/08/92	10:40		154	2140	xx----------------------	
01/08/92	10:45		216	2161	xx----------------------	
01/08/92	10:50		281	3019	xxx------------------------------	
01/08/92	10:55		185	1667	xx-----------------	
01/08/92	11:00		240	2505	xxx------------------------	
01/08/92	11:05		199	1835	xx-------------------	
01/08/92	11:10		233	1800	xxx-----------------	
01/08/92	11:15		192	2170	xx----------------------	
01/08/92	11:20		199	2063	xx--------------------	
01/08/92	11:25		397	1626	xxxx--------------	
01/08/92	11:30		264	2989	xxx----------------------------	
01/08/92	11:35		421	1595	xxxxx-------------	
01/08/92	11:40		192	1595	xx----------------	
01/08/92	11:45		230	1664	xxx----------------	
01/08/92	11:50		199	1711	xx----------------	
01/08/92	11:55		390	1650	xxxx--------------	
01/08/92	12:00		161	3567	xx------------------------------------	

Polsi S per sortir, o qualsevol tecla per continuar

Figure 4. Day oriented chronogram

LLISTAT GLOBAL DE TRAMES D'ENTRADA AL HOST. COOB 92   10/02/92   Pàgina   1
                                                         Total registres   176419

(R:SIR / O:SICO / F:SIFO / G:SIGO)					
Codi Aplicació	Data interval	Hora interval	Bytes Longitud Trama	Nombre Trames Rebudes	Missat/seg. KBytes/seg.
F	25/07/92	09:00	———		
			— 200 B	172.5	0.048 Mis/s
					0.009 KB/s
F	25/07/92	10:00	———		
			— 200 B	154.5	0.043 Mis/s
					0.008 KB/s
F	25/07/92	11:00	———		
			— 200 B	154.5	0.043 Mis/s
					0.008 KB/s
		. . .			
G	15/08/92	09:00	———		
			— 1024 B	225.8	0.063 Mis/s
			— 2048 B	539.8	0.150 Mis/s
					0.363 KB/s
G	15/08/92	10:00	———		
			– 1024 B	225.8	0.063 Mis/s
			– 2048 B	539.8	0.150 Mis/s
					0.363 KB/s
G	15/08/92	11:00	———		
			— 1024 B	225.8	0.063 Mis/s
			— 2048 B	539.8	0.150 Mis/s
					0.363 KB/s
		. . .			
O	30/07/92	19:00	———		
			— 150 B	720.0	0.200 Mis/s
			— 300 B	1359.6	0.378 Mis/s
					0.140 KB/s
O	30/07/92	20:00	———		
			— 150 B	1110.0	0.308 Mis/s
			— 300 B	3062.3	0.851 Mis/s
					0.294 KB/s
O	30/07/92	21:00	———		
			— 150 B	970.0	0.269 Mis/s
			— 300 B	3308.3	0.919 Mis/s
					0.309 KB/s
		. . .			
R	01/08/92	09:30	———		
			— 150 B	5951.9	1.653 Mis/s
			— 5120 B	5.5	0.002 Mis/s
			— 7168 B	2.2	0.001 Mis/s
					0.254 KB/s
R	01/08/92	10:00	———		
			— 150 B	10111.6	2.809 Mis/s
			— 5120 B	124.1	0.034 Mis/s
			— 7168 B	101.8	0.028 Mis/s
					0.782 KB/s
R	01/08/92	11:00	———		
			— 150 B	6565.1	1.824 Mis/s
			— 5120 B	82.2	0.023 Mis/s
			— 7168 B	65.3	0.018 Mis/s
					0.508 KB/s

Figure 5. Global packet extraction report

4.3 January-August 1991: Early testing

The goals of this period were:
- To test and verify the installations.
- To refine the planning of the pending installations.
- To verify and test the developed application software.

These goals were attained by the organisation of a test with the current version of the application software. So, during approximately two weeks in Summer 1991 a test was organised with a third of the sports and a hundredth of the communication resources and people involved. However, it was of the highest level because the computer system was giving service to 13 World or European championships of several sports. It was mainly a functional test of the application software and of the computer and communication infrastructure, but the information taken from it was used for the final sizing in the next phase.

To get information on what was happening inside the computer system to be used for the final sizing, a set of measuring tools was installed:
- CMF [4] to get information about the behaviour of the CICS.
- RMF [5] to get information about the resource utilisation.
- SLR [6] to get information about the general behaviour of the computer system.
- OMEGAMON for DB2 [7] to get information about the data base behaviour.

4.4 September 1.991 - June 1.992: Final development and testing

The goals of this phase were:
- To define exactly the needs of computer system and communication networks:
- To install this infrastructure
- To implement and test the application software using this infrastructure.

To attain these goals the following techniques were used:
- Conventional capacity planning to size the CPU needs.
- TPNS [8] simulation to test and tune the system performance with all the applications in execution and also to test the functionality and integrity of the system applications under heavy load.
- Real tests of the behaviour of specific installations.
- Analytical simulation to test the system performance and different disk organisations under different workload scenarios.
- General rehearsals using a mixed workload of real users and TPNS simulated load.

Capacity planning. From the information collected during the test of the previous phase the final sizing of CPU needs was done using the USAGE (Understanding System Applications and Growth Environment). It consists of the following steps;
- To identify the workload applications.
- To compute the transaction throughput of each application during the peak hour.
- To compute the CPU consumption per each one of the application transactions.
- To multiply the previous values to obtain the CPU consumption per application.
- To add the CPU consumption of all the applications and increase it by the estimated overhead.
- To size the computer system in such a way that the CPU utilisation is not greater than 80% (in order to not to degrade the response time).

As on August 1 between 20:00 and 22:00 there were the maximum number of active terminals due to the maximum number of active sports, this period was chosen as the peak load.

Application identification. During the daily service (from 7:00 to 24:00) the on line applications can be grouped as follows:
- SIFO or AMIC was a client-server type application with centralised data base and dialogues prepared in PSs. It should be used by participants, trainers, coaches, journalists, Olympic officers and COOB staff. Its main subapplications were:
  * Queries about event calendar, results, medals, participants, biographies and general information.
  * Electronic mail and information boards per user groups.
  * Biography interface with SICO.
- SIGO was an application with centralised data base and data presentation driven by the host. It should be used by the operative COOB staff. Its main subapplications were:
  * Accreditation of the Olympic family, COOB staff, volunteers, etc.
  * Olympic family housing.
  * Transportation between Olympic villages and venues.
- SIR was an application with data bases distributed in the LANs and consolidated in the host. It should be used by the COOB staff. Its main subapplications were:
  * Result data base updating (to be consulted by AMIC).
  * Reception of result pages and their distribution.
- Batch applications: The only critical batch process during the day hours is the route generation of the transportation application.

Throughput estimation.
- SIFO or AMIC: The throughput estimation was based in two sources. Considering the number of attendees and the number of terminals, it was considered that the number of queries should be 2.5 the number of queries in Seoul'88. The value obtained was in accordance with one come from the load simulator. The following estimations were done for the three components of this set of applications:
  * Queries: The Seoul'88 extrapolation gave 27720 queries during the peak hour equivalent to 7.7 queries/s. Each query required in average 3 interaction with the host with a 40 s interval between them. The transaction throughput should be 7.7 * 3 = 23.1 tr/s. This throughput should require 23.1 tr/s * 40 s/tr = 920 terminals.
  * Mail: The Seoul'88 extrapolation predicted 5843 functions in the peak hour equivalent to 1.62 funct/s. The number of generated transactions was variable depending on the function type (read or send). The measurements of July 1991 suggested an average of 2.3 tr/funct with a mean interval of 50 s. In consequence the throughput should be 1.62 * 2.3 = 3.7 tr/s generated from 3.7 tr/s * 50 s/tr = 188 terminals. The estimated total number of terminals was 920 + 188 = 1108 according with the value obtained by other estimations.
  * Biography questions from SICO: During the peak hour there were an estimation of 600 journalist open terminals. Estimating a query per terminal each 10 minutes and that each query generates just one transaction, the estimated throughput should be 600 term * 1 tr / (10 min * 60 s/min) = 1 tr/s.
- SIGO: The throughput estimation was based in the analysis of the number of terminals devoted to its applications during the peak hour and of the characteristics

of each application. The following estimations were done for the two main components of this set of applications:

* Accreditation: This application consisted in three parts: Reception, Accreditation and Reclamation. However as the peak load of this application was estimated for the days before the Games, the workload used for this capacity planning was the one estimated for the considered peak hour of August 1; in consequence it was necessary to consider the estimated activity at that hour. Reception: 10 terminals * 6 funct/min * 1 tr/funct * 0.1 / 60 s/min = 0.1 tr/s. Accreditation: 47 terminals * 1 funct/1.5 min * 3 tr/funct * 0.6 / 60 s/min = 0.94 tr/s. Reclamation: 27 terminals * 1 funct/2.5 min * 3 tr/funct * 0.3 / 60 s/min = 0.32 tr/s.

* Transportation: This application consisted in one part on line and another of critical batch (concurrent with the real time applications). The estimated throughput of the on line part of the application was 11 terminals * 4 tr/min / 60 s/min = 0.73 tr/s. The critical batch to be executed at the peak hour had to be completed in 40 minutes.

The characteristics of other SIGO applications were bad known at that time.

- SIR: The peak load of this set of applications was the considered one. The volumes generated by the 220 sportive events at the peak hour were:

* Information reception: 660 input messages/hour from reception of results; 440 pages/ to be printed at the EPH (Electronic Page Holder) and 220 structures/hour to update the data base.

* Information broadcast: 440 pages/hour to be distributed to a great number of receptors estimating in average 12540 pages/hour.

CPU consumption estimation. The measurement of the CPU consumption per transaction was done in September 1991 by means of the monitors CMF [4] for the CICS, OMEGAMON [7] for the DB2 and RMF [5] for the MVS. The CPU consumption per transaction was the addition of the corresponding consumption given by CMF and OMEGAMON and the overhead by means of the RMF. The estimated capacity of each one of the six processors of the IBM 9021-720 was 120 Mips.

- SIFO or AMIC: The measured values were:
* Queries: 0.94 millions of instructions per transaction (MI/tr).
* Mail: 3.43 MI/tr.
* Biography queries from SICO: 1.1 MI/tr.

In consequence the estimated total CPU consumption for this set of applications was:

Queries	23.1 tr/s * 0.94 MI/tr = 21.7 Mips
Mail	3.7 tr/s * 3.43 MI/tr = 12.7 Mips
SICO	1.0 tr/s * 1.10 MI/tr = 1.1 Mips
Total AMIC	27.8 tr/s * 1.30 MI/tr = 35.5 Mips

- SIGO: The measured values were:
* Accreditation: Reception: 4.6 MI/tr; Accreditation: 6.6 MI/tr; Reclamation: 6.6 MI/tr
* Transportation: On line: 2.0 MI/tr. Batch: 420 s of CPU.

In consequence the estimated total CPU consumption for this set of applications was:

Reception	0.10 tr/s * 4.6 MI/tr =	0.5 Mips
Accreditation	0.94 tr/s * 6.6 MI/tr =	6.2 Mips
Reclamation	0.32 tr/s * 6.6 MI/tr =	2.1 Mips
On line transportation	0.73 tr/s * 2.0 MI/tr =	1.5 Mips
Batch transportation	420 s / (40 min * 60 s/min) * 22 Mips =	3.9 Mips
Total SIGO		14.2 Mips

- SIR: The measured values were:
  * Information reception: 1.1 MI/input message; 1.8 MI/input page; 4.4 MI/updating structure.
  * Information broadcast: 2.6 MI/distributed page; 3.5 MI/distributed copy.

In consequence the estimated total CPU consumption for this set of applications was:

Input message	660 mess/h / 3600 s/h * 1.1 MI/mess =	0.20 Mips
Input page	440 mess/h / 3600 s/h * 1.8 MI/mess =	0.22 Mips
Updating structure	220 mess/h / 3600 s/h * 4.4 MI/mess =	0.27 Mips
Distributed page	440 mess/h / 3600 s/h * 2.6 MI/mess =	0.32 Mips
Distributed copy	12540 mess/h / 3600 s/h * 3.5 MI/mess =	12.19 Mips
Total SIR transactions		13.20 Mips
Internal SIR overhead	13.2 Mips * 0.08 =	1.06 Mips
Total SIR		14.26 Mips

Total CPU consumption. In addition to the direct transactional and batch loads there were other CPU consuming tasks (network services, spool, monitoring, etc.). During the test of July 1991 it was observed that this overhead consisted in two parts:
- a constant amount of 2.4 Mips
- a variable quantity approximately equal to the 40% of the CPU devoted to CICS (included DB2), batch and TSO applications.

It was assumed that during the Games the overhead would keep the same structure.
In consequence the total need of CPU during the peak hour would be:

AMIC	35.5 Mips
SIGO	14.2 Mips
SIR	14.3 Mips
Subtotal	64.0 Mips
Fixed overhead	2.4 Mips
Variable overhead 0.4 * 64.0 =	25.6 Mips
Total CPU consumption	92.0 Mips

The CPU utilisation should be no greater than 80%. In consequence the computer system should be able to give 92 / 0.8 = 115 Mips. As the IBM 9021-720 with six processors was able to give 120 Mips, the choice seemed reasonable.

TPNS simulation. To get information about the behaviour of the system when the applications were used by number of terminals similar to one of the peak hour, several simulations using TPNS [8] were done. Their goals were from the performance view point:
- To test the system performance.
- To do load tests during several hours to know the appropriate size of the files whose size grows with the activity (logs, insertions, etc.).
- To determine the exact amount of needed main memory in the computer system.

- To tune the system choosing the appropriate parameters and file distribution on the disks.
- To establish the appropriate number of buffers for each file.
- To obtain the transaction profile (CPU consumption, file and data base calls, disk accesses, etc.) in order to observe the effect of the changes in programs, operating system parameters, file distribution, etc.

and from the functional view point to detect the inter transaction locking in order to correct them or to modify the data base index implementation. This last point allowed to detect and correct application or DB2 misconceptions, that appear only when the system was running under heavy load of all the applications simultaneously.

TPNS description. The Terminal Program Network Simulator is a software allowing the simulation of a set of terminals connected locally or by means of some teleprocessing network. It sends messages to the host applications with the same timing as they were real ones. In its best way of working it needs a computer to simulate the terminal network in addition to the one that is executing the load.

A set of scenarios can be defined corresponding to different script (dialogue sequence between the terminal and the host). A terminal can execute randomly a set of scenarios.

Script structuring. The dialogue between a human driven terminal and an application can be traced, formatted and filed in a library by means of a TPNS utility. For each interaction this library element consists of the input and output messages in order to allow the later testing by the TPNS of the correct execution of the application program. Two kinds of scripts were used:
- The lineal script: The script so generated is directly executable. However it always accesses the same date during the execution. Normally it is more interesting to randomly access the records as it happens in the structured scripts.
- The structured script: It is a lineal script in which some fields of the message sent to the host are replaced by variables. Each time that a message is sent, the variables are replaced by values taken randomly or sequentially from a table of allowed values. This process is easy to set in the first interaction of the dialogue. However in the following ones it is necessary to take into account the previous interaction in order to keep the coherence of the data. This process had a variable difficulty depending on the characteristics of the application. For example:
  * If the next query is composed by several fields of the previous answer, to structure it is very easy as it happened in the SIGO application of accreditation.
  * If the answer has a simple repetitive structure, as for example an item list, and one of them is randomly chosen to create the next query, it is also easy to structure the script. This was the case of the AMIC application of electronic mail.
  * If the preparation of the next query requires a complex logic because the terminal is intelligent or because it is used by a person, the script structuring is not so easy. However it can be described by means of the TPNS language. This was the case of the AMIC application to query the information stored in the data base.

The scenario definition was based on the information given the load simulator modified by the experience obtained during the events of July 1991.

Real tests of the behaviour of specific installations. All the installations were tested for both verifying the hardware and software functionality, specially when there was some geographical constraint, and analysing the behaviour of all the elements. In this way some minor deficiencies were corrected.

Analytical simulation. The information obtained from the TPNS was very useful. However, it is a heavy tool and it was just possible to test the apparently more critical combinations of scenarios and configurations (mainly due to the disk space allocation of the files whose space increased with the time). To diminish the risk of a disagreeable surprise on the system performance (mainly response time) an exhaustive test of all the combinations was done by means of an analytical model built in QNAP2. This model with the general structure of a central server had two submodels. The first one was an open BCMP model with interest in the CPU behaviour (represented by a processor sharing server); the disks were represented by FIFO exponential queues with load independent service time. The second one was an open diffusion model in which the disk path contention (computed in a one-step iteration by the model itself) was taken into account in the disk behaviour (represented by a FIFO queue with hyperexponential service time with load dependent mean value); the CPU was represented also by a FIFO queue.
The data used for these simulations were coherent with those of the TPNS simulations. The results for those cases studied by both tools were reasonably similar. Hundreds of cases were executed and no inconvenient situation was discovered.

General rehearsals. During the two months immediately before the Olympic Games two general rehearsals were organised. Their goal was to test the system functionality, the organisation behaviour (people directly concerned with the use of critical equipment, as for example those in charge to enter the results, participated in these tests) and the functionality of the back-up systems (simulating incidents of different gravity levels). The workload of these tests was partly generated from the human driven terminals and partly simulate by means of the TPNS to attain the level estimated during the Games.

5 July-August 1992: Operation

The computer system run without major problems during its life period, the two weeks before the Games and two weeks during the Games. In order to get information about what was happening inside the system the same set of monitors was connected permanently. Also they were used to detect any kind of problem and, if it was possible, to correct it. Fortunately, or perhaps due the good work done, just some minor functional and performance problems appear. The information obtained by the set of mentioned performance measurement tools allowed to detect these problems that were easily corrected.
Figures 6, 7 and 8 show some of the performance measurements of the computer system during the Olympic Games.

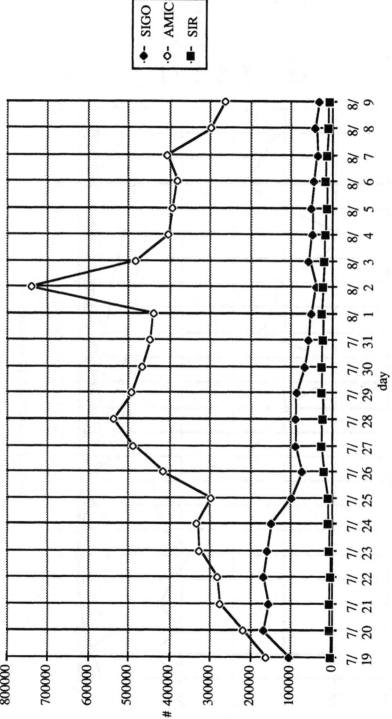

Figure 6. Number of transactions per day

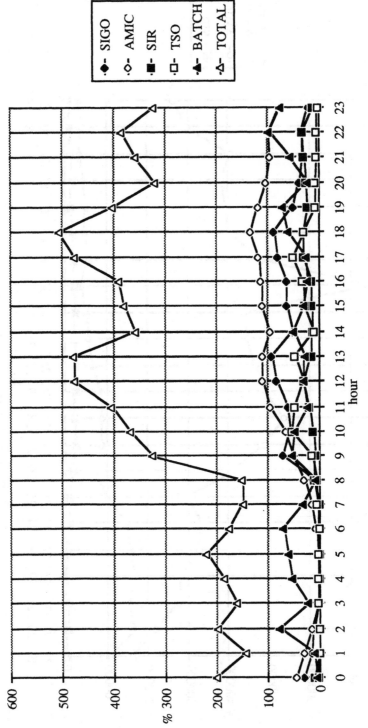

Figure 7. CPU consumption on July 29

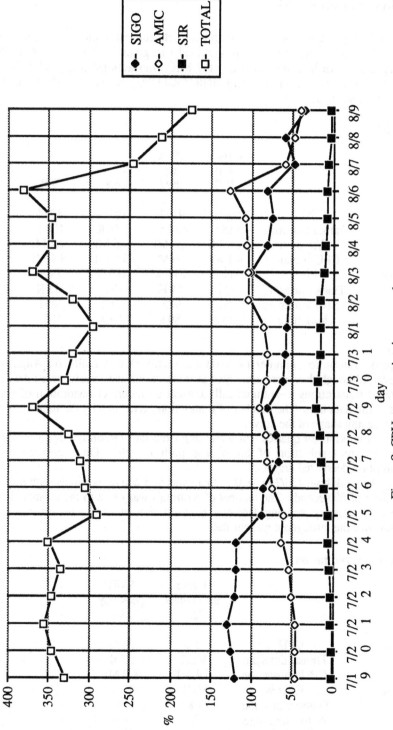

Figure 8. CPU consumption in user tasks

6 Reality vs. estimation

In the following sections some comparisons between estimation and reality are exposed. Considering that the typical workload was given by AMIC, the peak day was July 28, according the number of AMIC transactions executed (536882 transactions) or July 29 according the AMNIC total CPU consumtion (91%). The last one has been chosen as peak day.

6.1 AMIC

		Games start 26/7	Peak day 29/7	6/8	7/8
Average	CPU %	73 %	91 %	128 %	57 %
	Transactions/day	415 K	474 K	381 K	408 K
	CPU/tr queries	1.5 MI	2.0 MI	4.6 MI	1.1 MI
	CPU/tr mail	4.1 MI	3.6 MI	3.5 MI	3.4 MI
Peak hour	CPU %	117 %	134 %	189 %	74 %
	Transactions/h	32 K	44 K	30 K	33 K
	Active terminals	786	985	712	952
	Think time	88 s	80 s	87 s	104 s

Remarks.
- The peak transaction number (738703 transactions) was really on August 2. However, although they were accounted as transactions, they really were internally generated transactions to automatically correct an incorrect restoration of some tables in the data base. The number of transactions generated by the user followed the decreasing shape of the curve.
- CPU consumption for queries was increasing from the start until August 6, due to some incorrect selection of the access path in DB2. Once corrected the consumption decreased drastically.
- During the peak hour (28/7 at 18:00), there was 985 active terminals with a mean think time of 80 seconds. The number of terminals was close to the estimated one in the CPU capacity planning. However the think time was twice the estimated; in consequence the throughput was the half.

6.2 SIGO: Accreditation

		Before the Games 22/7	AMIC peak day 28/7
Average	CPU %	43 %	24 %
	Transactions/day	82 K	37 K
	CPU/transaction	6.4 MI	8.0 MI
Peak hour	CPU %	80 %	41 %
	Transactions/hour	7 K	4 K
	Active terminals	73	44
	Think time	45 s	83 s

6.3 SIGO: Others

		Before the Games 22/7	AMIC peak day 28/7
Average	CPU %	33 %	29 %
	Transactions/day	65 K	42 K
	CPU/transaction	6.2 MI	8.5 MI
Peak hour	CPU %	50 %	52 %

6.4 SIR

		Games start 26/7	AMIC peak day 28/7
Average	CPU %	11 %	20 %
	Input documents/day	2 K	4 K
	Distributed documents/day	50 K	82 K
	CPU/transaction	2.6 MI	2.8 MI
Peak hour	CPU %	19 %	34 %

6.5 Comparison

A comparison of the CPU consumption in Mips for the peak day appears in the following table:

	Planned	Reality
AMIC	35.9	26.8
SIGO Accreditation	8.8	7.4
SIGO others	1.5	10.2
SIR	14.3	6.0
CICS terminals		3.8
Total real time	60.1	3.8
Batch	3.9	12.0
TSO		6.2
MVS services	28.0	28.8
Total CPU	92.0	101.2

Remarks:
- The planned total of the real time load is quite close to the reality. However, the underestimation due to the lack of information about the rest of the SIGO applications and the CICS consumption controlling terminals was compensated by the less consumption of AMIC and SIR.
- The estimation on the MVS services (overhead) was perfectly acceptable.
- Nobody expected a such high consumption of TSO and batch. However as the system was giving good performance the policy concerning the use of these workloads was relaxed.

7 Conclusion

Although any estimations were not completely accurate, the project of the Barcelona Olympic Games computer system can be considered as successful by its performance, by the respect of the time scheduling and, what is very surprising, by the respect of the budget.

From the performance point of view, its success was the consequence of the cooperation of the simulation, modelling and measurement tools used along the project.

Bibliography

[1] Simulog: "QNAP2" Reference Manual Version 9.0

[2] Bux, W; Truong, H. L.: "Mean Delay Approximation for Cyclic Service Systems". Performance Evaluation. August 1983.

[3] Bux, W.: "Token-Ring Area network and tjeir Performance". Proceedings of the IEEE, 77-2. February 1989.

[4] IBM: "CICS Monitoring Facility". CICS/ESA V.3. Customisation Guide: SC33-0665.

[5] IBM: "Resource Measurement Facility". MVS/ESA RMF V.4. General Information: GC28-1028.

[6] IBM: "Service Level Reporter". SLR V.3. General Information GH19-6529.

[7] CANDLE: "OMEGAMON for DB2". Reference Manual.

[8] IBM: "TPNS. Teleprocessing Network Simulator". TPNS V.3. General Information: GH20-2487.

Performance Tools on Intel Scalable High Performance Computing Systems[1]

Thomas Bemmerl[2] and Bernhard Ries

Intel European Supercomputer Development Center (ESDC)
Dornacher Str. 1, 85622 Feldkirchen, FRG, mail: {thomas, bernhard}@esdc.intel.com

Abstract. A massive breakthrough of scalable high performance computing (SHPC) systems depends strongly on the ease of programming. Programming environments have to be provided wich enable the user to program and use SHPC systems in an efficient and transparent manner. A special challenge within these parallel programming environments are performance tuning tools. The paper gives an industrial perspective on the development of performance tools for SHPC systems with a focus on the ParAide programming environment for Intel's Paragon (tm) supercomputer. An integrated approach that provides an extensible and consistent programming environment for monitoring and visualizing the performance of parallel applications is presented. From the experiences made in the development of the Paragon performance toolset, future research directions are derived.

1   Motivation

Many compute intensive applications relevant for the human society require computing performance far beyond the performance of today's fastest supercomputers. These fields of applications include, but are not restricted to, fluid dynamics, weather forecast, image processing, design and simulation, databases and artificial intelligence. Modern scalable high-performance computing systems provide the high potential for sustained performance that is needed to solve these Grand Challenge computing problems. The Paragon architecture is already used in several projects to solve Grand Challange projects. Paragon systems are also used for early application demonstrators in Intel's latest research, development and demonstration program toward TeraFLOPS performance.

To fully understand the performance behavior of SHPC systems and to operate the machines near the high end of their performance range, new performance data-collection, analysis and visualization tools are needed. Some of the most challenging problems that have to be solved by integrated performance environments are:

---

1. Parts of this paper have already been published in [8].
2. Thomas Bemmerl also directs the Chair for Operating Systems at RWTH Aachen University of Technology, e-mail: thomas@lfbs.rwth-aachen.de

- **System and Application Performance:** Different categories of users require different types of performance tools. System administrators are interested in observing the overall behavior of the system while application programmers need to focus on a specific application run. Thus, there is a need for both system and application performance analysis tools.

- **Massive Amounts of Data:** Monitoring the performance of massively parallel systems or applications can lead to very large amounts of performance data. Different techniques must be devised for the performance environment to be able to handle or reduce the amount of data that needs to be processed.

- **Multiprogramming:** The current trend towards multiprogramming in operating systems for scalable high-performance computers introduces new problems for application performance tools. In order to be an aid in tuning applications the tools must be able to distinguish system effects from application bottlenecks.

- **Perturbation and Intrusion:** Monitoring the performance of parallel applications necessarily introduces perturbation. The performance environment must strive to minimize the amount of intrusion and should try to filter out monitoring effects or provide the user with a measure of the intrusiveness of the measurements.

- **Flexibility:** The diversity of programming models available on scalable high-performance computing systems and the lack of programming model standards make it necessary for the tools to be highly flexible. Tools should be able to present the performance data in the context of the programming model being used but it should not be necessary to design a completely new performance environment for every new programming model.

This paper describes how these inherent problems are addressed in the integrated performance monitoring tools of the ParAide programming environment developed by Intel and initially targeted for use on Intel Paragon systems. It consists of an instrumentation front-end, a distributed monitoring system and a set of graphical and command-oriented tools that can be used to analyze performance data. On the basis of this implementation future research on integrated tool environments is conducted within Intel's TeraFLOPS research, development and demonstration program. Potential areas of research on performance tools are discussed at the end of the paper.

2 Methodology and Basic Research

Most state-of-the-art performance tuning tools for SHPC systems are based on the concept of interactive parallelization. With this interactive parallelization methodology, the user can experiment easily with different versions of his parallel program. The interactive performance tools give the programmer the opportunity to switch quickly between parallelization strategies and evaluate the quality and efficiency of different versions of the codes parallelized.

The field of performance monitoring and performance visualization for parallel machines has received a lot of attention lately. This has led to a large number of monitoring environments with different capabilities. The Pablo [10] project has implemented a system in which performance tools are built from individual modules that can be easily interconnected and reconfigured. The Pablo team has also developed a self-documenting trace-format that includes internal definitions of data types, sizes and names. The TOPSYS environment [3] is based on a distributed monitoring system that is used by multiple tools, including an on-line performance visualizer. The IPS-2 system [11] takes a hierarchical approach to performance visualization and integrates both system and application based metrics. ParaGraph [6] is a tool that provides a large number of different displays for visualizing the performance of parallel application. Intel has integrate some results of these research projects into it's performance monitoring environment and will likely continue to exploit the results of this research.

3 The Paragon System

3.1 Hardware Architecture

The Paragon [4] hardware architecture consists of processing nodes based on the Intel i860XP microprocessor. Each processing node has one i860 (tm) XP application processor, a second i860XP used for message-protocol processing and a Network Interface Controller (NIC) that connects the node to the two-dimensional interconnection network. Each node board also contains hardware support for performance monitoring in the form of a special daughterboard that provides hardware counters. In addition, the daughterboard implements a 56-bit global counter that is driven from the global 10 MHz clock from the backplane.

3.2 Operating System and Programming Model

The operating system and programming model supported by the target architecture have to be taken into account during the design of performance monitoring tools since the performance tools should be able to present the performance data in the context of the programming model being used.

The Paragon OSF*/1 AD operating system [12] is based on the Mach 3.0 microkernel technology and an extended OSF*/1 server. Each node in the system runs a microkernel that provides core operating system functions. Other operating system services are handled by the OSF/1 server. The operating system provides a single system image with standard UNIX functionality. Multiprogramming is fully supported by the operating system - multiple users may log on to the machine and run applications at any time.

Partitions with different scheduling characteristics can be defined within the Paragon architecture. The service partition normally uses conventional UNIX scheduling with dynamic load balancing. Applications in the compute partitions are scheduled according to a modified Gang scheduling scheme.

The current Paragon implementation supports as it's primary programming model a message passing model that is compatible with the NX/2 programming model available on previous generation Intel machines [2].

4 Performance Monitoring Architecture

4.1 The Components

Figure 1 gives an overview of the Paragon Performance Monitoring Architecture. It consists of the following parts:

- The Tools Graphical Interface (TGI) is used as a front-end for instrumenting the parallel application and to visualize the performance data produced by the performance tools.

- The Instrumentation Front End is used to parse user requests for performance instrumentation of a parallel application. It generates the commands that are then sent to the distributed monitoring system through an RPC interface.

- The Distributed Monitoring System consists of the Tools Application Monitor (TAM), the performance library that is linked with the application program, and the event trace servers. The TAM is responsible for instrumenting the application with jumps to the performance libraries. The libraries produce trace data that are sent to the event trace servers. The servers perform post-processing tasks and write the performance data to file or send them directly to the application performance analysis tools.

- The application performance analysis tools interpret the trace data they receive from the event trace servers. Currently, three tools have been implemented. A modified version of ParaGraph allows detailed graphical animation of processor activity and message passing. Profiling tools (prof and gprof) can be used to examine performance at the statement and procedure level.

- The System Performance Visualizer (SPV) is designed to monitor the overall system usage by displaying parameters such as CPU utilization, bus usage and network traffic. Since this tool is not application oriented, there is no need for instrumentation. The data are partly generated by the hardware performance monitor, partly by the Mach kernel.

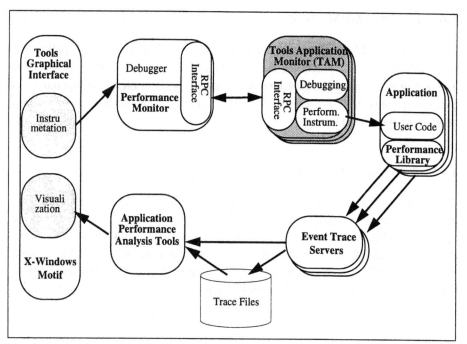

Fig. 1. Performance Monitoring Architecture

4.2 Common Graphical Interface

The Tools Graphical Interface serves two purposes. On the one hand, it is used for loading, starting and instrumenting an application. On the other hand, it provides a common interface through which the user can operate the performance tools and display performance data. The fact that all tools share a common user interface that is based on the Motif style guide standard makes the functionality of the tool environment easily and obviously available to the user.

4.3 Instrumentation Front End

The instrumentation front end parses commands it receives from the graphical or command-line interface and generates the commands that are then sent to the distributed monitoring system to instrument the application. In the first implementation, instrumentation is restricted to specifying a monitoring context and a tool (prof, gprof, ParaGraph) for which instrumentation is to be performed.

4.4   Distributed Monitor

To overcome the difficulties associated with monitoring massively parallel applications with large numbers of processes, the Paragon performance monitoring environment uses a distributed monitoring system that is able to perform high-level operations in parallel. It consists of a network of TAM processes, a performance library and a network of event trace servers.

4.4.1   Tools Application Monitor

The Tools Application Monitor is made up of a network of TAM processes arranged in the form of a broadcast spanning tree with one TAM process for every node used by the application. This makes it possible to broadcast monitoring requests to all nodes in an efficient manner. The instrumentation front-end communicates with the TAM network through an RPC interface based on the Mach Interface Generator (MIG). Communication within the TAM network also occurs by message-passing over Mach ports.

4.4.2   Performance Library

To instrument an application for performance measurements, the TAM modifies the application code to vector to a performance library. To minimize the data collection perturbation, the performance library is automatically linked with every application so that event traces can be gathered without incurring the overhead of context switches. The performance data that can be captured include event traces and counters as well as profiling information.

4.4.3   Event Trace Server

The Event Trace Servers perform post-processing tasks on the performance data they receive from the performance library. One event trace server services a number of performance libraries. The exact ratio of servers to performance libraries is configurable and depends on the type of performance measurement being performed. In the normal case, the output of the event trace servers consists of one or many trace files that can be interpreted by the application performance analysis tools. Plans also exist for having the trace servers may forward their output directly to a tool for on-line visualization of an application's performance. The Paragon performance monitoring environment supports event traces in the Pablo Self-Defining Data Format[10]. The SDDF approach standarizes the interfaces to the trace file not the physical trace file itself.

4.5 Application Performance Analysis Tools

The application performance analysis tools interpret the performance data gathered by the distributed monitoring system and report them back to the user. They either have a command-line interface that can be invoked from the graphical interface or directly use the functionality of the tools graphical interface to present graphical output. Currently, three application performance tools have been implemented.

4.5.1 ParaGraph

ParaGraph [6] is one of the most widely used tools for analyzing the performance of parallel applications. It can be used in a post-mortem fashion to visualize the performance data contained in a trace file generated during an application run. ParaGraph was originally written for use with the PICL Portable Instrumented Communication Library which runs on a variety of message passing parallel architectures. In the mean time, many groups have adapted ParaGraph to other programming models and architectures. The widespread use of ParaGraph was the prime motivation for providing the tool as part of the Paragon performance monitoring environment.

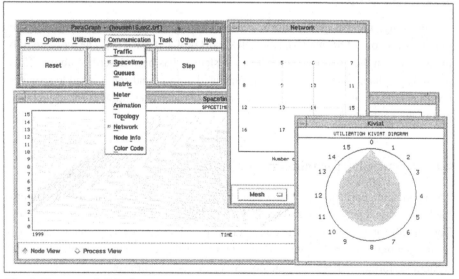

Fig. 2. Paragraph Displays

ParaGraph is a graphical tool that provides a variety of displays to visualize the performance of a parallel application. The user can choose as many displays as will fit on the screen from three different types of displays. The utilization displays are concerned primarily with depicting processor utilization, communication displays can be used to visualize interprocessor communication, and the task displays provide a way of relating the events in the trace to source locations in the parallel programs. Figure 2 shows some of the displays generated by ParaGraph for a Householder transformation algorithm executed on 16 nodes. The general functionality of the tool has been left unchanged and will not be described any further. However, a number of important changes and enhancements have been made to the tool.

- ParaGraph has been integrated into the Tools Graphical Interface. Thus, the tool now has a standardized Motif interface that is easy to learn and consistent with the rest of the Paragon tool environment.

- ParaGraph now supports the major features of the Paragon/NX programming model such as synchronous, asynchronous and interrupt-driven message passing and probe operations.

- The instrumentation front-end allows the user to specify a monitoring context in an on-line fashion. This makes it possible to focus on subsets of nodes and parts of an application's code without the need to recompile.

- ParaGraph now supports traces in the Pablo Self-Defining Data Format [10] instead of the PICL trace format.

4.5.2 Prof and Gprof

For profiling, the Paragon performance monitoring environment provides slightly modified versions of the well-known UNIX tools prof and gprof [5]. The tools are fully integrated with the distributed monitoring system as has been described above. Thus, the tools graphical interface and the instrumentation front-end can be used to specify a profiling context in the usual way, which results in instrumentation of the appropriate part of the application.

4.6   System Performance Visualization

The application performance analysis tools described above allow the user to monitor the behavior and performance of parallel applications at different levels of granularity. They try to filter system effects out as far as possible. However, in many cases it is useful to monitor overall system usage without focusing on a specific application. The System Performance Visualization tool (SPV) serves this purpose.

SPV is a graphical tool that allows the user to visualize the Paragon front panel on a workstation but it also provides zooming capabilities that make it possible to display more detailed information such as CPU, mesh and memory bus utilization values. Colors that can be configured according to the user's wishes are used to represent different utilization values. Figure 3 shows some of the displays supported by SPV. The following displays are available:

- The CPU display shows the CPU utilization of all the nodes. Mesh utilization is not displayed.

- The Mesh display is a visualization of the Paragon front panel which shows both CPU and mesh utilization.

- The Values display adds the actual utilization values to the Mesh display.

- The Node display provides the greatest level of detail. It shows symbolic representations of the Paragon node boards (CPU, message passing processor, memory, bus and Network Interface Controller) and the mesh and displays the utilization values for all components.

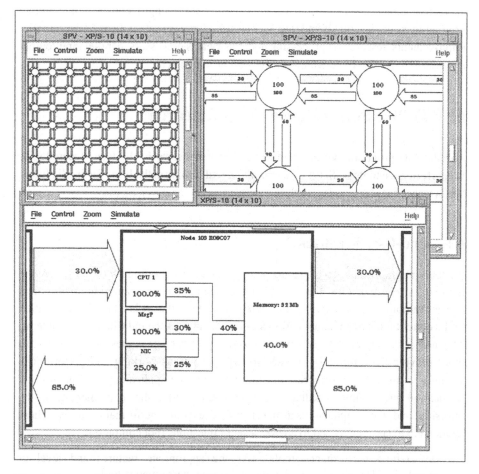

Fig. 3. System Performance Visualization

5   The TeraFLOPS Program

The TeraFLOPS technology program is a collaborative effort between Intel Corp. and the Advanced Research Project Agency ARPA. The primary objective of the technology program is: Develop early in the second half of this decade the technolygy that will deliver sustained one TeraFLOPS SHPC systems at the price of today's vector supercomputers.

The program will produce subsequent generations of system architectures demonstrated in prototype systems between now and 1996. On the basis of the success of the Touchstone program which delivered the technology to build machines with hundreds of GFLOPS, the Paragon architecture will be used as the starting point for many new research activities. The research agenda of the new program follows an evolutionary top-down R&D methodology. The necessary R&D activities include work-load characterizations of Grand Challenge problems, improvements of parallel programming environments and the OSF*/1 AD operating system. In addition compiler improvements, research on processing nodes as well as processors, parallel input/output and research in interconnection networks will be conducted.

6   Future Performance Tools Research

The ParAide programming environment incorporates some of the latest technologies for performance tools on SHPC systems and it will also form the basis for Intel's future performance tools research.

6.1   End-User Feedback

The current graphical state-of-the-art perfomance tools are offered for the first time in industrial strength quality. The ParAide integrated parallel programming environment is one example. The next step in improving the state of the art is a detailed evaluation of the current tools by real application users on the basis of Grand Challenge problems. This necessity may be derived from application programmer quotes as "Current tools don't provide the information needed to get applications working and performing". Therefore a major area of research in the future will be application experiments and application programmer feedback on the functionality, the implementation, the intrusion and the graphical user interfaces of existing performance tools. These application experiments need to answer a set of questions (not complete):

- Do exsiting performance tools offer the appropriate features and level of granularity? Do we need application and system performance tools?

- Are available (mostly graphical user interfaces) the appropriate man-machine interface?

- How strongly should the various performance tools be integrated with other tools of a parallel programming environment? What is the priority of performance tools compared to other programming tools?

6.2 Missing Technology Base

Another characteristic of the current state-of-the-art tool environments for SHPC systems is the missing technology base. In particular performance tools depend on:

- The rapid evolution of new programming models, as there are message passing models like NX/2, PVM, PARMACS, various flavours of data parallel programming models (Fortran D, HPF, Vienna Fortran, etc.) and virtual shared memory programming models. The current tools in most cases only concentrate on one of these programming models, e.g. the ParAide environment on NX/2 message passing

- The rapid evolution of the underlying architecture, i.e. the processing node models and operating system structures. No common interfaces between the various system components are yet defined, which makes the adaptation of the performance tools to new architectural changes difficult and resource intensive.

- The ever increasing number of processing elements in future SHPC systems. Therefore scalability is an ever important issue. While current performance tools scale up to hundreds of processing nodes, the tools of future generation machines need to scale up to thousands of nodes. This requires improvements on the data capturing, data compression and the display side.

6.3 Potential Research Areas

These problems of integrated performance tools due to the missing technology base and the missing end-user feed back need to be addressed in future research. Areas that will be investigated in the research agenda of the TeraFLOPS program may include the following:

- **New tools:** The design of the performance monitoring architecture makes it relatively easy to add new tools to the environment. We plan to add tools that help relate performance bottlenecks to locations in the application program. In many cases, simple table of numbers may be more effective than sophisticated graphical displays. Thus, a good candidate for addition to the tool environment is a critical path analyzer such as the one implemented within the IPS-2 system [11]. As new programming models emerge, the tools will have to be modified to support these programming models. Only small adjustements of the current toolset have to be made to support implementations of PVM, PARMACS, MPI. For support of non-message passing programming models like HPF and shared virtual memory, more basic redesigns of the current environments are needed. Different display and instrumentation techniques are required to support shared memory programming. In addition, higher level abstraction techniques in the performance tools will be needed to support the data parallel directives and data distributions in HPF.

- **Basis for Tool Development:** It is not feasible to cover all the needs users have for performance tools in one single, but very complex tool. The Pablo system [10] is very flexible in that it allows tools to be configured out of existing modules. We investigate to integrate parts of our tool environment with the techniques proposed and demonstrated by the Pablo software. This will allow users to add new tools and displays to the existing performance monitoring environment. It will also make it possible to run different modules of the performance analysis tools in parallel to fully utilize the power of the parallel machine for performance analysis. Selective instrumentation techniques can then be used to reduce the amount of performance data generated by tracing only the information needed for the displays the user has selected.

- **Tool Infrastructure:** We also plan to further modularize the tool environment. Through this concept a smooth evolution of the integrated tool environment on the basis of evolving architectural concepts as well as evolving programming models will be possible. Primary objective of this work is to reduce the complexitiy of the tools environment, in particular when adaptation to new programming models and architectures is a strong requirement. Currently, some of the features needed for performance analysis are implemented within the debugger (e.g. instrumentation, symbol table lookup). It would be desirable to implement these parts as separate modules that can be joined to form either performance or debugging tools. This example demonstrates, that an object oriented structure of the integrated tool environment with clearly separated functionalities of each component will contribute most to an advanced tool infrastructure concept. Modules like parallel software monitor, symbol table management, debugging engine, event trace engine, profiling engine, source browser have to be defined. The interfaces, data structures and information flow between these components has to be specified. Concepts known from sequential toolsets like Fuse, PCTE, ToolTalk will be investigated and evaluated.

Paragon and i860 are trademarks of Intel Corporation. Other products mentioned are the trademarks or registered trademarks of the manufacturers or marketers of the products with which the marks are associated.

7   References

[1]   D. Breazeal, R. Anderson, W.D. Smith, W. Auld, and K. Callaghan. *A Parallel Software Monitor for Debugging and Performance Tools on Distributed Memory Multicomputers.* In *Proceedings of the Supercomputer Debugging Workshop 1992*, Dallas, Texas, October 1992. Los Alamos National Laboratory.

[2]   D. Breazeal, K. Callaghan, and W.D. Smith. *IPD: A Debugger for Parallel Heterogeneous Systems.* In *Proceedings of the ACM/ONR Workshop on Parallel and Distributed Debugging*, pages 216-218, Santa Cruz, CA, May 1991.

[3]   T. Bemmerl. *The TOPSYS Architecture.* In H.Burkhart, editor, *CONPAR 90 - VAPP IV*, volume 457 of *Lecture Notes in Computer Science*, pages 732-743. Springer-Verlag, Berlin, Heidelberg, New York, September 1990.

[4]   Intel Supercomputer Systems Division. *Paragon XP/S Product Overview.* Intel Corporation, 15201 N.W. Greenbrier Parkway, Beaverton OR 97006, 1992.

[5]   S.L. Graham, P.B. Kessler, and M.K. McKusick. *gprof: a Call Graph Execution Profiler.* *ACM SIGPLAN Notices*, 17(6):120-126, June 1982.

[6]   M.T. Heath and J.A. Etheridge. *Visualizing the Performance of Parallel Programs.* *IEEE Software*, 8(5):29-39, September 1991.

[7]   P. Pierce. *The NX/2 Operating System.* In *Proceedings of the 3rd Conference on Hypercube Concurrent Computers and Applications*, pages 384-391. ACM, 1988.

[8]   B. Ries, R. Anderson, W. Auld, D. Breazeal, K. Callagham, E. Richards, W. Smith. *The Paragon Performance Monitoring Environment.* In *Proceedings of Supercomputing '93*, pages 850 - 859, Nov. 1993, Portland.

[9]   D.T. Rover, M. B. Carter, and J. L. Gustafson. *Performance Visualization of SLALOM.* In *Proceedings of the Sixth Distributed Memory Computing Conference*, pages 543-550, Portland, Oregon, May 1991. IEEE, IEEE Computer Society Press.

[10]   D. A. Reed, R. D. Olson, R. A. Aydt, T. M. Madhyastha, T. Birkett, D. W. Jensen, B. A. Nazief, and B. K. Totty. *Scalable Performance Environments for Parallel Systems.* In *Proc. of the Sixth Distributed Memory Computing Conference*, pages 562-569, Portland,Ore, April 1991. IEEE.

[11]   C.Q. Yang and B.P. Miller. *Critical Path Analysis for the Execution of Parallel and Distributed Programs.* In *Proceedings of the 8th International Conference on Distributed Computing Systems*, pages 366-375, San Jose, CA, June 1988. IEEE.

[12]   R. Zajcew, P. Roy, D. Black, C. Peak, P. Guedes, B. Kemp, J. LoVerso, M. Leibensperger, M. Barnett, F. Rabii and D. Netterwala. *An OSF/1 UNIX for Massively Parallel Multicomputers.* In *Proceedings of the Winter 1993 USENIX Technical Conference*, pages 449-468, San Diego, CA, January 1993. The USENIX Association.

Modelling Aspects of Model-Based Dynamic QoS Management by the Performability Manager

Leonard J.N. Franken†, Raymond H. Pijpers†
Boudewijn R.H.M. Haverkort‡

†PTT Research
P.O. Box 15000, 9700 CD Groningen, The Netherlands
E-mail: {l.j.n.franken,r.h.pijpers}@research.ptt.nl
‡University of Twente,
Department of Computer Science Tele-Informatics and Open Systems
P.O. Box 217, 7500 AE Enschede, The Netherlands
E-mail: b.r.h.m.haverkort@cs.utwente.nl

Abstract. The *Performability Manager* (PM) is a distributed system component which maintains the application-requested Quality of Service (QoS) by dynamically reconfiguring ANSAware-based distributed applications, using a model-based optimization procedure. The PM receives information about the ANSAware-based application from a distributed monitoring process based on JEWEL and DEMON. With this information, and using predefined stochastic Petri net (SPN) models of ANSAware applications, the PM automatically constructs an overall SPN performability model which is subsequently used for the determination of the provided QoS. Based on the analysis results, the PM can decide to initiate on-line system reconfigurations, if such is needed for maintaining the requested QoS. ANSAware provides facilities for these dynamic reconfigurations.

In this paper we focus on the modelling aspects of model-based dynamic QoS management by the performability manager. We present an ANSAware-based experimental distributed environment in which the modelling and evaluation aspects are totally automated. We also show the feasibility of the proposed PM by presenting some operational results.

1   Introduction

For modern distributed systems it is important to be able to realize and maintain a requested Quality of Service (QoS). This QoS can degrade for several reasons: the addition of new applications, the updating of applications, the change of workload (new users) or the occurrence of failures and repairs.

QoS is difficult to define in general. Most importantly, it describes the *user-perceived performance* [22, 28, 34, 35]. The QoS can be divided in the subjective and the objective QoS. The subjective QoS is user oriented, and hard to quantify and measure. The objective QoS can be measured. We will always refer to the objective QoS. The objective QoS is related to or can be transformed into the subjective QoS, but this is not a one-to-one relation.

Service Performance Parameters (SPPs) is the generic term for provider visible performance parameters [8]. These are quantitative parameters which indicate how well the system (service) is performing. Between the objective QoS parameters and the SPPs there exists a one-to-one mapping [22, 28]. The SPPs can be measured at the service, and they ultimately determine the QoS, but they do not describe the QoS in a way that is meaningful to users (the subjective QoS).

As the QoS describes the user-perceived performance, the separate evaluation of performance, reliability and availability during system design, implementation and maintenance is not sufficient. The mutual influence of these aspects is recognized by the QoS and demands for modelling and evaluation techniques which can handle the combined aspects [24].

In [9] we introduced the *Performability Manager* (PM), a distributed system component which maintains the application-requested QoS by dynamically reconfiguring a distributed system, using a model-based optimization procedure. The PM does so by creating alternative configurations using (or manipulating) a graph-oriented model of the current configuration of the distributed system. The performability models of the alternative configurations are automatically generated from a graph model of the system obtained via DEMON [21], a library of predefined SPN model components, and parameterized via a monitoring process with JEWEL [19, 26]. By carefully selecting an alternative configuration, based on performability evaluation, an alternative (new) configuration can be decided upon. The alternative configuration can dynamically be effected using dynamic reconfiguration [18] as supported by the ANSAware computing platform underlying the distributed application.

For the operation of the performability manager, *model creation* and *evaluation* is both crucial and difficult. An overall performability model must be created, at run-time, out of model components. Apart from that, the performability manager must also be able to create alternative configurations for the current configuration. The performability manager uses performability analysis because that type of analysis provides us with the means to model and evaluate distributed systems with respect to their QoS [7, 9, 14, 23, 24].

In the realization of the PM the following questions arises:

1. how to detect QoS degradations of the current configuration;
2. how to create an alternative configuration;
3. how to determine the QoS of the alternative configuration.

These questions can be answered by either modelling (and evaluation) and/or monitoring. In this paper we will address both modelling (and evaluation) as well as monitoring aspects of *performability or QoS management* as done by the PM. The modelling aspects are important for configuration creation and configuration evaluation. Monitoring is important for parameterization of the models and to detect QoS degradations. We discuss these modelling and monitoring aspects by means of an experimental distributed environment realized using ANSAware.

This paper is further organized as follows. Section 2 presents work related to dynamic reconfiguration and performability management. In Section 3 we present the experimental environment and the ANSAware computational model of our application. Section 4 describes the graph-oriented modelling of distributed environments and shows how our experimental environment can be described using the proposed graph-oriented model. In Section 5 we present the creation of a performability model using predefined stochastic Petri net models of the system components and the graph-oriented model of the experimental environment. The monitoring of the experimental distributed environment is discussed in Section 6, whereas parameterization and first results on measurements are presented in Section 7. Finally, in Section 8, we discuss implementation and operational issues of the presented modelling techniques, discuss our ongoing research and set out lines for future research.

2 Related work

The performability manager maintains the required QoS by dynamic reconfigurations. This requires that facilities for dynamic reconfiguration should be available in the distributed system. Such facilities would include access, concurrency, federation, location, migration and replication transparency [27]. These facilities can be realized at several levels in a distributed system: at the operating system level, at the middle-ware level, think of computing platforms or configuration languages or at the application level itself. The current trend in distributed systems is to provide these facilities at the middle-ware level, i.e. by computing platforms. These platforms allow for heterogeneous distributed systems, which are transparent to the application programmer. Examples of such middle-ware facilities are the configuration languages Gerel, Conic, Argus, Rex, Darwin [18], and the computing platforms ANSAware [1] and DCE [29].

Computing platforms and configuration languages provide the user with the functionality for dynamic or/and static (re-)configurations. Examples of the use of these facilities for qualitative configuration management are for example described by Cole and Dean in [18]. Cristian presents in [18] an approach for a so-called availability manager, which guarantees the availability of the applications using replicated components. A performability manager extends this functionality by also addressing performance aspects.

Most of the effort in the area of dynamically reconfiguring distributed systems has been put in supplying facilities to perform the reconfiguration rather than on reconfiguration management to guarantee a desired level of QoS. The performability manager is therefore designed to guide reconfigurations by using a model-based optimization procedure. The goal of the reconfigurations is to maintain the QoS as requested by the application users. Performability evaluation will be used in the optimization procedure. A similar, but less general and less "automatic" approach towards resource control has been proposed by Lee and Shin [20, 32].

Further related to our work is the area of optimal system design [11, 15] and dynamic load balancing [3, 33]. Closely related is also the area of task allocation [31]. An example of this has been presented by Bowen [4] in which heuristic and linear-program solution for optimal process allocation in heterogeneous distributed systems are compared. Hariri [12] presents an algorithm which takes care of optimizing reliability and communication delay. The above approaches, however, are all focussed on single aspects of performance or reliability and not on their combination as we propose.

Another area related to our work is network management. From this area Kheradpir *et al* [16] proposes model based network management to manage the *end-to-end* network performance and robustness (dependability). They advocate a model based solution for future telecommunication systems to manage the QoS.

3 An ANSAware-based distributed environment

The application in our distributed environment is realised using ANSAware. In ANSAware the *computational model* is the model used for creation of applications and application components. This computational model is both object-oriented and client/server oriented. From this computational model we want to come to a *performability model*, which will be discussed in later sections.

In Section 3.1 we present a distributed application which will be used throughout this paper to clarify the different models. Section 3.2 describes ANSAware and the ANSAware computational model. Section 3.3 presents the experimental application using the ANSAware computational model.

3.1 An ANSAware-based number translation service

In this section we describe the (telephone) number translation service (NTS) as provided in intelligent networks [2, 10, 36]. In the sequel we will refer to this application as the IN/ANSA application.

End-Users are submitting requests or tasks for the application with a certain rate. Since we do not have real users, we mimic the user behaviour by a so-called *Generating Component* (GC). The GC generates the calls for the NTS. The NTS is provided by the following application components (see also Figure 1):

1. *The Selection Component* (SC): this component selects a service using the contents of the requests it receives (number translation service in this example).

2. *The Number Translation Component* (NTC): this component receives requests for number translations. The NTC sends a request to a database component for the required number and to a billing component for the creation of a bill. The number received from the database is returned to the SC.

3. *The DataBase Component* (DBC): this component receives requests for specific numbers. It will fetch the number from disk and return the number to the component which requested the number.

4. *The Billing Component* (BC): this billing component receives requests for the preparation of a billing record.

The *Management Component* (MC) does not belong specifically to the NTS, but provides the PM with the necessary "buttons to push" for performing a reconfiguration. The MC uses ANSAware facilities to perform necessary reconfigurations. The other components (SC, NTC, DBC and BC) are components of the application and can be controlled by the MC.

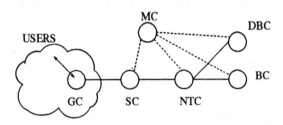

Fig. 1. The experimental ANSAware application

For the experimental application we use a small distributed system consisting of three SUN SPARC workstations connected by an Ethernet as depicted in Figure 2. The workstations run UNIX and, on top of that, ANSAware. Of course, more heterogeneous environments are possible as well, e.g.,using both SUNs and PCs.

application		
ansa	ansa	ansa
unix	unix	unix
sun 1	sun 2	sun 3
ethernet		

Fig. 2. The distributed system

Within this experimental distributed environment we use two monitors, DEMON and JEWEL. DEMON, the Distributed Environment MONitor [21], is used to visualize the structure of the experimental distributed environment. JEWEL [19] is used to do performance measurements in the experimental distributed environment.

3.2 ANSAware

ANSAware is a suite of programs which allows users to write applications suitable for heterogeneous distributed environments (see also Figure 2, although the possible heterogeneity is not directly apparent there). It is based on ISO-ODP, an emerging ISO standard for Open Distributed Processing [27]. ANSAware essentially consists of an infrastructure placed on top of the operating system, and provides a uniform, technology-independent platform upon which applications can be executed. The infrastructure permits interworking between applications running on remote and dissimilar machines. Several management applications, performing functions identified as important in ODP, are provided for the user's convenience. ANSAware provides a uniform view of a multi-vendor world, allowing system builders to link together components or existing software with minimal changes and overhead.

The basic building block of ANSAware is a *service*. Components that use a service are called *clients*. Components that provide a service are called *servers*. Services are provided at *interfaces*: an interface is a unit of service provisioning. This is also depicted in Figure 3. The ANSA computational model permits an object to be both client and server. A component or object described purely in terms of the way it provides and uses services, is referred to as a computational object. A client can invoke an operation or service at the interface of a server object in two different ways:

1. by interrogation, in which the invoking client waits for the server to perform the operation and return the result (similar to a remote procedure call);
2. by announcement, in which the invoking client does not wait for the server to perform the operation and no result is returned (remote process spawn).

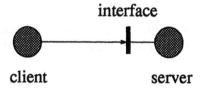

Fig. 3. A client object and a server object with its interface

The location of the computational objects or the type of machine they execute on can be changed at run-time: the ANSAware infrastructure enables a flexible configuration of application components and provides a uniform way of accessing them.

3.3 The computational model of IN/ANSA

In our experimental distributed environment the computational objects are the application components of the distributed system. One or more application components or computational objects make up a distributed application. This is depicted in Figure 4. Each computational object has been implemented as a process. All invocations for the experimental application are announcements, except for those between the NTC and the BC and those between the NTC and the DBC; these are interrogations.

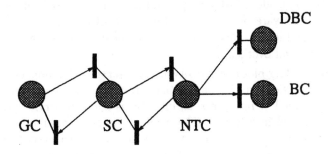

Fig. 4. The experimental application described in a computational form

4   A graph-oriented description of distributed environments

Because a reconfiguration must be performed with great care, the PM uses a model of the current distributed environment to prepare a reconfiguration. The performability manager uses this model to create alternative configurations and as a basis to derive a performability model.

Section 4.1 presents the graph-oriented approach to describe distributed environments. In Section 4.2 the IN/ANSA application is described using the graph-oriented approach.

4.1   Graph-oriented structure model of a distributed environment

In this section we propose a graph-oriented model suitable for the creation of alternative configurations and which allows for the transformation of these alternative configurations into performability models. For the graph-oriented model we look at a distributed environment at four levels (see also Figure 5) [9]:

| Task level | Tasks |
| Application level | Applications |

System level	System parts and network
Management level	Management

A similar distinction in levels has been proposed by Calzarossa *et al.* for workload modelling in network-based environments [5]. Each level will be described as consisting of components and their relations. These relations can be between the different levels as well as between the components at one level (mutual relations).

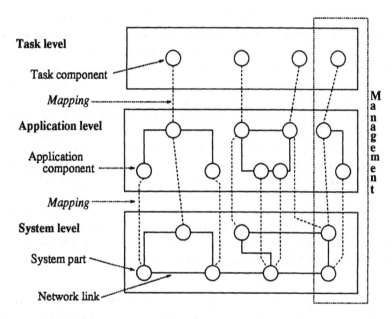

Fig. 5. A graph oriented model of a distributed environment

Task. At task level, tasks are viewed as *task components* t_d. A task is a certain amount of work, initiated by a user. For each application j we define the set of task components $T^j = \{t_1, \ldots, t_o\}$, $T^j \in TC$ with $TC = \{T^1, \ldots, T^n\}$ the set of all tasks for all applications. The task components have a relation with the application level. We assume that there are no mutual relations between task components.

Application. At application level, applications are viewed as composed of *application components* $a_i \in AC$, with $AC = \{a_1, \ldots, a_m\}$ the set of all application components. The application components have besides their relations to the task and system level, a mutual relation which represents the communication between the application components. Therefore, at application level we view an application as a structure of application components and their mutual relations (communication). We can describe a single application structure using a graph

$A^j = \langle VA^{\,j}, EA^{\,j} \rangle$ with $VA^{\,j} \subseteq AC$ and $EA^{\,j} \subseteq VA^{\,j} \times VA^{\,j}$. The complete set of (created) application structures which is active in the current distributed system is denoted by $\mathcal{A}$. Thus, we have $A^j \in \mathcal{A}$, and $\mathcal{A} = \{\langle VA^{\,j}, EA^{\,j} \rangle | j = 1, \cdots, n\}$, where n is the number of applications.

System. At system level we have a structure of the system level components, i.e. system parts $s_k \in SC$ with $SC = \{s_1, \ldots, s_w\}$ and the network links, $n_l \in NL$ and $NL = \{n_1, \ldots, n_t\}$ which provide the mutual relations between the system parts. We can describe the distributed system structure as a graph $S = \langle VS, ES \rangle$, with $VS \subseteq SC$ and $ES \subseteq VS \times VS$ and $F : ES \to NL$.

At the system level we also define *paths*. A path is a finite sequence of network links (arcs or edges) between any two system parts (nodes), i.e.,a finite sequence of links in which the terminal node (system part) of each link coincides with the initial node of the following link [6]. We define P as the set of all paths on S, p_{kl} is the set of paths between the system parts s_k, s_l and $p_{kl_v} \in p_{kl}$, where $p_{kl_v} = \{(s_k, s_h)(s_h, s_f) \ldots (s_g, s_l) | (s_k, s_h), \ldots, (s_g, s_l) \in ES \}$. Every path $p_{kl_v} \in P$ can be mapped on a set of links of S, by the function $F : P \to ES$.

The system parts have besides their mutual relation, a relation to the application level. The definition of a path will be used when assigning the communication between the application components to the system level. If no path exists between two system parts then these system parts are not connected.

Management. The management level is orthogonal to the other levels. It consists of management tasks, applications and systems, i.e.,the MC component and the monitors. They are structured as described above and are composed of components with the already mentioned relations and mutual relations.

Mapping. The *mapping* is the logical allocation of higher levels to lower levels, i.e.,the *allocation* of application components onto system parts and the *routing* of the communication over the network links. This is also reflected in Figure 5 where the mutual relations are represented by thick lines and the relations between levels (the mapping of the levels) by dotted lines. The allocation of tasks to application components is a special case of allocation. Are the other allocations subject to change, by migration etc., the task components are always allocated to the same application component of an application.

We can describe a mapping function M^j for each A^j. The allocation and routing of a mapping can be described in a more formal way. We will now elaborate more on the description of the allocation and routing.

Allocation. The allocation consists of two types of allocations: the allocation of task components to application components and the allocation of application components to system parts. We start with the former. We describe the allocation of task components to application components by the matrix $Z^j = T^j \times VA^{\,j}$:

$$z^j_{d,i} = \begin{cases} 1, & \text{if the } t_d \in T^j \text{ is allocated on } a_i \in VA^{\,j}, \\ 0, & \text{otherwise.} \end{cases}$$

For the allocation of application components $a_i \in A^j$ on system component $s_k \in S$ the mapping can be described by $ma^j : VA^j \to VS$. We can derive such a function for all $VA^j \in \mathcal{A}$.

For the allocation of $a_i \in VA^j$, and $VA^j \in \mathcal{A}$ on system component $s_k \in VS$, we define a parameter $x^j_{i,k}$ as follows:

$$x^j_{i,k} = \begin{cases} 1, & \text{if the } a_i \in VA^j \text{ is allocated on } s_k \in VS , \\ 0, & \text{otherwise,} \end{cases}$$

For A^j we can create an allocation matrix $X^j = VA^j \times VS$, where $X^j[i,k] = x^j_{i,k}$. The matrix X^j represents the allocation part of the mapping M^j of A^j and still leaves the routing to be solved.

Routing. We now present a way of describing the routing for an application A^j. From an application point of view the communication between application components is described by $(a_i, a_j) \in EA^j$. The routing of the communication of $(a_i, a_j) \in EA^j$ on a path between $(s_k, s_l) \in ES$ can be described by $mr^j : EA^j \to ES$. We can derive such a function for all $EA^j \in \mathcal{A}$.

For the routing of (a_i, a_j) of EA^j on the path $p_{kl_v} \in P_{k,l}, v = 1, \dots, n$ and $P_{k,l} \in P$ we define a parameter $y^j_{(a_i,a_j),p_{kl_v}}$, where:

$$y^j_{(a_i,a_j),p_{kl_v}} = \begin{cases} 1, & \text{if } (a_i, a_j) \in EA^j \text{ is routed on } p_{kl_v} \in P_{k,l}, \\ 0, & \text{otherwise.} \end{cases}$$

Every path $p_{kl_v} \in P$ can be projected on a set of physical links of S, i.e.,by the earlier derived function $F : P \to ES$. Thus $p_{kl_v} = \{es_g, \dots, es_h | es_g, \dots, es_h \in ES \}$ and $es_g = \{(s_k, s_r) | s_k, s_r \in VS \}$. Using this notation allows us to create a routing matrix $Y^j = EA^j \times ES$, where $Y^j[(a_i, a_j), p_{kl_v}] = y^j_{(a_i,a_j),p_{kl_v}}$. Y^j represents the routing part of the mapping M^j of application A^j.

Overall mapping. The mapping is determined by the set of routing vectors $Y = \{Y^1, \dots, Y^n\}$, and the set of allocation vectors $X = \{X^1, \dots, X^n\}$ and $Z = \{Z^1, \dots, Z^n\}$, thus $M^j = \{Z^j, X^j, Y^j\}$. We define mapping as follows: $\mathcal{M} = \bigcup_{\forall A^j}^{\mathcal{A}} M^j$.

Using the above view we can state that a *configuration* of a distributed environment consists of the structures and mapping of the distributed system, i.e.,the constellation of components, their physical interconnection and their mapping on each other... As a consequence, a *reconfiguration* is the changing of the structure or the mapping. For the creation of alternative configurations we intend to use application placement procedures as proposed in [4, 11, 30, 31, 33]. Formally, the distributed system configuration, Ω, is a function of the task components, the application and system structure and the mapping, i.e.,$\Omega = F(TC, \mathcal{A}, S, \mathcal{M})$.

4.2 A graph model of IN/ANSA

In this section we describe our experimental environment using the graph notation as presented above. First we start with the presentation of the available components for our distributed application (see also Figure 6):

$$TC = \{T^1\}$$
$$AC = \{\text{GC}, \text{SC}, \text{NTC}, \text{DBC}, \text{BC}\}$$
$$SC = \{sun2, sun3\}$$
$$NL = \{n1\}$$
$$P = \{p_{2,3_1}\}$$

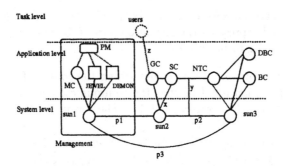

Fig. 6. The graph-oriented view of the distributed environment

In this description we left out the aspects of the management level, because they are not assumed to have influence on the performance and therefore do not contribute to the performability model to be created. *sun1* (see Figure 6), has been reserved for the management level activities. The user component is included in the GC as said before, still we present it also as a separate component at task level for parameterization reason to become clear below.

Below we describe the task, application and system level of the environment using the notation presented in the previous section:

$$T^1 = \{\text{USERS}\} \qquad\qquad S = \langle VS, ES \rangle$$
$$A = \{\langle VA^{\,1}, EA^{\,1}\rangle\} \qquad\qquad VS = \{sun2, sun3\}$$
$$A^1 = \langle VA^{\,1}, EA^{\,1}\rangle \qquad\qquad ES = \{(sun2, sun3)\}$$
$$VA^{\,1} = \{\text{GC}, \text{SC}, \text{NTC}, \text{DBC}, \text{BC}\} \qquad p_{2,3_1} = \{(sun2, sun3)\}$$
$$EA^{\,1} = \{(\text{GC}, \text{SC}), (\text{SC}, \text{NTC}), (\text{NTC}, \text{DBC}), (\text{NTC}, \text{BC})\}$$

The only aspect that still needs to be described is the mapping $\mathcal{M}$ of the different levels. Because there is only one application, A^1, the mapping remains simple, $\mathcal{M} = \{M^1\}$. For the mapping we use the allocation and routing matrices, $M^1 = \{Z^1, X^1, Y^1\}$.

$$Z^1 = \begin{pmatrix} 1 & 0 & 0 & 0 & 0 \end{pmatrix}, \quad X^1 = \begin{pmatrix} 1 & 0 \\ 1 & 0 \\ 0 & 1 \\ 0 & 1 \\ 0 & 1 \end{pmatrix}, \quad Y^1 = \begin{pmatrix} 0 \\ 1 \\ 0 \\ 0 \end{pmatrix}.$$

This mapping in combination with the application, task and system level components results in a configuration as presented in Figure 6; we can see that the application components NTC, BC and DBC are allocated on one system part and therefore do not use any communication paths as shown by Y^1.

We can change the mapping and thereby create an alternative configuration. For the experiments in Section 7 we used the following two alternative mappings, M^2 and M^3. In the first alternative configuration (represented by $M^2 = \{Z^2, X^2, Y^2\}$) we moved all the application components to the same system part, $sun2$. For this configuration the application does not use the communication paths. In the second alternative configuration (represented by $M^3 = \{Z^3, X^3, Y^3\}$), replicated components are used, i.e.,NTC', DBC' and BC'. For this alternative configuration we also have to change (expand) the set of application components and the application graph. These replicated components are treated as independent components. The corresponding allocation matrices then have the following form:

$$Z^2 = \begin{pmatrix} 1 & 0 & 0 & 0 & 0 \end{pmatrix}, X^2 = \begin{pmatrix} 1 & 0 \\ 1 & 0 \\ 1 & 0 \\ 1 & 0 \\ 1 & 0 \end{pmatrix}, Y^2 = \begin{pmatrix} 0 \\ 0 \\ 0 \\ 0 \end{pmatrix}.$$

$$Z^3 = \begin{pmatrix} 1 & 0 & 0 & 0 & 0 \end{pmatrix}, X^3 = \begin{pmatrix} 1 & 0 \\ 1 & 0 \\ 0 & 1 \\ 0 & 1 \\ 0 & 1 \\ 1 & 0 \\ 1 & 0 \\ 1 & 0 \end{pmatrix}, Y^3 = \begin{pmatrix} 0 \\ 1 \\ 0 \\ 0 \\ 0 \\ 0 \\ 0 \end{pmatrix}.$$

In Figure 6, the relations between the management level and the application components have not been shown, but each of the application components actually has a relation with all the management components.

5 A performability model of IN/ANSA

For the evaluation of an alternative configuration we need a performability model. Therefore, an alternative configuration Ω_{alt} is transformed into a performability model Ψ_{alt}. We do this by replacing each component of T^1, A^1 and S by a predefined stochastic Petri net (SPN) model. We use the mapping, the application and system graphs to create the overall performability model. The resulting performability models are both flexible and relatively easy to solve by current day software tools [13]. In this paper we will deal with the performance aspects of the model only.

In Section 5.1 we present the generic SPN modelling of user, application and system components. The performability model of the experimental distributed environment is presented in Section 5.2.

5.1 The SPN models used to realize the performability model

In this section we present a generic way to transform each component of the distributed environment into an SPN sub-model. We start with the application level, then the system level and finally present how the users are modelled using SPN.

For each operation or service provided by an application component a SPN model component is predefined. In such an SPN model a service is represented by a timed transition. The invocation of a service at the interface by a client is represented by putting a token in the corresponding "service-input place". Resources must be allocated (e.g.,an CPU) and the operation can be performed (the timed transition). In Figure 7 we see (at the right hand side of the arrow) the SPN representation of one operation of a computational object (shown at the left ahnd side of the arrow) or application component. The output of the timed transition, i.e. the operation, is an announcement or an interrogation to another operation (see Figure 8). With an interrogation invocation as output the component will await an answer and continue operation.

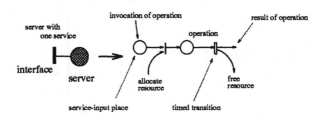

Fig. 7. The SPN representation of an ANSAware service provision

The duration of an operation is represented by a timed transition. These transitions represent the work demanded from the resource, for example the CPU busy time ($1/\mu_{i_v,k}$, the average service time of operation v of application component i on system component k) if the CPU is the scarce resourse. We can estimate these parameters by running and monitoring the component in isolation (one component on a single workstation).

The communication between components can be represented in a similar way as the operations. Per (remote) operation, or communication between two application components allocated to different system parts, a network link must be allocated. The duration of a communication operation is also represented by a timed transition. In this case a timed transition represents the communication time ($1/\mu_{i_v,j,n_i}$, the average communication time of application component j to

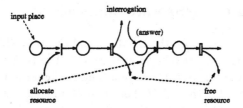

Fig. 8. The SPN representation of an announcement or interrogation operation

application component i for operation v using network link i) per invocation of an operation per network link (see [9]).

The generation of requests by the USERS is modelled as a Poisson arrival process, represented by a single timed transition.

5.2 The performability model of the IN/ANSA environment

In this section we discuss the SPN performability model of the IN/ANSA application. Tools for SPN analysis normally only allow finite state space models. This does not correspond to the experimental environment. However, we can approximate an open model by closed model with a very large customer population. The average *request rate* Λ for an application A^j is modelled by the task component USERS $\in T^j$. In Figure 9 the SPN representation of the distributed environment, using configuration M^1, is given using the predefined SPN models.

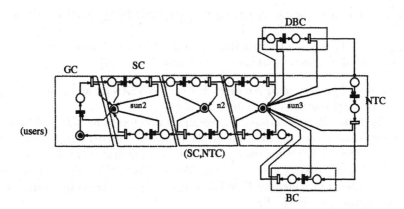

Fig. 9. The SPN model of configuration M^1

For the evaluation of the model we need the labels for the components as presented in [9]. These labels are the transition rates (service rates) of the timed transitions in the SPN model.

6 Monitoring of ANSAware applications

Two different monitoring tools, the DEMON and the JEWEL tool, monitor the experimental environment introduced in Section 3. The DEMON tool [21] monitors and visualizes the functional behaviour and configuration of the ANSAware components (as defined in the set AC) on the system nodes (as defined in the set SC). These can be used to provide the performability model with configuration information. The JEWEL monitoring tool [19] extracts performability indices from the ANSAware environment and visualizes them for each component on a graphical display. The performability indices are used to detect a decrease of QoS and to parameterize the performability model.

In order to provide the monitoring tools with the information needed, the ANSAware application components has to be instrumented with additional code for both monitoring systems. Instrumentation for the DEMON tool is performed automatically by a pre-compiler designed and implemented at PTT Research [17]. Instrumentation for the JEWEL monitoring tool is performed in a generic manner using the ANSAware operations as a reference point to detect relevant events. The implementation of the invocation of an operation is embraced by the two events: *request* and *confirm*. These events are detected by JEWEL and used to derive the turnaround time of an operation. The implementation of the operation is also embraced by two events: *indication* and *response*, these are detected by JEWEL and used to derive the service time of an operation, see Figure 11. A detailed prescription of generic instrumentation for ANSAware is provided in [25].

7 Experiences with monitoring, modelling and evaluation

A performability model of a distributed application can automatically be constructed guided by three input sources (see also Figure 10):

1. *A library of predefined SPNmodels.* For each ANSAware and system component a model has to be available in a library .
2. *Configuration determination.* The configuration has to be obtained from the system to construct the model from the predefined model components in the library. The DEMON monitor provides this configuration information.
3. *Performability indices determination.* We use the performability monitoring measurements provided by JEWEL to determine the transition rates of the timed transitions in the SPN.

The method of tuning the performability model, provided in [9], obtains the transition rates for the SPN from the requirements of the components and the capacities of the system nodes. A major drawback of this method is the required *a priori* determination of the requirements and capacities. Because the source code of the ANSAware components is processed by several pre-compilers and linked with library functions, exact requirements of the components with respect to processing workload, communication workload, memory access, etc.

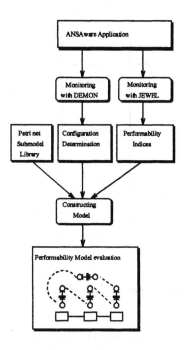

Fig. 10. The performability model is constructed from three input sources

are hard to assess. Capacities of the system nodes may be exactly specified by the manufacturers, but mechanisms like memory caching or disk access cause dynamically changing capacities of the system nodes. Therefore, we have used a more practical approach to parameterize the performability model, guided by the measurements provided by the JEWEL monitoring tool. The transition rates can be obtained by measuring the service times of the individual components. In Figure 11 a timing diagram is depicted containing the monitored time-stamps of the events: request, confirm, indication and response.

The service times can be derived from these measurements under minimal load. No queueing will occur under minimal load, so the residence time of a component will be equal to the service time of that component decreased by the residence times of the interrogation operations invoked during the service provisioning and the encountered communication delays:

$$T_j = \frac{1}{R_j - \sum_{i \in K} R_i - \sum_{i \in K} C_i}$$

where T_j is the transition rate of the sub-model of component j, R_j is the average turnaround time of component j and K is the set of operations invoked (as an interrogation) by component j. R_i is the average turnaround time of operation i and C_i the average communication delay to component i. As an example consider

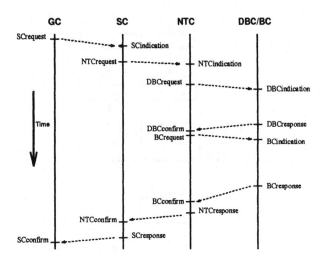

Fig. 11. The timing diagram for one configuration

the service time of the NTC in Figure 11. The service time of component NTC can be derived from the residence time of NTC (NTCresponse - NTCindication) decreased by the residence times of DBC and BC and the communication delays (differences between request and indication and the differences between response and confirm). In this way service times and communication delays can be derived from the measurements depicted in the diagram. A drawback of this method is that for each combination of components and system nodes a measurement under minimal load has to be done to obtain the residence time without queueing. A major advantage of this method, however, is the higher level of abstraction maintained, i.e. the capacities of the system nodes and the requirements of the components are implicitly incorporated.

We now discuss some comparative results from the modelling and monitoring. The IN/ANSA application has been monitored using different (alternative) configurations. The performability model has been parameterized with statistics (averages) over the measurements, obtained by monitoring the different configurations under minimal load. This leads to one set of parameters applicable for all configurations. The SPNP implementation of the performability model has been verified with the performability indices actually measured by the JEWEL monitor under various workloads.

In Table 1 the monitored results for the three different configurations are presented in comparison with the values calculated by SPNP. We see that the model results, under minimal load, come very close to the actually measured values. Notice that these results are obtained using a very simple performance model, only taking into account application components and CPU possession.

Finally, we compare the measured results with the model evaluation results under higher load. Note that the models parametrization is the same as for the

config.		Time in ms.		
		Monitored	SPNP	Difference
M^1	Turnaround time of GC	112	113	+0.9%
	Turnaround time of SC	91	88	-3.4%
	Turnaround time of NTC sun3	76	73	-4.1%
M^2	Turnaround time of GC	157	157	0%
	Turnaround time of SC	132	132	0%
	Turnaround time of NTC sun2	103	103	0%
M^3	Turnaround time of GC	138	135	-2.2%
	Turnaround time of SC	115	110	-4.5%
	Turnaround time of NTC sun2	106	103	-2.9%
	Turnaround time of NTC' sun3	73	73	0%

Table 1. Evaluation SPNP model with monitoring results under minimal load for configurations M^1, M^2 and M^3.

minimal load case. The configurations investigated are M^1, M^2 and M^3. Due to scheduling strategies of ANSAware the approximations for the turnaroundtime of the "internal" components, i.e., NTC, SC, BC and DBC are not comparable to the SPNP results. More important for the performance, however, is the QoS provided to the user, i.e.,the turnaroundtime of the complete application, which is equal to the turnaround time of GC. We therefore address this measure.

The results of the SPNP model and the measurements are graphically depicted in Figure 12. For each configuration eight monitoring sessions were conducted for different workloads (λ). The results of the SPNP model are reasonably good (less than 10% error) when the load is low to moderate. When the load increases, however, the monitoring results differ substantially from the results calculated by SPNP. The workload range of our interest is the moderate range were the turnaround time does not exceed the requested QoS. If the QoS is violated (or the turnaroundtime has increased significantly) the performability manager is triggered and runs the Performability Model for alternative configurations.

Furhter research is necesary to estimate the level of confidence we can put in our models.

8 Discussion and future work

In this paper we focused on the modelling aspects of model-based dynamic QoS management by a PM. We presented an ANSAware-based experimental distributed environment in which the modelling and evaluation aspects are totally automated. We proposed a generic modelling strategy in which the structure of the client/server and the computational model of the ANSAware computing platform are used. This structure allows for a generic transformation of the com-

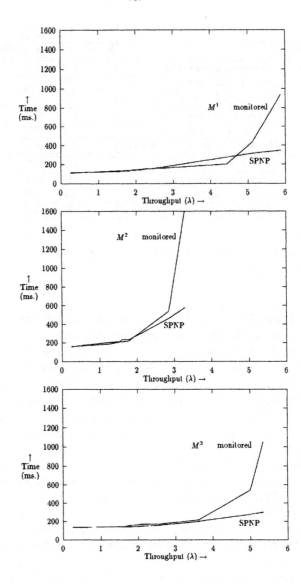

Fig. 12. Evaluation SPNP model with monitoring results for configurations M^1, M^2 and M^3.

putational models into performability models using predefined SPN models. The PM receives information about the ANSAware-based application from a distributed monitoring process based on JEWEL and DEMON. With this information, and the SPN model library of ANSAware applications, the PM automatically constructs an overall SPN performability model which is subsequently used for the determination of the provided Quality of Service (QoS).

The modelling as presented is based on our general view on distributed systems. This view only bears four levels of which we modelled three, which might not be sufficient. For example the scheduling activities, i.e. the use of threads and tasks, in ANSAware are not modelled at the moment which introduces inaccuracies. Because ANSAware is implemented on top of the system level (and even partly resides in the management level), the introduction of a middle-ware level might be necessary.

We showed the feasibility of the proposed PM by presenting some operational results. More work, however, will be necessary to make the PM fully operational. The process of automatically creating alternative configurations, selecting the best and make it operational is not completely automated yet. Currently we are working on proper mapping algorithms for the creation of the alternative configurations.

The required on-line and therefore necessarily fast evaluation of the created SPN models also requires further study. Currently we are experimenting with MVA algorithms (thereby ignoring the "simultaneous resource possession" aspects) and the use of closed-form solutions for the SPNs [25].

The monitoring process is realised using two monitoring tools. In a future environment the use of one monitoring tool is preferred because of the interference of the monitoring process with the monitored applications. The current experience with generic monitoring shows satisfying results which makes it applicable for further use.

In this paper we mainly adressed pure performance issues of the performability manager. The use of replicated components and the evaluation of the models w.r.t. "real" performability measures, i.e.,including dependability aspects, will be subject of further study. We also intend to use the performability manager as a conceptual framework for the study of resource control issues in multimedia conferencing systems.

References

1. APM Ltd., Cambridge, U.K. *ANSA : An Engineer's Introduction to the Architecture*, November 1989.
2. R.L. Bennett and G.E. Policello II. Switching Systems in the 21st Century. IEEE Communications Magazine: *Feature Topic: Toward The Global Intelligent Network*, 31(3):24–30, March 1993.
3. Y. Berders and P. Dickman, editors. *Workshop on Dynamic Object Placement and Load Balancing in Parallel and Distributed Systems*. The Sixth European Conference on Object-Oriented Programming, ECOOP'92, 1992.

4. N.S. Bowen, C.N. Nikolaou, and A. Ghafoor. On the Assignment Problem of Arbitrary Process Systems to Hetrogeneous Distributed Computer Systems. *IEEE Transactions on Computers*, 41(3):257–273, March 1992.

5. M. Calzarossa and G. Serazi. Workload Charaterization: A Survey. *Proceedings of the IEEE*, 81(8):1136–1150, August 1993.

6. B. Carré. *Graphs and Networks*. Clarendon Press, Oxford, 1979.

7. N.M.van Dijk, B.R. Haverkort, and I.G. Niemegeers. Guest editorial: Performability Modelling of Computer and Communication Systems. *Performance Evaluation*, 14(3-4):61–78, February 1992.

8. ETSI. Network Aspects (NA); General aspects of quality of service and network performance in digital networks, including ISDN. Technical Report ETR 003, ETSI, 1990.

9. L.J.N. Franken and B.R.H.M. Haverkort. The Performability Manager. IEEE Network: The Magazine of Computer Communications *Special Issue on Distributed Systems for Telecommunications*, 8(1), Januari 1994.

10. J.J. Garrahan, P.A. Russo, K. Kitami, and R. Kung. Intelligent Network Overview. IEEE Communications Magazine: *Feature Topic: Toward The Global Intelligent Network*, 31(3):30–38, March 1993.

11. A. Gersht and R. Weihmayer. Joint Optimization of Data Network Design and Facility Selection. *IEEE Journal on Selected Areas in Communications*, 8(9):1667–1681, December 1990.

12. S. Hariri and C.S. Raghavendra. Distributed Functions Allocation for Reliability and Delay Optimization. *Proceedings of the Fall Joint Computer Conference (IEEE)*, pages 344–352, 1986.

13. B.R. Haverkort and K.S. Trivedi. Specification and Generation of Markov Reward Models. *Discrete-Event Dynamic Systems: Theory and Applications*, 3:219–247, 1993.

14. B.R.H.M. Haverkort. *Performability Modelling Tools, Evaluation Techniques, and Applications*. PhD thesis, University of Twente, 1990.

15. K. Kant. *Introduction to Computer System Performance Evaluation*. McGraw-Hill, Inc., 1992.

16. S. Kheradpir, W. Stinson, J. Vucetic, and A. Gersht. Real-Time Management of Telephone Operating Company Networks: Issues and Approaches. *IEEE Journal on Seclected Areas in Communications*, 11(9):1385–1403, December 1993.

17. H. Korte. Visualising ANSAware Programs with EXP93. Technical report, PTT Research, the Netherlands, unpublished, June 1993.

18. J. Kramer, editor. *Proceedings of the International Workshop on Configurable Distributed Systems*. Computing Control Division of the Institution of Electrical Engineers, IFIP, Imperial College of Science, Technology and Medicine, IEE, March 1992.

19. F. Lange, R. Kroeger, and M. Gergeleit. JEWEL: Design and Implementation of a Distributed Measurement System. *IEEE Transactions on Parallel and Distributed Systems*, 3(6):657–671, November 1992.

20. Y.H. Lee and K.G. Shin. Optimal Reconfiguration Strategy for a Degradable Multimodule Computing System. *Journal of the ACM*, 34(2):326–348, April 1987.

21. MARI Computer Systems Ltd. *DEMON V3.0 User's guide and Reference manual*, 1993.

22. L. Mejlbro. QOSMIC-Deliverable D1.3C: QoS and Performance Relationships. Deliverable QOSMIC R1082, RACE, 1992.

23. J.F. Meyer. Performability Evaluation of Telecommunication Networks. In Network Teletraffic Science for Cost-Effective Systems and ITC-12 Services, editors, *M. Bonatti*, pages 1163–1172. IAC, Elsevier Science Publishers B.V. (North Holland), 1989.

24. J.F. Meyer. Performability: a Retrospective and some Pointers to the Future. *Performance evaluation*, 14(3-4):139–156, Februari 1992.

25. R.H. Pijpers. Performability Monitoring and Modelling of ANSAware Environments. M.Sc. thesis, University of Twente, the Netherlands, December 1993.

26. R. Pooley and J. Hillston, editors. *Proceedings of the Sixth International Conference on Modelling Techniques and Tools for Computer Performance Evaluation*. University of Edinburgh, Athony Rowe Ltd, Chippenhame, Wiltshire, September 1992.

27. Project JTC1.21.43. Reference Model for Open Distributed Processing. Draft Recommendation X.901: Basic Reference Model of Open Distributed Processing Part 1: Overview and Guide to use reference SC21 N7053, , 1993-1-28.

28. QOSMIC. General Aspects of Quality of Service and System Performance in IBC. Deliverable RACE D510, RACE, 1991.

29. W. Rosenberry, D. Kenney, and G. Fisher. *Understanding DCE*. O'Reilly & Associates, Inc, 1992.

30. S. M. Shatz, J. Wang, and M. Goto. Task Allocation for Maximizing Reliability of Distributed Computer Systems. *IEEE Transactions on Computer*, 41(9):1156–1168, December 1992.

31. S.M. Shatz and J. Wang, editors. *Tutorial: Distributed Software Engineering.* IEEE Computer Society, Press, 1989.

32. K.G. Shin, C.M. Krishna, and Y. Lee. Optimal Dynamic Control of Resources in a Distributed System. *IEEE Transactions on Software Engineering*, 15(10):1188–1197, October 1989.

33. N.G. Shivaratri, P. Kreuger, and M. Singhal. Load Distributing for Locally Distributed Systems. *IEEE Computer*, 25(12):33–44, December 1992.

34. International Telecommunication Union. General Characteristics of International Telephone Connections and Circuits. Red Book Fsc. II.1, CCITT, 1985.

35. International Telecommunication Union. Telegraph and Mobile Service and Quality of Service. Blue Book Fsc. II.4, CCITT, 1989.

36. Studygroup XI. Q.1200, Draft recommendations. Technical report, CCITT, 1991.

Waiting time distributions for processor sharing queues with state-dependent arrival and service rates

Jens Braband*

Institut für Mathematische Stochastik
Technische Universität Braunschweig

Abstract

Several variants of multiple server queues with Poisson input, exponentially distributed service demands and processor sharing discipline are considered, in which the arrival rate and the service capacity may depend on the actual number of customers in the system. These queues are approximated by a sequence of models featuring a new natural discipline called Random Quantum Allocation (RQA) operating in discrete time. This approach can be used for the numerical approximation of waiting and response time distributions for processor sharing queues. Numerical examples are provided for open and closed M/M/N processor sharing queues. In particular the effect of the number of parallel processors on the response time distribution under the condition of fixed total service capacity is discussed.

Keywords: Processor sharing, multiple server queues, waiting time distributions.

1 Introduction

Queueing models with processor shared service have attracted considerable interest as models of time sharing computer systems. If there are n customers present in a single server processor sharing (PS) model each customer receives service at an equal fraction of $\frac{1}{n}$ of the total service rate. The main performance measure of such a computer model is a customers response or waiting time either conditioned on the state of the system at his arrival and/or his total service demand or under stationary conditions.

While it is commonly agreed that a model is solved if some transform of its waiting or response time distribution can be given [16], it has recently been pointed out [1], that the inversion of e. g. Laplace transforms is a nontrivial numerical task. It seems as if most results derived in terms of transforms do not solve the practical problem of actual calculation of the waiting or response time distributions because only a successful few results in this direction have been reported.

---

*The main part of this research was carried out at the Technische Universität Braunschweig. The author is now with Siemens AG, Bereich Verkehrstechnik, Ackerstraße 22, D–38126 Braunschweig

Explicit analytical results, which can be exploited for the numerical calculation of the stationary waiting or response time distribution, exist only for simple M/M/1–PS models [7, 9, 12]. While several variants of single server PS models have been studied, including

- M/M/1–PS with a finite population of customers e. g. [9]

- M/M/1–PS with limited multiprogramming e. g. [2, 13]

- gated M/G/1–PS e. g. [3, 14]

- M/M/1–PS with state-dependent arrival rate e. g. [8]

- head–of–the–line–PS e. g. [11]

the few results on waiting or response time distributions are either not explicit or cannot be used for the numerical computation of waiting or response time distributions.

Relatively little is known about multiple server PS models (see the survey [16]). Multiple server PS models are of particular interest because the usual way to increase the performance of a computer system is to use a faster processor. But as faster processors become very expensive, a natural way to overcome this bottleneck is to use parallel processors.

The common approach to PS queueing models is via an approximating sequence of Round Robin (RR) models in discrete time, see for example [7] for the M/M/1–PS model or [15] for the M/G/1–PS model. From the practical point of view this is a realistic and natural approach because PS models are idealizations of RR queues. But if we are mainly interested in PS models, RR models have two major drawbacks: Firstly in the PS model there is no overt queueing at all, and secondly the description of RR models is much more complicated.

A natural approach in the above sense is the Random Quantum Allocation (RQA) model [6], which in contrast to the RR model gives up the ordering of the customers in the queue. The basic idea is to select the next customers to be served at random. For the M/M/1–RQA it has been shown [6], that the RQA approach results in an improved numerical approximation of the corresponding PS model and that the convergence of the waiting time distributions of the RQA model to the waiting time distributions of the PS model is very fast.

This paper is organized as follows: We define a general multi–server M/M/N–RQA model which serves as an approximation for the corresponding general M/M/N–PS model and show how the waiting time distributions for the RQA model can be determined numerically. We choose this model because there exist no results [16] about the waiting time distributions in this general M/M/N–PS model. Examples of important classes of PS models covered by this general model are given. We show that for these models much more detailed numerical results on waiting or response time distributions can be obtained than by analytical techniques. In particular we present numerical results for open and closed M/M/N–PS models.

2   A general M/M/N–RQA model

Consider a multi–server system with an unbounded waiting room for customers. Let $\lambda_n = \lambda(n)$ and $\sigma_n = \sigma(n)$ be nonnegative real– respectively integer–valued functions. λ_n represents an arrival rate and σ_n a service capacity which depend only on the number n of customers in the system. The system operates in discrete time as follows:

1. Arrivals and departures of customers occur only at multiples tq, $t \in \mathbb{N}$, of the time slice length $q > 0$.

2. There is at most one arrival at tq. The probability of an arrival $\lambda_n q < 1$ dependends only of the number n of customers in the system before the arrival.

3. If a new customer arrives at tq no service at all is provided during the next time slice $[tq, (t+1)q)$.

4. If there are $n \leq \sigma_n$ customers already in the waiting room at tq and there is no arrival, all customers are being served in the next time slice by one server. If there are $n > \sigma_n$ customers present without a new customer arriving, σ_n customers are selected at random for service during the next time slice.

5. Each customer that has been served during the time slice $[tq, (t+1)q)$ leaves the system with probability $\mu q < 1$ after his service, independent of all other events in the system.

The general M/M/N–RQA model is obviously a discrete approximation for the corresponding general M/M/N–PS model with Poisson input stream with intensity λ_n, state-dependent service capacity σ_n and customers having exponentially distributed service demands with mean $\frac{1}{\mu}$. The weak convergence of the waiting time distributions in the general M/M/N–RQA model to the corresponding waiting time distributions in the general M/M/N–PS model can be proved along the same lines of reasoning as in [15], for details see [5].

We remark that the denial of service immediately after the arrival of a customer simplifies the numerical evaluation of the waiting time distributions significantly. Other possible choices are to treat arriving customers in the same way as waiting customers or to serve the new customer immediately after arrival. We remark that in general the service behaviour of the model can be changed for a limited number of time slices immediately after the arrival of new customers without changing the weak convergence properties.

3   Waiting time distributions for general M/M/N–RQA models

Consider a tagged customer with a remaining service demand of k time slices. Let $p_{k,m}(n)$ be the probability that the tagged customer remains in the system for exactly $k + m$ time slices, conditioned on his remaining service demand k and the number n of competing customers in the system (not including the tagged customer).

Theorem: *The conditioned waiting time distribution $\{p_{k,m}(n), m \in \mathbb{N}\}$ of a tagged customer in the general M/M/N–RQA model is completely determined by the following set of recursive equations:*

$$p_{0,m}(n) = \delta_{m,0}$$

$$p_{k,m}(n) = \lambda_{n+1}q p_{k,m-1}(n+1) + (1-\lambda_{n+1}q)\left(\sum_{i=0}^{n} \mu_i^n p_{k-1,m}(n-i)\right),$$

$$n+1 \leq \sigma_{n+1}$$

$$p_{k,m}(n) = \lambda_{n+1}q p_{k,m-1}(n+1) \tag{1}$$

$$+(1-\lambda_{n+1}q)\left\{\frac{\sigma_{n+1}}{n+1}\sum_{i=0}^{\sigma_{n+1}-1} \mu_i^{\sigma_{n+1}-1} p_{k-1,m}(n-i)\right.$$

$$\left. + \left(1-\frac{\sigma_{n+1}}{n+1}\right)\sum_{i=0}^{\sigma_{n+1}} \mu_i^{\sigma_{n+1}} p_{k,m-1}(n-i)\right\}, n+1 > \sigma_{n+1}$$

$$p_{k,m}(-1) = p_{k,-1}(n) = 0$$

where $\mu_i^n = \binom{n}{i}(\mu q)^i(1-\mu q)^{n-i}$ denotes the probability that among n customers that have been served i leave the system and δ the Kronecker symbol.

Proof: We can derive the recursions by conditioning on the events that occur during the next time slice (see the details given in the description of the model in the last section).

The first formula states only that the tagged customer leaves the system immediately after receiving the last service time slice. The second formula stands for the case where all customers are being served during the next time slice, while (1) represents the case where we have $n+1$ customers (including the tagged one) are competing for service. Note that the tagged customer cannot leave the system before his fixed service demand is completed and that the maximal number of the customers who can leave the system depends on whether the tagged customer is being served during the time slice or not. ∎

The conditioned waiting time distributions for the general M/M/N–RQA model can be evaluated numerically via (1) with reasonable computational effort for moderate values of q. The computational effort can be further reduced by the introduction of further approximations, e. g.

- For small time slice length q we have $\mu_i^n = o(q)$ for $i \geq 2$ and a reasonable approximation for this case is $\mu_0^n = 1-\mu n q$, $\mu_1^n = \mu n q < 1$ and $\mu_i^n = 0$ for $i \geq 2$.

Practical application shows that this approximation is already very accurate for moderate values of q, e. g. $q = 0.01$ in most applications.

In the same manner recursions for the waiting time distribution $\{p_m(n), m \in \mathbb{N}\}$, conditioned on the number n of competing customers only, can be found. The only difference is that now the tagged customer himself is allowed to leave the system after each service with probability μq. As an example we give the recursion for the waiting time distribution $\{p_m(n), m \in \mathbb{N}\}$ modified by the additional approximation proposed above. Using the abbreviation $\varphi_{n+1} = \min\{n+1, \sigma_{n+1}\}$, which simply stands for the number of customers which can be served during the next time slice, we have

$$p_0(n) = (1 - \lambda_{n+1}q)\frac{\varphi_{n+1}}{n+1} \tag{2}$$
$$\{\mu q + (1 - \mu q)\left((\varphi_{n+1} - 1)\mu q p_0(n-1) + (1 - (\varphi_{n+1} - 1)\mu q)p_0(n)\right)\}$$

$$p_m(n) = \lambda_{n+1}q p_{m-1}(n+1) \tag{3}$$

$$+(1 - \lambda_{n+1}q)\left\{ \frac{\varphi_{n+1}}{n+1}(1 - \mu q)\left((\varphi_{n+1} - 1)\mu q p_m(n-1) + (1 - (\varphi_{n+1} - 1)\mu q)p_m(n)\right)\right.$$

$$\left. + \left(1 - \frac{\varphi_{n+1}}{n+1}\right)(\varphi_{n+1}\mu q p_{m-1}(n-1) + (1 - \varphi_{n+1}\mu q)p_{m-1}(n))\right\}, m > 0,$$

$$p_m(-1) = p_{-1}(n) = 0.$$

We remark that due to the first approximation all sums that appeared in (1) are reduced to just two terms, which is a significant advantage for the analysis and numeric evaluation of the recursions. We notice that the corresponding recursion for the response time distribution $\{\hat{p}_m(n), m \in \mathbb{N}\}$ can be obtained in a simpler form

$$\hat{p}_1(n) = (1 - \lambda_{n+1}q)\frac{\varphi_{n+1}}{n+1}\mu q \tag{4}$$

$$\hat{p}_m(n) = \lambda_{n+1}q\hat{p}_{m-1}(n+1) \tag{5}$$

$$+(1 - \lambda_{n+1}q)\left\{ \frac{\varphi_{n+1}}{n+1}(1 - \mu q)\left((\varphi_{n+1} - 1)\mu q\hat{p}_{m-1}(n-1)\right.\right.$$

$$\left.\left. +(1 - (\varphi_{n+1} - 1)\mu q)\hat{p}_{m-1}(n)\right)\right.$$

$$\left. + \left(1 - \frac{\varphi_{n+1}}{n+1}\right)(\varphi_{n+1}\mu q\hat{p}_{m-1}(n-1) + (1 - \varphi_{n+1}\mu q)\hat{p}_{m-1}(n))\right\}, m > 1,$$

$$\hat{p}_m(-1) = \hat{p}_{-1}(n) = 0,$$

where m now stands for the discrete response time, which, in contrast to the waiting time, is reduced during each time slice by one.

Stationary response or waiting time distributions can be approximated by averaging the distributions given above by the stationary distribution $\{\pi_k, k = 0, 1, \ldots\}$ of customers in the general M/M/N–PS model which is known from the theory of birth–and–death processes:

$$\pi_k = \pi_0 \prod_{i=0}^{k-1} \frac{\lambda_i}{\sigma_{i+1}\mu}$$

provided that

$$\pi_0 = \frac{1}{1 + \sum_{k=1}^{\infty}\prod_{i=0}^{k-1}\frac{\lambda_i}{\sigma_{i+1}\mu}} < \infty.$$

The computation of the stationary distribution of customers in the general M/M/N–RQA model is more difficult.

We remark that the waiting time distributions for general M/M/N–PS models are unknown [16], except for the classical case M/M/1–PS [7], where their Laplace–Stieltjes transforms have been found. Additionally an integral representation for the distribution function of the equilibrium response time has been given [12]. For the M/M/N–PS model implicit representations for the waiting time distributions and their moments have recently been derived [4].

We give some examples of models which are covered by our general model:

1. **M/M/N–PS model:** $\lambda_n = \lambda$ and $\varphi_n = \min\{n, N\}$ gives the open multi–server M/M/N–PS model.

2. **Closed $\langle T/M/M/N\rangle$–PS model:** $\lambda_n = \lambda \cdot \max\{(T-n), 0\}$ and $\varphi_n = \min\{n, N\}$ results in the closed multi–server M/M/N–PS model with a fixed number T of customers.

The approach presented above can easily be adapted to handle

1. **Limited batch arrivals:** Let $\{b_i, i = 1, \ldots, B\}$ be the distribution of the number of new customers arriving in a batch, then for example the first part of (1) has to be changed into

$$p_{k,m}(n) = \lambda_{n+1} q \sum_{i=1}^{B} b_i p_{k,m-1}(n + i) + \ldots$$

General batch arrival distributions ($B = \infty$) destroy the recursive structure of (1).

2. **Customers with unlimited service demands:** If there are always C customers present in the system with unlimited demand for service, then this behaviour can be approximated by the recursions

$$p_{0,m}(n) = \delta_{m,0}$$

$$p_{k,m}(n) = \lambda_{n+1} q p_{k,m-1}(n + 1) + (1 - \lambda_{n+1} q) \left(\sum_{i=0}^{n} \mu_i^n p_{k-1,m}(n - i) \right),$$

$$n + C + 1 \leq \sigma_{n+1}$$

$$p_{k,m}(n) = \lambda_{n+1} q p_{k,m-1}(n + 1)$$

$$+ (1 - \lambda_{n+1} q) \left\{ \frac{\sigma_{n+1}}{n + C + 1} \sum_{i=0}^{\sigma_{n+1}-1} \mu_i^{\sigma_{n+1}-1} p_{k-1,m}(n - i) \right.$$

$$\left. + \left(1 - \frac{\sigma_{n+C+1}}{n + 1} \right) \sum_{i=0}^{\sigma_{n+1}} \mu_i^{\sigma_{n+1}} p_{k,m-1}(n - i) \right\},$$

$$n + C + 1 > \sigma_{n+1}$$

$$p_{k,m}(-1) = p_{k,-1}(n) = 0$$

where additionally μ has to be substituted by

$$\hat{\mu} = \mu \frac{n + 1}{n + C + 1}.$$

This appoximation results in the appropriate state-dependent reduction of the system capacity (for the tagged customer by reducing his probability to be selected for service and for the other customers by increasing their service demand) by the additional customers with unlimited service demand.

The RQA approach can obviously be extended to cover the case of gated PS models or models with limited multi–programming capacity.

4   Numerical examples for M/M/N–PS models

All numerical computations in this section have been carried out on a PC with Intel 80486 processor in single precision and on the mainframe computer IBM 3090J where the results have been checked with quadruple precision.

We compute the distribution of the equilibrium response time V in the M/M/1–RQA model via (4), (5) and the stationary distribution of customers in the M/M/1–RQA model. For the M/M/1-PS model an integral representation for the equilibrium response time distribution is known [12], which can be evaluated numerically. For time slice length $q = 0.01$ respectively $q = 0.001$ the absolute differences between the RQA approximation and the numerical results from [12] in our examples were less than 1.5×10^{-3} respectively 2×10^{-4}, the plots of the response time distributions cannot be distinguished in the scale of the following figures. We show in figure 1 and 2 the complementary response time distribution $P(V > t)$ for M/M/1-RQA and M/M/2–RQA models for the same load in logarithmic scale. In particular the parameters are $\lambda = 0.1, 0.2, \ldots, 0.9$ and $\mu = 1$ respectively $\mu = \frac{1}{2}$.

The largest absolute error between the distribution functions for the RQA approximations with $q = 0.01$ and $q = 0.001$ in the numerical examples below was less than 1.2×10^{-3}. The results show that the response time increases with the number of processors and the effect is larger for small load. For high load the response times are practically identical because almost no processor is idle at any time.

We now compute conditioned waiting time distributions for a $\langle T/M/M/1 \rangle$–RQA model for small time slice length q obtained from (1). For this model only results on moments or stationary response time distributions [9] or approximations for response time distributions are known.

The following figures show numerical examples for $\mu = 1$, $T = 10$, and $\lambda = 0.1$ dependending on the time slice length q, the service demand $k = \lceil \frac{\tau}{q} \rceil$ and the number n of active customers. Figure 3 shows for $q = 10^{-4}$ the distribution functions obtained by RQA approximation for $\tau = 1$ and $n = 0, \ldots 9$. At least for $n = 0, 1, 2$ we notice jumps for $t = n$. It is clear, that there must be jumps in the waiting time distributions, because for each fixed number of customers n and each fixed service demand τ it may happen with positive probability, that during the service of the tagged customer no new arrivals or departures of customers occur. In this case the waiting time of the tagged customer is constantly $\tau \frac{n}{\varphi_n}$.

We now focus our attention on the part of the waiting time distributions which admits a density function. To approximate this part via the $\langle 10/M/M/1 \rangle$-RQA model, we scale the discrete waiting time distribution $\{p_{k,m}(n), m \in \mathbb{N}\}$ by the factor $\frac{1}{q}$. Because the corresponding waiting time distributions in the $\langle 10/M/M/1 \rangle$-PS model

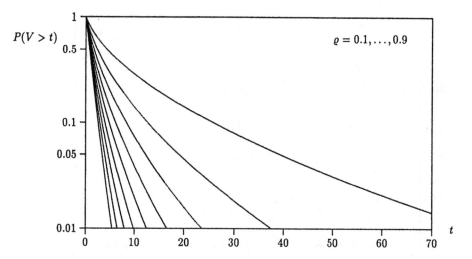

Figure 1: Response time distributions for the M/M/1–RQA model

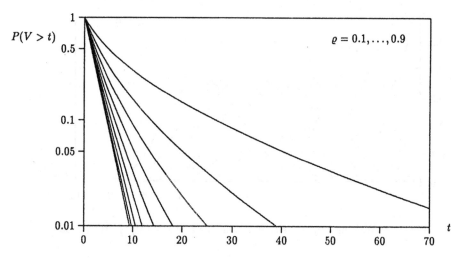

Figure 2: Response time distributions for the M/M/2–RQA model

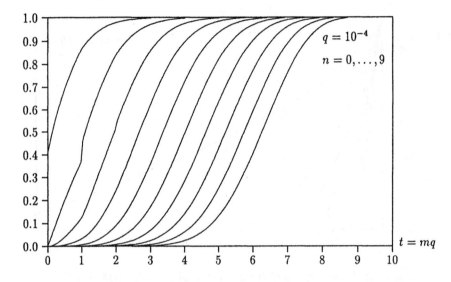

Figure 3: Waiting time distributions for a $\langle 10/M/M/1 \rangle$–RQA model

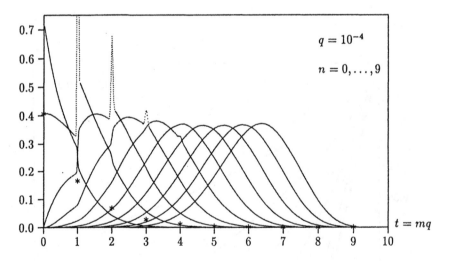

Figure 4: Waiting time density functions for a $\langle 10/M/M/1 \rangle$–RQA model

are always mixtures containing a discrete component, we have peaks at the points having positive probability.

Figure 4 shows the distributions from figure 3 in this form. The discrete parts of the distributions have been marked according to their probability with stars ($*$).

Figure 5 illustrates for fixed $n = 5$ and $\tau = 1$ the relation between the discrete waiting time distribution and the time slice length $q = 10^{-i}$, $i = 1, 2, 3$. The results for $q = 0.01$ can already serve as a satisfactory approximation for the waiting time distribution in the corresponding PS model.

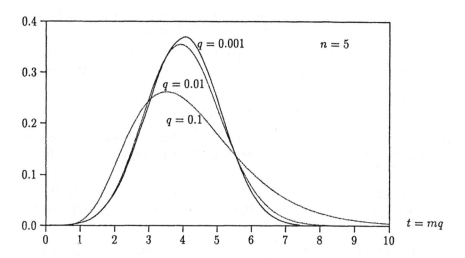

Figure 5: Waiting time density functions for $\langle 10/M/M/1\rangle$–RQA models dependent on the quantum size q

We note that especially for large τ or n we obtained a good numeric agreement of our results and the normal approximations suggested for the $\langle T/M/M/1\rangle$–PS model [10]. But for small load, small service time demands or a small number of jobs we obtain distributions of a totally different form (see figure 4).

The universality of application, the simplicity of programming and the numerical robustness are advantages of the RQA approximations (1), because already for moderate size of $q = 0.01$ we obtain satisfactory and for small $q = 0.001$ we obtain very good approximations for the waiting time distributions of the corresponding PS model.

5 Conclusion

We showed in this paper how the RQA approach can be used for the approximation of waiting and response time distributions for general M/M/N–PS models. The weak convergence of the waiting and response time distributions of the general M/M/N–RQA model to the corresponding quantities of the general M/M/N–PS model is known. Because of this fact the efficient numerical computation of the waiting and response time distributions of the general M/M/N–RQA model allows the determination of the waiting and response time distributions of the general M/M/N–PS model with high accuracy, which seems to be sufficient at least for practical purposes.

Acknowledgement: The author is indebted to R. Schaßberger for several hints and valuable discussions on the subject.

References

[1] J. Abate and W. Whitt. The Fourier series method for inverting transforms of probability distributions. *Queueing Systems*, 10:5–88, 1992.

[2] B. Avi-Itzhak and S. Halfin. Response times in M/M/1 time–sharing schemes with limited number of service positions. *J. Appl. Probab.*, 25:579–595, 1988.

[3] B. Avi-Itzhak and S. Halfin. Response times in gated M/G/1 queues: The processor–sharing case. *Queueing Systems*, 4:263–269, 1989.

[4] J. Braband. Waiting time distributions for M/M/N processor sharing queues. Preprint, Technische Universität Braunschweig, Braunschweig, 1992. Submitted to *Stochastic Models*.

[5] J. Braband. *Wartezeitverteilungen für M/M/N–Processor–Sharing–Modelle*. Dissertation, Technische Universität Braunschweig, Braunschweig, 1992. (in german).

[6] J. Braband and R. Schaßberger. Random quantum allocation: A natural approach to M/M/N processor sharing queues. In B. Walke and O. Spaniol, editors, *Messung, Modellierung und Bewertung von Rechen- und Kommunikationssystemen*, 130–142, Berlin, 1993. 7. ITG/GI Fachtagung, Springer Verlag.

[7] E. G. Coffman, R. Muntz, and H. Trotter. Waiting time distributions for processor–sharing systems. *J. Ass. Comp. Mach.*, 17:123–130, 1970.

[8] C. Knessl, B. J. Matkowsky, Z. Schuss, and C. Tier. Response times in processor-shared queues with state–dependent arrival rates. *Stochastic Models*, 5:83–113, 1989.

[9] D. Mitra. Waiting time distributions from closed queueing network models of shared–processor systems. In F. J. Kylstra, editor, *Performance '81*, pages 113–131, Amsterdam, 1981. North Holland.

[10] J. A. Morrison. Conditioned response–time distribution for a large closed processor–sharing system in very heavy usage. *SIAM J. Appl. Math.*, 47:1117–1129, 1987.

[11] J. A. Morrison. Head of the line processor sharing for many symmetric queues with finite capacity. Technical report, AT&T Bell Laboratories, Murray Hill, NJ 07974, 1992.

[12] J. A. Morrisson. Response time distribution for a processor-sharing system. *SIAM J. Appl. Math.*, 45:152–167, 1985.

[13] K. M. Rege and B. Sengupta. Soujourn time distributions in a multiprogrammed computer system. *AT&T Techn. J.*, 64:1077–1090, 1985.

[14] K. M. Rege and B. Sengupta. A single server queue with gated processor–sharing discipline. *Queueing Systems*, 4:249–261, 1989.

[15] R. Schaßberger. A new approach to the M/G/1 processor sharing queue. *Adv. Appl. Prob.*, 16:202–213, 1984.

[16] S. F. Yashkov. Processor–sharing queues: Some progress in analysis. *Queueing Systems*, 2:1–17, 1987.

Queueing Models of Parallel Applications: The *Glamis* Methodology

Henk Jonkers

Delft University of Technology, Faculty of Electrical Engineering
P.O. Box 5031, 2600 GA Delft, The Netherlands
E-mail: h.jonkers@et.tudelft.nl

Abstract. In the development of efficient parallel applications, reliable performance predictions are essential. However, many performance modelling formalisms, such as queueing networks, are not directly suitable for modelling parallel applications, while for other formalisms the analysis is too expensive. We present a methodology for performance modelling of parallel processing systems (*Glamis*), based on extended queueing networks, aiming to overcome these problems. The methodology yields reliable performance predictions for a class of parallel machines and programs at relatively low (polynomial time) analysis cost. Additional reductions of analysis cost are obtained by exploiting inherent replications in parallel systems.

1 Introduction

Performance predictions of a parallel program running on a parallel machine can be of great importance in a number of situations. A first application is the decision which one from a range of parallel machines is most suitable for a given application. Another application is the optimisation of program and machine parameters, possibly even compile-time optimisations.

A central issue in performance modelling is the trade-off between the expressive power of a modelling formalism (and, related to this, the accuracy of the performance predictions) and the efficiency of the analysis. For instance, simple complexity analysis of a parallel algorithm gives fast results, but these will often not be very accurate because certain machine influences are disregarded (e.g. resource contentions). On the other extreme of the trade-off, unrestricted timed Petri net models are able to express almost everything, but the analysis cost is in general exponential to the net size.

We present a performance modelling methodology with associated analysis techniques based on queueing networks, which enables us to predict the performance of as wide as possible a class of parallel machines/programs, while keeping the analytical complexity within acceptable limits. Some previous case studies [16, 17] showed that queueing models have the potential to yield reasonably accurate performance predictions of parallel applications for acceptable analysis costs. Therefore, we believe that they offer a good foundation for a general performance modelling methodology for parallel systems. However, standard queueing networks are in many cases not sufficiently powerful to capture all characteristics of parallel applications. Therefore, some extensions will be defined, especially at the program model level. Care has to be taken that these extensions do not lead to an unacceptable increase in analysis cost. For this reason, analysability is taken as a starting point: we first restrict ourselves to a class of models that is known to have an efficient (product-form, or at least polynomial time) solution; then we try to extend this class in such a way that the analysis efficiency is maintained. This in contrast to many other approaches, in which modelling power is taken as a starting point, i.e. a modelling formalism is selected that can express all aspects of a certain class of problems. As a result, often a lot of effort is needed afterwards to reduce the analytical complexity.

The remainder of this paper is organised as follows. In Sect. 2 an account is presented of existing performance modelling techniques, with the emphasis on the problems that occur when using them for parallel applications. Section 3 introduces the *Glamis* methodology, and describes the associated analysis methods. In Sect. 4, the features of the methodology are demonstrated by a case study. Finally, Sect. 5 presents some conclusions and recommendations for future research.

2   Performance Modelling of Parallel Applications

An important aspect in which parallel applications differ from traditional sequential applications is the presence of more and different kinds of *synchronisations*. Two types of synchronisation can be distinguished: *condition synchronisations* and *mutual exclusion*. The former typically occurs at the program level (e.g. barrier synchronisations, send/receive synchronisations), while the latter occurs both at the software level (critical sections) and the hardware level (resource contentions; these also appear in sequential applications). Mutual exclusion can be nested at both levels, and even across the levels. At the hardware level, this is known as *simultaneous resource possession*, which takes place in, among others, circuit-switched interconnection networks. An example of nesting of mutual exclusion across the levels is resource usage within a critical section. The main differences in expressive power of the different modelling formalisms are found in the types of synchronisation that can be expressed. Basic task graph models of parallel programs include condition synchronisations but no mutual exclusion. Product-form queueing networks, on the other hand, describe (non-nested) mutual exclusion but cannot express condition synchronisations. In Petri net models all types of synchronisation, including nested mutual exclusion, can be expressed.

The various existing performance modelling formalisms can be classified in *deterministic* and *probabilistic* formalisms. Also a combination of both is possible. In deterministic models all quantities, such as timing parameters and loop bounds, are fixed. This corresponds to many deterministic parameters in a real parallel system, e.g. constant execution times of floating-point operations. When very quick (linear-time or constant time) performance predictions are required (e.g. when used for compile-time optimisations), at the expense of some accuracy, deterministic models can be necessary. An example of a method yielding such fast performance approximations is *serialisation analysis* associated with the PAMELA modelling paradigm [13].

In probabilistic models some degree of uncertainty exists, which is incorporated in the models as stochastic quantities. Making stochastic assumptions for quantities which in reality are fixed (e.g. assuming an exponentially distributed service time while in reality the time is fixed) leads to errors in the predictions. However, often this error is limited, while probabilistic modelling offers many advantages.

One advantage of probabilistic modelling is that the assumption of certain time distributions (usually a negative exponential distribution) enables the solution of many models that would otherwise be analytically intractable. Another major advantage of using probabilistic modelling is that aspects of parallel applications can easily be taken into account which are inherently non-deterministic. Examples are mutual exclusion (often it is uncertain which one of several concurrent processes first obtains a resource or enters a critical section), data-dependencies (e.g. branching) and the the influence of the memory hierarchy (e.g. cache behaviour). Also, probabilistic modelling enables us to *generalise* over classes of machines and programs.

2.1 Probabilistic Performance Modelling Techniques

Many probabilistic performance modelling techniques have been developed. Their analysis is either based on the solution of the underlying Markov chain, or on the existence of a product-form solution. The latter is preferable, because the complexity of this analysis is polynomial while the Markov state space generally grows exponentially with the system size. An example of a formalism which is usually solved by means of the underlying Markov chain are *Stochastic Petri Net* models [2, 20], although recently product-form solutions for subclasses of Petri net models have been reported [10, 12]. Apart from the analytical complexity, another drawback of formalisms such as Petri nets and Markov chains is that the structural correspondence between the real system and the model is not always obvious, as opposed to e.g. queueing models. Although several techniques have been investigated to reduce the Markov state space [7, 18, 21, 27], both during and after Markov chain construction, the worst-case complexity is still exponential. An advantage of Petri nets is that both machine aspects and program aspects can easily be expressed. For this reason, Petri nets are often chosen as the modelling formalism in unified parallel application modelling environments [4, 11, 29].

Separable queueing networks are an example of models with a product-form solution, and are therefore a popular formalism to model (traditional) computer systems. A queueing network is separable if it obeys certain criteria [6]. The most widely used product-form solution methods for queueing networks are the *convolution algorithm* [8] and *mean value analysis* (MVA) [23]. At the expense of some accuracy faster results can be obtained using the iterative Bard/Schweitzer approximation of MVA [5, 26]. If a queueing network is separable, parts of the model can be aggregated into one flow-equivalent queueing centre with a population-dependent service rate, while the solution remains exact [9]. In case of a non-separable network aggregation yields approximate solutions. In general, the approximations are more accurate if the sub-model and the rest of the system are loosely coupled.

2.2 Extensions to Queueing Models

Standard queueing theory is generally sufficient to obtain reasonably reliable performance predictions for traditional uniprocessor computer systems. However, as mentioned before, the modelling power of queueing networks often runs short for parallel systems due to their inability to express condition synchronisations.

Different approaches have been suggested to incorporate condition synchronisations in queueing models. One approach is the combination of task graph program models with queueing network machine models. Examples are Thomasian and Bay's method [28] and Mak and Lundstrom's method [19]. The former approach is based on Markov chains, while the latter approach uses MVA but is restricted to series-parallel task graphs. A drawback of these methods is also that not only the service times but also the loop bounds are stochastic values (based on routing probabilities, and hence geometrically distributed; the response times have an approximately exponential distribution), which, when combined with condition synchronisations, can lead to a considerable over-estimation of execution times when in reality loop bounds are fixed (see [17] for an example).

Another solution is the combination of queueing networks with Petri nets. [3, 25]. Aggregation techniques are used to incorporate the queueing submodels in a higher level Petri net model (or the reverse). A hybrid queueing/Petri net approach was considered in [17], and appeared to yield predictions that only marginally improved the simple MVA predictions.

3 The *Glamis* Methodology

Glamis (*GeneraLised Architecture Modelling wIth Stochastic techniques*) stands for a combined parallel machine and program modelling formalism with associated analysis methods and tool support, based on separable closed queueing networks. The intention of *Glamis* is to provide means to model and analyse as wide as possible a class of parallel machines/applications, while keeping the analysis cost within acceptable limits. *Glamis* allows to make use of replications (symmetries) in the system to be modelled (e.g. processes with an identical behaviour at the program level, or identical switches or memory banks at the machine level), which can significantly reduce the analysis cost through aggregation. It also results in models that are easily *scalable*.

In principle, *Glamis* applies to all parallel architectures of which the essential features can be captured in a separable queueing model. This class basically includes all machines with a packet-switched interconnection network. However, some circuit-switched networks, that can be shown to be (approximately) equivalent to packet-switched networks, can also be handled.

As for programs, we will restrict ourselves to the class of programs which we will call *SPS-programs*. Their task graphs (*SPS-graphs*) consist of a sequence of *parallel sections*, which feature an implicit barrier synchronisation (a serial section is just a parallel section with only one parallel task). The class of SPS-graphs is a subclass of the well-known class of *series-parallel* graphs (SP). Some frequently occurring types of parallel computation (e.g. divide-and conquer and software pipelining) have a task graph which closely resembles an SPS structure. It might be possible to approximate the completion time of programs with approximately an SPS-structure (which we will call *ASPS-programs*, and their task graphs *ASPS-graphs*) by first transforming the program to a nearly equivalent SPS-program. The situation that can most easily be analysed is when all tasks in a parallel section are equivalent, i.e. their service demand on the various system resources is the same. However, in practice also parallel sections with non-equivalent parallel tasks often occur. In order to be able to obtain reliable performance predictions for these sections, a novel fixed-point algorithm was developed.

3.1 Machine Models

In *Glamis*, a queueing model of a machine is defined in terms of a number of building blocks, each block representing a single service centre or a number of equivalent service centres. Service centres are equivalent if the service demand of a job on each of the centres is the same. In order to keep the models separable, we will only consider queueing centres with an exponentially distributed service time (infinite servers can have any service time distribution).

We will use a tuple notation to define a machine model. Every item in the tuple represents a queueing model building block. A general block is denoted by $Q_k^m(D)$, where m is the number of service centres in the block, k is the number of servers in every service centre, and D is the service demand of a job on each of the service centres for *one visit* to the block. For convenience, an index 1 can be omitted. In this way, many elements frequently appearing in queueing models can be described:

- A single queueing centre with one server and service time S, denoted by $Q(S)$.
- A single queueing centre with k servers (multiple-server) and service time S, denoted by $Q_k(S)$.
- A block of m single-server queueing centres, each with the same service demand per visit D, denoted by $Q^m(D)$ (e.g. m parallel queues with service time S and an equal probability of being visited, denoted by $Q^m(S/m)$, or a sequence of m queues with service time S, denoted by $Q^m(S)$).
- A delay centre or infinite server with service time S, denoted by $Q_\infty(S)$; we will use the shorthand $I(S)$ for this (note that a block of m delay centres $Q_\infty^m(S)$ is identical in behaviour to a single delay centre $I(S)$).

Since the queueing models are separable, the exact way in which service centres are interconnected is irrelevant for the performance predictions. Only the service time and the number of visits to the centre need to be known. This is the reason why tuples can be used to define a machine model, and the order of the blocks in the tuple is irrelevant (except for the correspondence with a workload model of a program, as will be described in the next subsection).

Although the tuple notation is very compact, for illustrative purposes a graphical representation of a machine model can be useful. Figure 1 shows the graphical representations of the above-mentioned building blocks, which closely match the traditional queueing network representations.

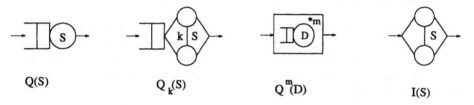

$$Q(S) \qquad Q_k(S) \qquad Q^m(D) \qquad I(S)$$

Fig. 1. Graphical representations of machine model building blocks

As an example, consider a machine with n processors connected to n shared memory banks by a packet-switched multistage interconnection network (MIN) of $\log_2 n$ stages (n is assumed to be a power of 2). Every processor runs exactly one process. Therefore, work carried out within the context of the processors (e.g. arithmetic, register moves) can simply be modelled as a workload on a delay centre. A floating-point operation takes t_f time units, a memory bank access t_m time units, and the switching time of an interconnection switch is t_s time units. Assuming a uniform distribution of memory accesses over the banks, all memory banks experience the same workload, and the same applies to the output links of the interconnection switches. Therefore, the machine model is given by $\langle I(t_f), Q^{n \log_2 n}(t_s/n), Q^n(t_m/n)\rangle$, i.e. only three model building blocks while the number of machine resources is $2n + n \log_2 n$.

3.2 Program Models

A model of an SPS-program consists of a sequence of descriptions of its parallel sections. A program section is characterised by its workload on the system resources. Because the service demand for one visit to a building block is specified in the machine model, the workload is completely determined by the *visit count* of a job to every block. Visit counts can be non-integer numbers. This can occur e.g. when different branches are possible in a program.

The workload of a parallel section is denoted by the number of equivalent parallel jobs, followed by a colon and a tuple of visit counts to the building blocks of the machine model. If there are different classes of parallel jobs, a separate workload description is given for every class. The descriptions are separated by a plus-symbol. A workload model of an SPS-program simply consists of a sequence of parallel section workload descriptions, separated by semicolons.

A workload model of a program is machine-dependent, because the different machine resources must be known. In order to separate program models from machine models, which is necessary to obtain reusable program models, a definition of the programming interface in terms of workload is needed, i.e. a mapping from machine instructions to workload parameters. For example, given a packet-switched MIN-based architecture, as described in the previous subsection, the following shared-memory interface could be defined:

$$\text{LOAD, STORE} \quad \rightarrow \quad \langle 0, 1, 1 \rangle$$
$$\text{MULT, DIV, ADD, SUB} \quad \rightarrow \quad \langle 1, 0, 0 \rangle$$

In other words, for a memory access operation both (the queueing centres representing the) bus and one of the memory banks are visited once. For a floating-point operation, the delay centre representing local CPU work is visited once. The instruction mappings can be considered to be a part of the machine model. Given the following program fragment

> **forall** $i = 1 \ldots n$
>   **for** $j = 1 \ldots k$
>   **do begin**
>   LOAD; MULT; STORE
>   **end**

and the aforementioned mapping, the workload is defined as follows:

$$n : \langle k, 2k, 2k \rangle$$

3.3 Model Analysis

The main performance measure we are interested in is the *completion time* of a parallel program. At the end of a parallel section in an SPS-program, processes wait for each other's completion at a barrier. Therefore, the parallel sections can be analysed in isolation. Thus we can simply add all the section completion times T_i to obtain the total program completion time T.

The completion time of a parallel section is the amount of time between the initiation of the tasks and the termination of the last task. Although in reality a parallel program section is finite, i.e. a transient process, the completion time of a parallel section will be estimated using the steady-state response time of a job in the system. This approach introduces errors because in reality the system population decreases as a program section approaches the end (due to the barrier effect), while in steady-state analysis of a closed queueing network the population is assumed constant. This is mainly a problem when different classes of tasks in one parallel section occur. In case of only one class of parallel tasks, all tasks approximately terminate at the same time, so that the system load is nearly constant during the whole section, and the job response time is a good estimation for the section completion time. (Note that this assumes that the variance in the task completion times is small. Adve and Vernon [1] showed that this assumption is justified for the class of applications that we consider).

The three subsections of this section describe how the completion time of a parallel section can be predicted, respectively for a single class of jobs, two classes of jobs and more than two classes of jobs.

Sections with a Single Class of Jobs. The behaviour of all of the building blocks can be described by a flow-equivalent single-server queueing centre, possibly with a service rate depending on the queue length. Table 1 gives the rates for the different building blocks. The rates for the single server $Q(S)$, the multiple server $Q_k(S)$ and the infinite server $I(S)$ are straightforward. However, infinite servers will be treated separately rather than as a special case of a single server with a population-dependent rate, because this allows for a more efficient analysis and has the additional advantage that we do not have to restrict ourselves to a negative exponential service time distribution.

Table 1. Service rates of basic building blocks

Building block	Service rate $\mu(n)$
$Q(S)$	$1/S$
$Q_k(S)$	$\min(n,k)/S$
$Q^m(S)$	$\dfrac{n}{(m+n-1)S}$
$I(S)$	n/S

The service rate for a block of m equivalent queueing centres Q^m needs some explanation. The given rate can be derived using aggregation as described in Sect. 2.1 (a similar derivation is presented in [30]). Given a short-circuited system of m identical queueing centres, each with a service demand $D_i = S$ ($1 \le i \le m$). With a population of n jobs, the job response time at every centre is $R_i = (1 + Q_i(n-1))S$. Because the centres are identical, the n jobs will be distributed evenly over the centres, assuming a steady-state situation, i.e. $Q_i(n) = n/m$ for all i. Therefore, $R_i = (1 + (n-1)/m)S = ((m+n-1)/m)S$. The total system response time is $R(n) = \sum_{i=1}^{m} R_i = mR_i = (m+n-1)S$. The flow-equivalent service rate of the aggregate queueing centre, which is equal to the system throughput, is given by

$$\mu(n) = X(n) = n/R(n) = \frac{n}{(m+n-1)S}$$

Sections with Two Classes of Jobs. In case of two classes of jobs in a parallel section, ignoring the effect of a barrier synchronisation from the model is no longer acceptable. Since the response times for the two classes will, in general, be different, the jobs with the lower response time (the short jobs) will finish before the jobs with the higher response time (the long jobs), resulting in a decrease of the system load. Since the short jobs will experience a full system load during their whole execution, regular multiple-class MVA will yield a good estimate for their completion time. However, the completion time for the longer jobs, which equals the section completion time, will be over-estimated.

A more accurate prediction of the parallel section completion time can be obtained by adding a delay centre to the queueing network, representing the barrier synchronisation delay of the short jobs. The synchronisation delay equals the difference between the completion time of the long jobs and the completion time of the short jobs. However, the completion time of the long jobs is not known in advance, so that the value to be chosen for the synchronisation delay cannot be determined. This problem can be solved by using the following iterative method (which resembles Heidelberger and Trivedi's *method of complementary delays* [14], but in contrast to this method applies to sections with a constant completion time).

Consider a *Glamis* machine model with L blocks $\langle C_1, \ldots, C_L \rangle$, where C_i ($i \in \{1, \ldots, L\}$) can be any block $Q_k^m(D_i)$ or $I(D_i)$. A program consisting of one parallel section with two types of jobs imposes the following workload on the resources: $N_A : \langle V_{A1}, \ldots, V_{AL} \rangle + N_B : \langle V_{B1}, \ldots, V_{BL} \rangle$

Analysing the model yields two response time values, R_A and R_B. The response time $\min(R_A, R_B)$ of the short jobs is a good estimation for the actual completion time of these jobs. Construct an augmented machine model, consisting of the original model extended with a delay centre representing the barrier synchronisation delay of the short jobs: $\langle C_1, \ldots, C_L, I(D_{\text{synch}}) \rangle$

A first estimation of the synchronisation delay is $D_{\text{synch}} = |R_A - R_B|$. This is an over-estimation, because the response time for the long jobs is over-estimated. The new workload description is

$$N_A : \langle V_{A1}, \ldots, V_{AL}, 1 \rangle + N_B : \langle V_{B1}, \ldots, V_{BL}, 0 \rangle \qquad \text{if } R_A < R_B, \text{ or}$$

$$N_A : \langle V_{A1}, \ldots, V_{AL}, 0 \rangle + N_B : \langle V_{B1}, \ldots, V_{BL}, 1 \rangle \qquad \text{if } R_B < R_A$$

Analysing this model yields two new response time values R_A and R_B. For the short jobs, the synchronisation delay should not be included in the new response time value. Use the new response times to calculate the new estimation $D_{\text{synch}} = |R_A - R_B|$ for the synchronisation delay. Now, both the response time for the long jobs and the synchronisation delay are under-estimations of the real values. We analyse the model again, which yields new estimations, etc. the above-mentioned steps are repeated until $\min(R_A, R_B) + |R_A - R_B| - \max(R_A, R_B) < \delta$, where δ is the required precision. The analysis algorithm is summarised in Fig. 2 in pseudo-code.

$$(R_A, R_B) := MVA(\langle C_1, \ldots, C_L \rangle, N_A : \langle V_{A1}, \ldots, V_{A2} \rangle + N_B : \langle V_{B1}, \ldots, V_{B2} \rangle);$$
if $R_A < R_B$ then $i_A := 1$ else $i_A := 0$;
$i_B := 1 - i_A$;
repeat
    $D_{\text{synch}} := \text{abs}(R_A - R_b)$;
    $(R_A, R_B) := MVA(\langle C_1, \ldots, C_L, I(D_{\text{synch}}) \rangle,$
                    $N_A : \langle V_{A1}, \ldots, V_{A2}, i_A \rangle + N_B : \langle V_{B1}, \ldots, V_{B2}, i_B \rangle);$
    if $R_A < R_B$ then $R_A := R_A - D_{\text{synch}}$ else $R_B := R_B - D_{\text{synch}}$;
until $\text{abs}(\min(R_A, R_B) + D_{\text{synch}} - \max(R_A, R_B)) < \delta$;
return$(\max(R_A, R_B))$;

Fig. 2. The multiple-class analysis algorithm

Sections with More Classes of Jobs. The described method to predict the completion time of a section of non-identical parallel tasks can be generalised to more than two classes of jobs in a straightforward manner. In that case, for every job class except the one with the greatest response time a delay centre is defined representing the synchronisation delay. Every iteration now yields a new response time estimation for every class. Accurate results are only obtained for the jobs with the lowest response time (first step, without synchronisation delays) and for the jobs with the highest response time (i.e., the section completion time), this is the value to which the method converges. The complexity of MVA increases as the number of classes increases. However, in many practical applications load imbalances are captured with sufficient accuracy by only a limited number of classes (this applies for instance to cyclic load distributions, cf. the example in section 4 where two classes suffice).

4 Case Study: LU Factorisation

In this section the concepts of the *Glamis* methodology and the associated performance prediction methods are demonstrated, using LU factorisation on a shared-memory computer as a case study. The predictions are compared to detailed simulations. The simulation programs have been written in C, making use of the VOP concurrent simulation library [22].

Consider a parallel implementation of the LU-factorisation (without pivoting) of an $n \times n$ matrix [24], running on a shared-memory computer with p processors and m memory banks interconnected through a packet-switched single bus interconnection. Every processor processes n/p columns of the matrix (assuming $p|n$). The matrix columns are divided over the processors in an interleaved way. The algorithm (in pseudo-code) is shown in Fig. 3.

Figure 4 shows the task graph for a shared-memory implementation of the algorithm in case $p = 4$ and $n = 8$. The black tasks represent one of the processors

```
for k = 1 to n − 1
do begin
      scale(k);
      forall j = 1 to p
      do for l = 0 to n/p − 1
         do if j + l * p > k
            then update(k, j + l * p)
end
```

```
procedure scale(k)
begin
   LOAD; DIV;
   for i = k + 1 to n
   do begin
         LOAD;
         MULT; STORE
      end
end
```

```
procedure update(k, l)
begin
   LOAD;
   for i = k + 1 to n
   do begin
         LOAD; LOAD;
         MULT; SUB; STORE
      end
end
```

Fig. 3. LU factorisation algorithm

scaling a matrix column. The other tasks represent the processors updating their columns in parallel (the actual Gaussian elimination). Open tasks represent processors updating one column, while shaded tasks represent tasks updating two columns. Because of the load imbalance in the former three parallel sections, the multiple-class *Glamis* algorithm must be used for these, while for the latter four parallel sections the single-class *Glamis* algorithm suffices. The serial sections will not be further considered (except for the total execution time), because for these simple complexity analysis yields exact results.

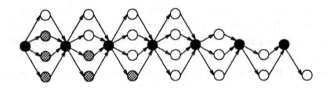

Fig. 4. Task graph for LU factorisation with $P = 4$ and $N = 8$

4.1 Machine Model

The queueing model representing the used architecture can be constructed in a straightforward manner. The floating-point arithmetic carried out locally within a processor is represented by a delay centre, with service time t_f being the duration of a single floating-point operation. The bus and each of the memory banks are represented by a single-server queueing centre, with respectively service time t_b (the bus access time) and t_m (the memory access time). We will assume a uniform distribution of memory references over the memory banks, so that the m memory banks can be represented by a single building block $Q^m(t_m/m)$. The resulting queueing model is shown in Fig. 5, or in tuple notation: $\langle I(t_f), Q(t_b), Q^m(t_m/m)\rangle$. The machine instructions for memory access (LOAD

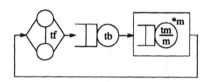

Fig. 5. Machine queueing model

and STORE) and floating point operations (MULT, DIV and SUB) are mapped to a workload description as indicated in Sect. 3.2.

4.2 Program Model

The given LU factorisation implementation consists of $n-1$ serial sections (the procedure **scale**) and $n-1$ parallel sections (p parallel processes each invoking the procedure **update** 0 up to n/p times). Every invocation of **scale**(k) results in $1+n-k$ floating-point operations and $1+2(n-k)$ memory accesses, while every invocation of **update**(k, l) results in $2(n-k)$ floating-point operations and $1+3(n-k)$ memory accesses. Applying the above-mentioned mapping of machine instructions to a workload model gives, for $p = 4$ and $n = 8$:

$1 : \langle 8, 15, 15\rangle;\ [1 : \langle 14, 22, 22\rangle + 3 : \langle 28, 44, 44\rangle];$
$1 : \langle 7, 13, 13\rangle;\ [2 : \langle 12, 19, 19\rangle + 2 : \langle 24, 38, 38\rangle];$
$1 : \langle 6, 11, 11\rangle;\ [3 : \langle 10, 16, 16\rangle + 1 : \langle 20, 32, 32\rangle];$
$1 : \langle 5, 9, 9\rangle;\quad 4 : \langle 8, 13, 13\rangle;\ 1 : \langle 4, 7, 7\rangle;\ 3 : \langle 6, 10, 10\rangle;$
$1 : \langle 3, 5, 5\rangle;\quad 2 : \langle 4, 7, 7\rangle;\quad 1 : \langle 2, 3, 3\rangle;\ 1 : \langle 2, 4, 4\rangle$

4.3 Results

First we consider the situation $p = 4$ and $n = 8$, i.e. two matrix columns are assigned to every processor. The other parameter settings are $m = 4$, $t_f = 0.200$,

$t_b = 0.025$ and $t_m = 0.200$ time units. The different parallel sections of the program are studied in isolation, because these are interesting from a performance modelling point of view. *Glamis* predictions are compared to simulation results (using deterministic delays) in Fig. 6. The sharp drop in the execution time at parallel section 4 is caused by the fact that in the former three parallel sections some of the processors update two matrix columns, while in the latter four parallel sections the processors update at most one matrix column. For comparative purposes the graph also shows section completion times obtained with regular MVA, assuming an equal distribution of the total workload over the processors. This leads to an under-estimation of the completion time of the former three sections, because their load imbalance is disregarded.

Section	Sim.	*Glamis*	Err.%
1	17.9	19.07	6.2
2	15.1	15.78	4.4
3	12.5	12.73	1.8
4	5.77	5.82	0.9
5	4.06	4.10	1.0
6	2.57	2.59	0.8
7	1.30	1.30	0.0
program	80.26	82.57	2.9

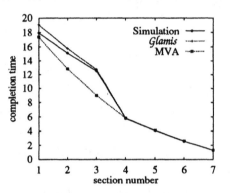

Fig. 6. Parallel section and program completion times: *Glamis* vs. simulation

Another interesting performance measure is the effect of the number of processors in terms of speedup, for a fixed matrix size. An experiment to study this effect was carried out for $n = 32$ and p ranging from 1 to 32 (the other parameter settings are the same as in the previous experiment). *Glamis* predictions versus simulation results are shown in Fig. 7. For comparative purposes complexity analysis results are included in the graph, showing the effect of the omission of resource contentions from the model. Complexity analysis does include the efficiency loss due to serial sections (Amdahl's law).

Both experiments show that the *Glamis* predictions approach the simulated values well. Although in many situations the absolute value of the execution times is slightly over-estimated (as a result of the probabilistic assumptions made), the relative influence of parameter choices and architectural choices is predicted correctly. The relatively small problem size and system size chosen in this example better illustrate the properties of the methodology, even though they might seem unrealistic. Especially in case of a larger number of processors the queueing model will become saturated and the bottlenecks will dominate the performance, which leads to misleadingly accurate results. Due to the scalability of the methodology, larger systems or applications will not pose any problems.

p	Sim.	*Glamis*	Err.%
1	1.00	1.00	0.0
2	1.81	1.77	-2.2
4	3.03	2.83	-6.6
8	4.26	3.94	-7.5
16	5.04	4.78	-5.2
32	5.43	5.30	-2.4

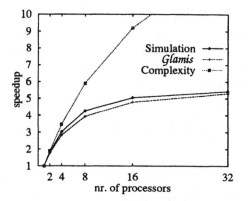

Fig. 7. Speedup for varying number of processors: *Glamis* vs. simulation

5   Conclusions

In this paper a general performance modelling methodology for parallel applications, *Glamis*, is presented. A new iterative algorithm is introduced to analyse parallel sections (with an implicit barrier synchronisation) with multiple classes of parallel tasks, while for parallel sections with one class of processes regular MVA suffices. A case study indicates that reliable performance predictions are obtained with our analysis methods, improving over regular MVA.

The class of programs that can be captured in *Glamis* models (for the moment SPS-programs) is a subset of the class of programs that can be handled by Mak's method [19] (SP-programs), which in its turn is a subset of the class of programs that can be handled by Thomasian's method [28] (general DAGs). However, *Glamis* has several advantages to compensate for that. Although *Glamis* makes probabilistic assumptions for e.g. service time distributions, certain other probabilistic assumptions, which have been shown to have a more serious effect on the accuracy (e.g. models with probabilistic loop bounds), are not needed. Other advantages of *Glamis* are the possibility to make use of replications in the system to be modelled, the separation of program models from machine models resulting in *reusable* models, and its suitability for tool support.

Future research includes the extension of the classes of parallel programs and parallel machines that can be handled. Special attention will be paid to the analysis of simultaneous resource possession, so that e.g. circuit-switched interconnection networks can be modelled. A default method to include simultaneous resource possession is the *method of surrogates* [15], but perhaps more efficient (approximate) solutions can be found. Furthermore we will consider the use of models with non-exponential service times in order to improve the prediction accuracy, without a significant increase of analysis cost. Another aspect that needs to be considered is the treatment of *data dependencies*, e.g. conditional statements. Finally, a more complete validation is required, including distributed-memory applications and a comparison of our predictions to measurements.

Acknowledgement

I am very grateful to Arjan van Gemund and prof. G.L. Reijns, for their support and many useful suggestions with respect to my research and the contents of this paper. I also wish to thank the referees for their valuable comments.

References

1. V.S. Adve and M.K. Vernon, "The influence of random delays on parallel execution times," in *Proc. 1993 ACM SIGMETRICS Conf. on Meas. and Mod. of Comp. Syst.*, May 1993.
2. M. Ajmone Marsan, G. Balbo and G. Conte, "A class of Generalized Stochastic Petri Nets for the performance analysis of multiprocessor systems," *ACM TOCS*, vol. 2, May 1984, pp. 93–122.
3. G. Balbo, S.C. Bruell and S. Ghanta, "Combining queueing networks and Generalized Stochastic Petri Nets for the solution of complex models of system behavior," *IEEE Tr. Comp.*, vol. C-37, Oct. 1988, pp. 1251–1268.
4. G. Balbo, "Performance issues in parallel programming," in *Proc. 13th Int. Conf. on Applic. and Theory of Petri Nets*, Sheffield, June 1992, pp. 1–23.
5. Y. Bard, "Some extensions to multiclass queueing network analysis," in *Performance of Computer Systems* (M. Arato, A. Butrimenko and E. Gelenbe, eds.), North-Holland, 1979.
6. F. Baskett, K.M. Chandy, R.R. Muntz and F.G. Palacios, "Open, closed, and mixed networks of queues with different classes of customers," *J. ACM*, vol. 22, Apr. 1975, pp. 248–260.
7. P. Buchholz, "Hierarchical Markovian models: Symmetries and reduction," in *Proc. 6th Int. Conf. Mod. Techniques and Tools for Comp. Perf. Eval.*, Edinburgh, Sept. 1992.
8. J.P. Buzen, "Computational algorithms for closed queueing networks with exponential servers," *Comm. ACM*, vol. 16, Sept. 1973, pp. 527–531.
9. K.M. Chandy, U. Herzog and L. Wu, "Parametric analysis of queueing networks," *IBM J. on Research & Development*, vol. 19, no. 1, 1975, pp. 36–42.
10. S. Donatelli and M. Sereno, "On the product-form solution for Stochastic Petri nets," in *Proc. 13th Int. Conf. on Applic. and Theory of Petri Nets*, Sheffield, June 1992, pp. 154–172.
11. A. Ferscha, "A Petri net approach for performance oriented parallel program design," *J. Par. and Distr. Comp.*, vol. 15, 1992, pp. 188–206.
12. D. Frosch-Wilke, "Exact performance analysis of a class of product-form stochastic Petri nets," in *UK Perf. Eng. Workshop for Comp. and Telecomm. Syst.*, Loughborough, July 1993.

13. A.J.C. van Gemund, "Performance prediction of parallel processing systems: The PAMELA methodology," in *Proc. 7th ACM Int. Conf. on Supercomputing*, Tokyo, July 1993, pp. 318–327.

14. P. Heidelberger and K.S. Trivedi, "Analytic queueing models for programs with internal concurrency," *IEEE Tr. Comp.*, vol. C-32, no. 1, Jan. 1983, pp. 73–82.

15. P.A. Jacobson and E.D. Lazowska, "Analyzing queueing networks with simultaneous resource possession," *Comm. ACM*, vol. 25, Feb. 1982, pp. 142–151.

16. H. Jonkers, "Queueing models of shared-memory parallel applications," in *UK Perf. Eng. Workshop for Comp. and Telecomm. Syst.*, Loughborough, July 1993.

17. H. Jonkers, "Probabilistic performance modelling of parallel numerical applications: A case study," in *Proc. Parallel Computing '93*, Grenoble, Sept. 1993.

18. G. Klas, "Net level aggregation using nonlinear optimization for the solution of hierarchical Generalized Stochastic Petri Nets in performance evaluation," in *Comp-Euro '92*, The Hague, May 1992.

19. V.W. Mak and S.F. Lundstrom, "Predicting performance of parallel computations," *IEEE Tr. on Par. and Distr. Syst.*, vol. 1, July 1990, pp. 257–270.

20. M.K. Molloy, "Performance analysis using stochastic Petri nets," *IEEE Tr. Comp.*, vol. C-31, Sept. 1982, pp. 913–917.

21. B. Plateau, J.M. Fourneau and K.H. Lee, "Peps: A package for solving complex Markov models of parallel systems," in *Proc. 4th Int. Conf. on ModTechn. and Tools for Comp. Perf. Eval.*, Palma, Sept. 1988, pp. 341–360.

22. R. Pulleman, "Simulation of VOP models," Tech. Rep. 92 TPD-ZP 938, TNO Institute for Applied Physics, Delft, Sept. 1992.

23. M. Reiser and S.S. Lavenberg, "Mean value analysis of closed multichain queueing networks," *J. of the ACM*, vol. 27, Apr. 1980, pp. 313–322.

24. Y. Robert, *The Impact of vector and Parallel Architectures on the Gaussian Elimination Algorithm*. Manchester University Press, 1990.

25. J.A. Rolia and K.C. Sevcik, "Fast performances estimates for a class of Generalized Stochastic Petri Nets," in *Proc. 6th Int. Conf. Mod. Techn. and Tools for Comp. Perf. Eval.*, Edinburgh, UK, 1992.

26. P. Schweitzer, "Approximate analysis of multiclass closed networks of queues," in *Proc. of International Conf. on Control and Optimization*, Amsterdam, 1979.

27. M. Siegle, "Using structured modelling for efficient performance prediction of parallel systems," in *Proc. Parallel Computing '93*, Grenoble, Sept. 1993.

28. A. Thomasian and P.F. Bay, "Analytic queueing network models for parallel processing task systems," *IEEE Tr. Comp.*, vol. C-35, Dec. 1986, pp. 1045–1054.

29. H. Wabnig, G. Kotsis and G. Haring, "Performance prediction of parallel programs," in *Proc. Workshop on Meas., Mod. and Eval. of Comp. and Comm. Syst.*, Aachen, Sept. 1993.

30. J. Zahorjan *et al.*, "Balanced job bound analysis of queueuing networks," *Comm. ACM*, vol. 25, Feb. 1982, pp. 134–141.

Automatic Scalability Analysis of Parallel Programs Based on Modeling Techniques

Allen D. Malony[1], Vassilis Mertsiotakis[2], Andreas Quick[3]

[1] Dept. of Computer and Information Science, University of Oregon, Eugene, OR 97403, USA
[2] University of Erlangen-Nürnberg, IMMD VII, Martensstr. 3, 91058 Erlangen, Germany
[3] Thermo Instruments Systems, Frauenauracher Str. 96, 91056 Erlangen, Germany

Abstract. When implementing parallel programs for parallel computer systems the performance scalability of these programs should be tested and analyzed on different computer configurations and problem sizes. Since a complete scalability analysis is too time consuming and is limited to only existing systems, extensions of modeling approaches can be considered for analyzing the behavior of parallel programs under different problem and system scenarios. In this paper, a method for automatic scalability analysis using modeling is presented. Initially, we identify the important problems that arise when attempting to apply modeling techniques to scalability analysis. Based on this study, we define the *Parallelization Description Language* (PDL) that is used to describe parallel execution attributes of a generic program workload. Based on a parallelization description, stochastic models like graph models or Petri net models can be automatically generated from a generic model to analyze performance for scaled parallel systems as well as scaled input data. The complexity of the graph models produced depends significantly on the type of parallel computation described. We present several computation classes where tractable graph models can be generated and then compare the results of these automatically scaled models with their exact solutions using the modeling tool PEPP.

1 Introduction

Implementing parallel programs for scalable parallel systems is difficult since the program's behavior could vary for different problem sizes and different system configurations. In order to implement portable and efficient programs which will also have good performance scalability, parallelization choices must be tested for many systems and problem testcases. Such empirical analysis is time consuming and is limited to existing parallel computer systems.

Modeling parallel programs with discrete event models like stochastic graph models [15] or stochastic Petri nets [1] is a well–known and proven method to analyze a program's dynamic behavior. It can be used to predict the program's execution time [16], and, by changing model parameters, help to understand the program's general performance behavior, to investigate reasons for performance bottlenecks, or to identify program errors.

When using modeling for scalability analysis, we desire to compute speedup values from the predicted runtimes of model instances for different numbers of processors or problem sizes. As done with performance monitoring, we also want to use modeling to analyze different parts of the program in order to obtain a detailed scalability profile [2, 11]. A significant advantage of modeling vs. monitoring is that model–based analysis is not restricted to existing systems and does not, necessarily, require access to existing systems for experimentation. Thus, we hope to be able, using modeling, to evaluate whether it is worth scaling a parallel machine and what the best scale of the system would be.

A systematic scalability analysis based on modeling techniques requires the creation of models for different configuration and topologies of the parallel system as well as for different problem sizes. However, model creation is a difficult problem – issues such as problem mapping and processor scheduling can quickly lead to large model complexity. One approach for automatic analysis might be to develop a model generator which automatically creates multiple "scaled" models by extending a basic "generic" model of the program to be analyzed. Adopting this idea, we have developed the *Parallelization Description Language* (PDL) for describing the structure of parallel programs, the parallelization scheme for each parallel program part, and various aspects of a program's runtime behavior.

Although this "generator" approach could be undertaken with different modeling techniques, we first target stochastic graph modeling for scalability analysis and automatic model generation. In several respects, scalability analysis using stochastic graph models is the most challenging one because of the close association between model complexity and solution tractability and accuracy. The key issue is to find model generation methods which produce approximately accurate models of scaled performance behavior but that do not exceed the solution capabilities of stochastic modeling tools. To evaluate the efficacy of our techniques, we have integrated methods for model generation and scalability analysis into our tool PEPP (*Performance Evaluation of Parallel Programs*) [4]. Based on the parallelization description language, PDL, a model generator for other model targets (like stochastic Petri net models) can be implemented in a similar manner.

The remainder of the paper is organized as follows. In section 2, the concept of model–based scalability analysis is introduced. The automatic creation of scalability models is addressed in section 3. Here, different parallel computation classes are described, and generic and scaled models of those classes are discussed. Our parallelization description language is presented in section 4. Finally, in section 5 it is shown how scalability analysis with stochastic graph models using our modeling tool PEPP can be carried out.

2 The Methodology

Modeling programs to be executed on parallel or distributed systems is too complicated a task to develop a model from scratch. For this reason Herzog proposed a "three step methodology" [9] to reduce modeling complexity. Instead of creating a single, monolithic model for each combination of workload, machine configuration, and load distribution, a *workload model* is developed independent of its implementation concerns. The *machine model* is also developed separately. The *system model* is then obtained by mapping the workload model onto the machine model. The combined model reflects the dynamic, mapped program behavior and shows how system resources are used (Figure 1).

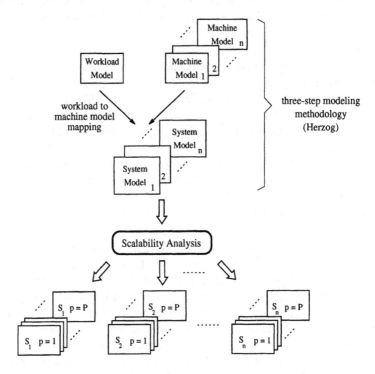

Fig. 1. Concept for Automatic Scalability Analysis

Our approach for scalability analysis uses this methodology as a basis, but extends it by another necessary step: a system model is enumerated to model various degrees of parallelism varying from $p = 1$ to $p = P$, where P is the maximum number of processors considered for scalability analysis. Since the number of processors of the parallel or distributed machine is not fixed in scalability analysis, the number of processors should not — in contrast to Herzog's approach — be set in the machine model. The mapping of the workload model onto the machine model must be realized for each number of processors separately by creating (ideally, automatically) *scaled models*. In our approach, the machine model only describes synchronization mechanisms and performance distributions for each important machine component.

Unfortunately, techniques that automatically generate scaled models must address several difficult issues. Parallelization schemes must be well understood in order for the automatic mapping of tasks to processors to occur. This requires some representational form to be defined that identifies parallelization characteristics for different classes of computations. The difficulty, however, is that some parallelization schemes, although simple, can significantly impact scaled graph model complexity — resulting in solution intractability — if exact execution behavior is modeled. Although modeling techniques have been developed that are "largeness tolerant" [17] (i.e., can deal to some extent with graph complexity), the process of creating a correct and accurate graph model is non–trivial. In order to overcome these model generation and evaluation problems, we should instead develop approximate models which can capture the correct behavior of the program, but that do not sacrifice analysis accuracy. Approximate model generation, however, is not easy due to problems such as task dependencies, scheduling, and synchronization. Furthermore, the generation of task density functions for approximate scaled models presupposes intimate knowledge of performance interactions between parallel tasks. Techniques must be developed for both graph structure approximations and execution time distribution approximations, keeping in mind, of course, that naive approximations can lead to invalid analysis results.

Finally, it is our aim to implement not only scalable, but also portable parallel programs. Therefore, our work should also address the use of different machine models in scalability study. By mapping the workload model onto n different machine models we obtain at least n different system models, each generating a scaled model set. Using this methodology, model–based scalability can be used to compare the scalability performance of different machines.

3 Scalability Models for Parallel Programs

Using stochastic graph models, the execution order of program activities, their runtime distribution, and branching probabilities can be represented. Besides modeling algorithmic properties, graph models can also be used to model the mapping onto a parallel machine, which is a prerequisite for scalability analysis. A parallel program is modeled by a graph $G = (V, E, T)$ which consists of a set of nodes V representing program tasks and a set of directed edges (arcs) $E \subset V \times V$ modeling the dependences between the tasks. To each program task v_i a random variable $T_i \in T$ is assigned which describes the runtime behavior of v_i ($T_i, i = 1, \ldots, n$, are assumed to be independent random variables).

Our goal with automatic scalability analysis is to make it possible for modeling tools to be applied to scaled versions of parallel programs where it is the number of processors or size of problem or both that are changing. The principal problem to solve is representational. That is, how scalability properties of a program – which might be known at different levels of detail and accuracy – are represented in a manner that a modeling tool can use.

We believe that this problem is best approached by considering different computation classes. In this section we consider several computation classes that are related to well-known parallel execution paradigms. We attempt to define the scaling behavior of these classes and to formulate how scaled models will be developed for them.

3.1 Parallelization of Independent Tasks

Perhaps the simplest parallel execution paradigm is one of n equivalent tasks executing on p processors. If $n > p$, then the tasks must be assigned to the processors. A trivial form of assignment is a dynamic one where each processor takes a task, executes it, and retrieves another until all tasks are completed. Given one processor, the tasks execute sequentially. Given $p \leq P$ processors, p tasks can be executing in parallel.

Let v_i be a task in the set of n tasks, V, that are to be executed. Suppose $f_i(t)$ represents the density function of the execution time of task v_i, for $1 \leq i \leq n$. Under the assumption of identically distributed tasks, $f_i(t) = f_j(t), \forall i, j \leq n$, let $f_T(t)$ represent the overall distribution density. In this case, the one processor (i.e., sequential) execution time is given by the convolution of all densities $f_i(t)$ resulting in the density $f_T(t) = \oplus_{i=1}^{n} f_i(t)$. The graph models in Figure 2(a) show two graph model versions of the sequential case.

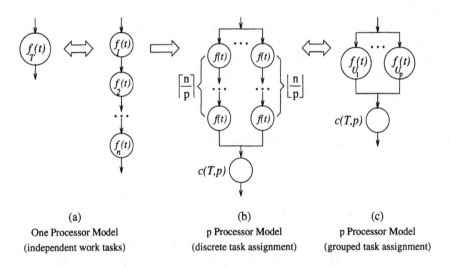

(a)
One Processor Model
(independent work tasks)

(b)
p Processor Model
(discrete task assignment)

(c)
p Processor Model
(grouped task assignment)

Fig. 2. Independent, Identically Distributed Tasks

From the single processor graph description, what is needed to allow a scaled analysis? In this case, very little. We know that the tasks must be assigned to at least one processor. Although it will be critically important to know the scheduling strategy in later cases, here only one strategy makes sense because the tasks are identically distributed. Knowing this, the modeling tool can easily generate the graph model shown in Figure 2(b), where the tasks are assigned to independent paths (representing processors) as equally as possible; each path has a subset of tasks, represented by U_1, ..., U_p. The tool can then reduce this graph to the simpler one shown in Figure 2(c). The important thing to note is that separate paths have the same mean execution time only when $n \bmod p = 0$. In any case, the mean completion time of the task set V will be determined by all subsets.

Even with this simple model, there are several issues to address. First, even though tasks are independent, it is often the case that their parallel execution can influence their execution time, due to resource contention in the target system. This contention will be a consequence of the machine model and the workload mapping. If we chose to model low-level contention overhead due to parallel interactions, the scaled models may become too complex to solve, even in this simple case. Another approach would be to model the accumulated delay caused by the interactions using a single distribution function based on the type(s) of interactions expected, the number of tasks, and the number of processors. With respect to the graph model, this would require an "interaction" or "contention" node to be added; this node is shown in Figure 2(b,c)) with the distribution function $c(T, p)$.

If the tasks are independent, but not identically distributed, $f_i(t) \neq f_j(t)$, performance will depend significantly on task load balance. Several static scheduling scenarios could be applied with accumulative "per processor" distributions computed by the convolution of the densities of tasks assigned to a processor, $f_{U_i}(t) = \oplus f_j(t), v_j \in U_i$. A dynamic scheduling scenario is more difficult to model, but a simple lower bound estimate is given by $\frac{\oplus f_j(t)}{p}$. As in the identically distributed case, a contention node is added at the end.

3.2 Parallelization of Dependent Tasks

Of course, any parallel program will have portions of the computation where the executing tasks are inter–dependent. The resulting models have a more complex structure than the models discussed in section 3.1 due to the synchronization arcs in the task graph. In general, dependencies can be quite arbitrary. In practice, certain dependent parallel computation classes are quite common in real–world applications. In the following, we discuss model scalability for some standard computation classes; Figure 3 shows some of the classes we will be considering.

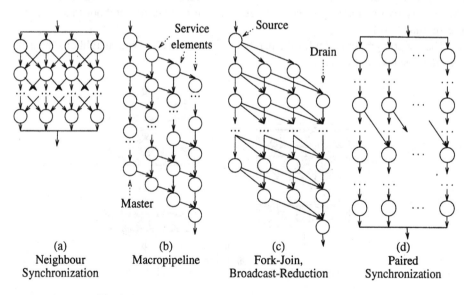

(a)	(b)	(c)	(d)
Neighbour Synchronization	Macropipeline	Fork-Join, Broadcast-Reduction	Paired Synchronization

Fig. 3. Parallel Computation Classes with Dependent Tasks

Before treating specific cases, it is instructive to consider what problems we might encounter. Computation classes are best defined by the pattern of task interaction; that is, dependency constraints. The problem size often translates into the number of tasks represented in the computation graph and the number of iterations of the basic graph structure (i.e., phases of the computation). The task density functions are rarely random: either they are related by the type of algorithm, or the same set of functions is used several times because the computation repeats. When generating a scaled model, we must try to determine some property of the computation class that allows us to transform the generic model, representing the detailed computation, to a tractable graph model.

Structurally, the scaled model should be of a form that tools like PEPP can analyze. The number of task nodes and the dependency structure of even basic computation graphs can overwhelm graph modeling tools. Thus, not only the size of the generic model, but also the dependency structure must be transformed. For instance, we would prefer that the task graph of the scaled model be a function of the number of processors, rather than the number of "scaled" generic tasks. Also, we want to use graph structures that can be reduced during model analysis. However, the trick will be to perform model scaling in a way that does not sacrifice modeling accuracy. Because performance scalability is intimately tied to parallel task interactions, reducing the detail at which these interactions are modeled in order to allow tractable solutions risks the loss of performance predictability.

Neighbor Synchronization. Many iterative solution methods for linear equation systems can be modeled by a neighbor structure (Figure 3(a)). The main characteristic of this computation class is that a processor can start the i-th iteration only after the $(i-1)$-th iteration has finished on its neighbor processors [10]. (Although Figure 3(a) shows only two neighbor processes, in general, the number of neighbor tasks can be greater than two.)

The generic graph model for the parallel computation class with neighbor synchronization is shown in Figure 4.

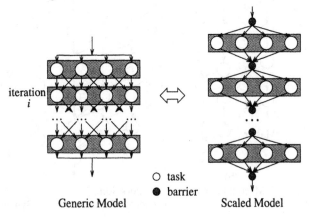

iteration i

○ task
● barrier

Generic Model   Scaled Model

Fig. 4. Scaling Neighbor Synchronization Models

If we were to represent, in the scaled model, each task and dependency in the generic model, the graph size and complexity would be unmanageable. However, it is easy to identify that the tasks at each iteration are independent and could be modeled by the techniques in section 3.1, but the neighbor synchronization has to be simplified. Our approach is to collapse the neighbor synchronization between iterations to a single barrier, as shown in Figure 4.

Although we can accurately capture the per iteration task execution times in the scaled model (albeit using a scaled independent task model), the synchronization approximation has certain model analysis consequences. In particular, because synchronization and computation cannot overlap in the scaled model, it will have poorer worst case behavior than expected for an exact model. However, this effect is reduced if the tasks at each iteration are identically distributed, or if the number of tasks at each iteration is significantly larger than the number of processors. The principal advantage of the scaled model is that it can be easily analyzed by graph modeling tools.

Macropipeline. If a task can be divided into subtasks, where the result of one subtask is the input to another subtask, the macropipeline computation paradigm (Figure 3(b)) results. The structure of a macropipeline is characterized by the number of service elements and the number of jobs. The execution time is determined by the interarrival time of new jobs and the runtime distribution of one pipeline stage. The macropipeline task graphs have also been referred to as mesh graphs [21] and are characteristic of wavefront computations.

Our scalability approach is similar to that for neighbor synchronization (Figure 5). We identify sets of independent tasks and separate their parallel execution in the approximate graph model by barrier synchronization. Many of the analysis ramifications are also the same. The graph structure is more complex, however, in that a different number of tasks are present at each graph stage. It is also possible that the tasks will be of different types, complicating the scaled submodel for independent tasks.

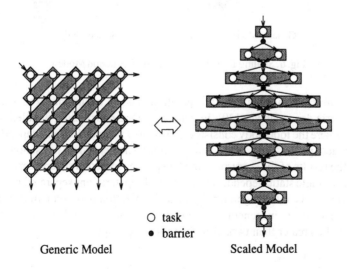

O task
• barrier

Generic Model   Scaled Model

Fig. 5. Scaling Macropipeline Models

Fork-Join, Broadcast-Reduction. Fork-join models are characteristic of computations where a source periodically generates jobs that spawn tasks to be completed with a drain node collecting results (Figure 3(c)). The graphs that result also have many similarities to graphs generated from computations that involve a sequence of broadcast and reduction operations. Such graphs are typical of linear algebra computations.

As an example, consider the generic LU-decomposition graph in Figure 6; the graph shown here is for a 6 x 6 matrix. Given a large matrix, the graph would consist of several thousands of nodes, making certain solution techniques computationally intractable [19]. However, we can transform the generic model to a simpler scaled model. Again, our standard technique can be applied in this case by identifying independent tasks at different iteration levels. However, because of the implicit fork-join nature of the computation, its explicit representation in the scaled model is less likely to lead to modeling inaccuracies.

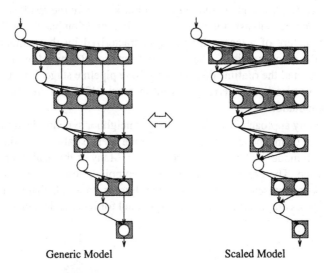

Generic Model　　　　　　　Scaled Model

Fig. 6. Scaling Fork-Join, Broadcast-Reduction Models

Paired Synchronization. Our last computation class is commonly found in parallel loops with static dependencies [20]. In general, we are considering parallelizable loop statements, where the loop body consists essentially of independent and dependent parts. The independent parts can be executed at any time. The dependent parts, however, have to wait for the results of the corresponding independent part and several earlier iteration steps. Often, a single static dependency spans two loop iterations separated by a constant number of iterations, resulting in a paired synchronization between two task nodes, as in Figure 7. However, in general, the number of static dependencies between loop iterations can be greater than two, as in the following loop:

```
do i = 1, n
    . . .
    a[i] = a [i-1] + a[i-5] + a[i-11]
done
```

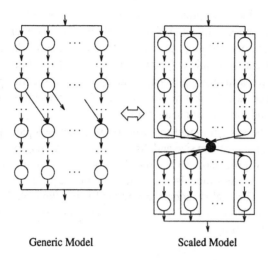

Generic Model   Scaled Model

Fig. 7. Scaling Paired Synchronization Models

Because the dependency constraints effectively separates an iteration into independent and dependent parts, we can scale the generic model by forcing a barrier synchronization before entering the dependent task nodes. Notice that we now have an option on how to parallelize the task sections. Shown in the figure is an assignment of processors to loop iterations.

For the scalability analysis of a program that contains independent and dependent working phases as described above, the following parameters are needed: the problem size (i.e. the number of iterations of the particular loop), the runtime distributions of the independent and dependent subtasks, and, finally, the scheduling strategy. The latter is especially important to enable the modeling tool to generate a model for any number of processors.

The presented list of computation classes is not complete. In order to analyze more general structures, a default parallelization scheme is provided, where the problem structure must be given in terms of a directed acyclic graph.

To conclude this section, the presented scaled models were gained from our experience in model evaluation using various bounding methods, since most of them apply modifications to the models until they are series–parallel reducible. In [13] different bounding methods are compared in order to obtain good scaled models.

4 The Parallelization Description Language PDL

In order to carry out scalability analysis of a parallel program based on modeling techniques, a general description of the program to be analyzed is needed, describing the different workload phases of the program's computation (workload model). We define the *Parallelization Description Language* (PDL) for describing how a program can be parallelized. We consider a program to consist of an arbitrary pattern of execution phases of the following types: sequential phase, parallelizable phase, and synchronization phase. Following the methodology presented in section 2, a workload model described in this manner can be combined with a machine model to build a system model, forming the basis for scalability analysis. The choice of a specific parallel architecture mainly influences a subtask's runtime, the mechanism and the duration of barrier synchronizations, and the communication times.

After creating the system model the user has to select two parameters needed for scalability analysis: the problem size and the maximal degree of parallelism, P (i.e., the maximum number of processors to be considered with this analysis). With these two input parameters, P system models are created automatically to carry out scalability analysis (Figure 8). The scaled models are derived using the techniques discussed in the previous section. (Note, if P is large, only certain numbers of processors may be selected for scalability study.)

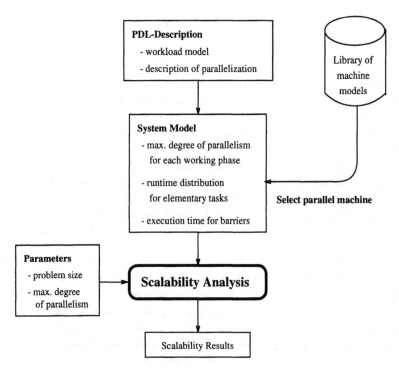

Fig. 8. Scalability Analysis Using PDL

4.1 Program Description

A description of a parallel program in PDL consists of the following three parts:

1. **Execution pattern**
   The execution order of the different work phases and synchronization phases must be specified in terms of a regular expression.
2. **Definition of the work phases**
   As the work phases represent the parallelizable parts of the program, the runtime distributions of the subtasks and the parallelization scheme for a particular work phase have to be specified.
3. **Concatenation of work phases**
   If no synchronization phase is between two parallelized work phases, they may be concatenated in different ways. The selected concatenation method must be described for each work phase which is not succeeded by a synchronization phase or a sequential phase (see section 4.3).

The specification of the program's name and the execution pattern form the first part of each PDL description. The name of the described program must be given for reference purposes; see Figure 9. After this, the work phases are described.

```
PROGRAM NAME IS 'prgA';
EXECUTION PATTERN IS  w₁– > sync– > w₂– > w₂– > w₃;
```

Fig. 9. Begin of a PDL–Description

4.2 Description of Work Tasks

In general, we distinguish between two types of work phases, namely work phases with a predefined parallelization scheme and work phases with an arbitrary parallelization scheme; see section 3.2. The definition of a work phase with a predefined parallelization scheme consists of the following elements:

- **Name of the work phase**
  For identifying the work phase in other parts of a PDL description a unique name must be assigned. This name is used for specifying the execution pattern (Figure 9) and the concatenation of two phases (Figure 12).
- **Parallelization scheme**
  Using PDL, the following six parallelization schemes (computation classes) are available (see section 3.2): INDEPENDENT, NEIGHBOR, NEIGHBOR-TORUS[4], MACROPIPELINE, FORK-JOIN, and PAIRED. For some of the parallelization schemes the partitioning of the data and the mapping of the partitions onto the processors must be given.
- **Runtime distributions**
  The user can choose between various parametric distributions like exponential, general Erlangian, or numerical distributions. Since the runtime distribution is

---

[4] The neighbor–torus class is a special case of the neighbor class were the tasks at the left and the right borders are also synchronized.

a property of each task and independent of the selected machine, the runtime
distribution must be specified in PDL. In order to represent the dependence on
the underlying machine, a runtime factor (RTFactor) can be given. This factor
is defined in the machine model and allows the comparison between different
parallel systems. The use of numerical distributions which may be obtained from
measurements makes the accurate modeling of real–world programs possible. Using
parameters obtained from monitoring renders performance prediction more relevant.

- **Problem size specification**

  This depends on the selected parallelization scheme. The problem size may be either
  a constant or a data structure which can be varied and whose size must be given
  by the user as a parameter. A variable data structure can be an array, a grid, or any
  other partitionable structure. In the case of a macropipeline or a fork–join model,
  the number of jobs and their interarrival times have to be specified. In Figure 10,
  both cases are shown.

- **Communication volume**

  For the generation of accurate system models, the amount of data to be transferred
  between processors must be known. By specifying the communication volume the
  underlying architecture can easily be changed. It must be specified in the machine
  model whether the data exchange is realized via message passing or via shared
  memory access. The time needed for data exchange is also specified in the machine
  model. In general, the duration is a function of the communication volume given in
  the PDL–description. Since the communication volume is dependent on the problem
  size, it has to be specified as a function of the problem size in case of a variable
  problem size.

```
WORK PHASE w₁ IS:
    PARALLELIZATION SCHEME IS INDEPENDENT;
    SUBTASK RUNTIME IS ERLANG(100,1*RTFactor);

    // constant problem size
    PROBLEM SIZE IS VECTOR A(100);
    COMMUNICATION VOLUME IS 100*SIZEOF(INT);
END;

WORK PHASE w₂ IS:
    PARALLELIZATION SCHEME IS NEIGHBOR;
    SUBTASK RUNTIME IS NUMERICAL(''dur_b.dis.$ARCH'');

    // variable problem size
    PROBLEM SIZE IS GRID B(height, width);
    COMMUNICATION VOLUME IS height*SIZEOF(FLOAT);
END;
```

Fig. 10. Description of Work Tasks with INDEPENDENT and NEIGHBOR Parallelization

The description of the three work phases from Figure 9, each with a predefined
parallelization scheme, is shown in Figure 10 and Figure 11. For the description of

work phases which do not match any of the predefined computation classes, PDL provides the parallelization scheme ARBITRARY (see section 3.2).

```
WORK PHASE w₃ IS:
    PARALLELIZATION SCHEME IS PAIRED;
    // i-th iteration depends on (i − 1)-th and (i − 5)-th iteration
    DEPENDENCY IS 1, 5;

    INDEPENDENT SUBTASK IS:
        RUNTIME IS ERLANG(5,1*RTFactor);
    END;

    DEPENDENT SUBTASK IS:
        RUNTIME IS DETERMINISTIC(5*RTFactor);
        COMMUNICATION VOLUME IS 1*SIZEOF(FLOAT);
    END;

    PROBLEM SIZE IS GRID C(height, width);
    COMMUNICATION VOLUME IS height*SIZEOF(FLOAT);
END;
```

Fig. 11. Description of a Work Task with PAIRED Parallelization

4.3 Concatenation of Work Phases

As already mentioned, sometimes it is desirable to model the execution of two work phases in series without a global synchronization between them. In this case, the user has to describe how to join the work phases. The following options are available: NEIGHBOR, NEIGHBOR-TORUS, TORUS, SERIAL, and BARRIER. Figure 12 shows how the concatenation of two work phases can be described in PDL.

```
CONCATENATION IS:
    // concatenation w₁, w₂ must not be specified,
    // since there is a synchronization between both phases
    w₂, w₂  : NEIGHBOR;
    w₂, w₃  : SERIAL;
END;
```

Fig. 12. Concatenation of Work Tasks

4.4 Flexibility of PDL

For flexible scalability analysis, the selected problem size and the underlying parallel machine should not be determined in the PDL–description. This has the advantage that it is not necessary to change the PDL–description each time the user wants to perform a new scalability analysis for different input data.

Therefore, PDL supports language constructs to describe program parts which are dependent on these input parameters on an abstract level. When creating scaled models the model generator replaces these variables by their actual input values. Such variables can be used to describe the problem size and the execution time of tasks in the PDL description. In Figure 10 the variables `height` and `width` are used to allow a flexible description of the problem size.

The abstract description of the communication volume in the same example (Figure 10) enables the abstraction from a real machine. Here the advantage of our modeling method becomes clear. Depending on the selected machine, the communication volume changes from 400 byte (size of an integer is 4 byte) to 800 byte, if an integer is represented by 8 bytes.

The above discussed examples demonstrate that the language PDL can be used to describe a wide range of parallel structures. The information given for each work phase is sufficient to create stochastic models for performance modeling. Since PDL is not dependent of any model type, a description in PDL can be used to generate arbitrary stochastic models like stochastic graph models or stochastic Petri net models. Since we have developed a modeling tool for efficient evaluation of stochastic graph models, we show in the next section, how a PDL–description can be used in combination with this tool to automatically create scaled stochastic graph models.

5   Automatic Scalability Analysis with the Tool PEPP

PEPP (*P*erformance *E*valuation of *P*arallel *P*rograms) [4] is a modeling tool for analyzing stochastic graph models. It provides various evaluation methods to compute the mean runtime of a modeled program. PEPP supports the analysis of real–world applications by applying efficient solutions methods including a series–parallel structure solver, an approximate state space analysis, and bounding methods to obtain upper and lower bounds of the mean runtime [8]. In order to model measured runtimes, numerical runtime distributions are allowed in all three cases. The following example illustrates how solutions to large graph models are calculated using bounding methods.

For systematic monitoring of parallel and distributed programs, PEPP implements the M^2–cycle methodology [14]. Here, a functional model of a program to be measured is used for event selection, automatic program instrumentation, and event trace evaluation. allowing the functional model to be extended into a performance model.

PEPP can be used for automatic scalability analysis by creating multiple stochastic graph models based on the program description given in PDL. After the maximal degree of parallelism P and the problem size for each variable used in PDL is specified, PEPP creates P different graph models. These models are then evaluated and speedup values are calculated from the predicted execution times. The results are presented in a speedup chart.

As done with performance monitoring, modeling can classify different parts of the program in order to obtain a detailed scalability profile (loss analysis [3, 2]). The relative influence of the different program phases on the program's execution time can be determined in the model. For this, the execution time of all program phases not considered should be set to deterministic runtimes with mean value 0. Using this technique, speedup values can be computed for only the selected program parts.

5.1 Example: Scalability Analysis of a Iterative Algorithm

In this section, we give an example for a model-based scalability analysis of a iterative algorithm which might be a part of a larger numerical computation. This example is influenced from an application running on the multiprocessor SUPRENUM [18]. The analyzed algorithm belongs to the neighbor computation class (see Figure 4). In order to model real program behavior, the execution of each task in our example is assumed as a Erlang–8 distribution. The domain to be calculated is a matrix consisting of 10 columns. If we assume that a column is the smallest unit of work that can be distributed among the processors, by default $P = 10$. The parallelization for 2 and 3 processors is depicted as an example in Figure 13 for two iterations. Each node in this figure represents a row iteration with rows numbered from left to right. Using two processors, there are inter–processor dependencies only between row 5 and 6. Using more processors, more inter–processor dependencies arise. The general model and the scaled model can be found in Figure 4.

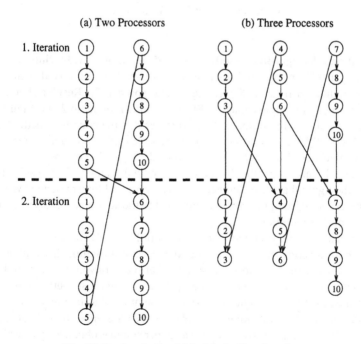

Fig. 13. Stochastic Graph Models for Two Iterations

This example shows the usefulness of model–based scalability analysis as well as the problems encountered when modeling parallel systems. Using PEPP, scalability analysis can be carried out in three different ways.

1. **Accurate modeling with state space analysis**[5]

Due to state space explosion, Markov analysis is not possible for more than 4 processors, even when reducing the number of iterations to 2 (Figure 13). For this case the state space is about 160,000 states, and model solution takes about 9 hours on a HP 715 workstation. Although PEPP solves graph models using an approximate state space analysis, model evaluation using state space analysis is not possible for a higher degree of parallelism (Table 1). It should be emphasized that the approximate state space analysis implemented in PEPP can deal with any runtime distribution (e.g. numerical distributions obtained from monitoring existing programs). Since evaluating large models (especially models with a high degree of parallelism causing a lot of task dependences) is hard, this method is not applicable in practice.

P	2 iterations	4 iterations	8 iterations
1	160	320	640
2	85	167.7	331.6
3	64.6	too high	
4	52.5	computations costs	

Table 1. Results of State Space Analysis

2. **Accurate modeling using bounding methods for model evaluation**

Bounding methods tolerate largeness because models are solved using series–parallel reduction instead of creating a state space [17]. Results obtained with bounding methods implemented in PEPP are shown in Table 2 and Figure 14. In [7] we have shown for various graph structures that the bounding methods implemented in PEPP are very accurate. In PEPP, three different bounding methods are implemented in order to select the best bound. Depending on the structure of the graph model, one or the other bounding method will yield the best result. It can be seen in the figure that the difference between upper and lower bound is very small. In some cases the computed bounds are equal to the state space analysis results.

3. **Approximate modeling**

Approximate models (like the model shown in Figure 4) can easily be evaluated if they have a series–parallel structure. In this case a runtime distribution can be calculated using the operators *series reduction* (i.e., convolution of two densities functions) and *parallel reduction* (i.e., product of two distribution functions) to reduce the graph to one single node. Here, generation of a state space is avoided.

Results obtained by evaluating our scaled models are upper bounds of the mean runtime. Kleinöder has proved [12] that an upper bound is obtained by inserting arcs. In this case, the added barrier synchronization arcs are causing higher execution times. The deletion of arcs leads to a lower bound, because execution constraints, with respect to the original model, may be violated.

The results obtained from evaluating our scaled models differ only slightly from the

---

[5] The presented numbers are modeled execution times and do not have a specific time unit

P	2 iterations	4 iterations	8 iterations
1	160	320	640
2	85 - 86.8	167.1 - 172.7	330 - 341.8
3	64.5 - 65.8	128.1 - 130.4	256 - 258.8
4	51.9 - 53.9	101.5 - 105.1	200 - 209
5	38.8 - 38.9	73.5 - 77.8	141 - 154
6	37.7 - 38.8	72.2 - 77.7	135 - 153
7	36.8 - 37.2	70.8 - 75.3	138 - 149
8	35.2 - 36.7	68.5 - 70.3	134 - 142
9	32.2 - 32.3	64 - 64.2	128 - 130
10	23.7 - 24.7	44.7 - 48.1	85.8 - 96.1

Table 2. Results of Bounding Methods

P	2 iterations	4 iterations	8 iterations
1	160	320	640
2	87.1	174.2	348.4
3	65.8	131.6	263.2
4	53.8	107.6	215.2
5	41.7	83.4	166.8
6	40.5	81	162
7	38.9	77.8	155.6
8	36.7	73.4	146.8
9	33	66	132
10	25.7	51.4	102.8

Table 3. Results of Approximate Modeling

results calculated with the sophisticated bounding methods implemented in PEPP (Table 3). Comparing the results of both methods (Table 2 and Table 3), we see that results obtained with approximate modeling estimate the runtime very well.

The speedup values presented in Figure 14 (right) verify the necessity of a systematic scalability analysis. To obtain these results with measurements, 30 measurements must be taken. For configurations of 2, 5, and 10 processors good speedup values are reached, since in these cases a good load balancing of the 10 tasks can be obtained.

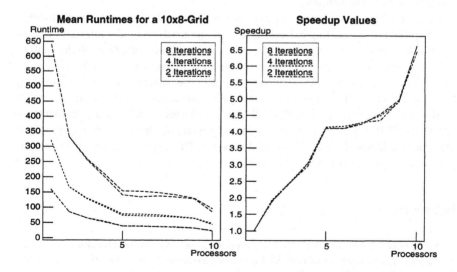

Fig. 14. Scalability Analysis Results

Note, in this example, communication delays are not considered. Using a model generator, this can easily be done by substituting each inter-processor arc by an arc, a node representing communication costs, and another arc. Communication delays depend on the underlying parallel system and are specified in the machine models. The communication volume for each data exchange is specified in PDL.

Using PDL, the problem size, the runtime distributions of all tasks and communication nodes, and the underlying synchronization mechanism can easily be varied. This is a prerequisite for a flexible and systematic scalability analysis. This first results of model–based scalability analysis show that our approach is well–suited to compare different scales of parallel systems and input data.

6 Conclusion and Prospect

Scalability analysis is an important issue when implementing parallel programs for scaled parallel systems. A systematic approach must be established wherein scaled performance can be estimated subject to the constraints of the analysis tool used. In this paper, we have presented an approach for scalability analysis based on stochastic graph modeling. There are several compelling reasons for a model–based approach from a performance evaluation standpoint, but the solution techniques must be efficient in order to return results in a timely manner.

By analyzing various computation classes, we have shown how scaled models can be created from a generic computation description in PDL and analyzed by an existing stochastic graph analysis tool, PEPP. Our results indicate that scalability analysis is possible with this approach and delivers performance predictions that are consistent with other solution techniques.

However, there are still many open issues to address. We have only briefly touched on how problem size scaling is handled or how the machine model interacts with the analysis. We are currently exploring these issue more thoroughly through the analysis of additional testcases. Another important problem is that we can handle only static computations, because the presented computation classes depend on knowing something about the structure of computation. Finally, we are investigating the integration of scalability model generation into PEPP. We believe that the model-based instrumentation support in PEPP may allow us to extrapolate a template of a generic model of programs from measurements of a few of its scaled versions. This appears particularly important when task density functions are unknown.

References

1. M. Ajmone Marsan, G. Balbo, and G. Conte. A Class of Generalized Stochastic Petri Nets for the Performance Evaluation of Multiprocessor Systems. *ACM Transactions on Computer Systems*, 2(2):93–122, May 1984.
2. F. Bodin, P. Beckman, D. Gannon, S. Yang, A. Malony, and B. Mohr. Implementing a Parallel C++ Runtime System for Scalable Parallel Systems. In *Proceedings of Supercomputing '93*, 1993.
3. H. Burkhart and R. Millen. Performance Measurement Tools in a Multiprocessor Environment. *IEEE Transactions on Computers*, 38(5):725–737, May 1989.

4. P. Dauphin, F. Hartleb, M. Kienow, V. Mertsiotakis, and A. Quick. PEPP: Performance Evaluation of Parallel Programs — User's Guide – Version 3.3. Technical Report 17/93, Universität Erlangen–Nürnberg, IMMD VII, September 1993.

5. J. Davies. Parallel Loop Constructs for Multiprocessors. Master's thesis, UIUC-CS, 1981.

6. Z. Fang, P. Yew, P. Tang, and C. Zhu. Dynamic Processor Self-Scheduling for General Parallel Nested Loops. In *ICPP-87*, pages 1–10, 1987.

7. F. Hartleb and V. Mertsiotakis. Bounds for the Mean Runtime of Parallel Programs. In R. Pooley and J. Hillston, editors, *Sixth International Conference on Modelling Techniques and Tools for Computer Performance Evaluation*, pages 197–210, Edinburgh, 1992.

8. F. Hartleb. Stochastic Graph Models for Performance Evaluation of Parallel Programs and the Evaluation Tool *PEPP*. Technical Report 3/93, Universität Erlangen-Nürnberg, IMMD VII, 1993.

9. U. Herzog. Formal Description, Time and Performance Analysis. In T. Härder, H. Wedekind, and G. Zimmermann, editors, *Entwurf und Betrieb Verteilter Systeme*, Berlin, 1990. Springer Verlag, Berlin, IFB 264.

10. U. Herzog and W. Hofmann. Synchronization Problems in Hierarchically Organized Multiprozessor Computer Systems. In M. Arato, A. Butrimenko, and E. Gelenbe, editors, *Performance of Computer Systems – Proceedings of the, 4th International Symposium on Modelling and Performance Evaluation of Computer Systems*, Vienna, Austria, Februar, 6–8 1979.

11. R. Hofmann, R. Klar, N. Luttenberger, B. Mohr, and G. Werner. An Approach to Monitoring and Modeling of Multiprocessor and Multicomputer Systems. In T. Hasegawa et al., editors, *Int. Seminar on Performance of Distributed and Parallel Systems*, pages 91–110, Kyoto, 7–9 Dec. 1988.

12. W. Kleinöder. *Stochastic Analysis of Parallel Programs for Hierarchical Multiprocessor Systems (in German)*. PhD thesis, Universität Erlangen–Nürnberg, 1982.

13. A.D. Malony, V. Mertsiotakis, A. Quick. Stochastic Modeling of Scaled Parallel Programs. Technical Report, Universität Erlangen–Nürnberg, IMMD VII, 1994.

14. A. Quick. A New Approach to Behavior Analysis of Parallel Programs Based on Monitoring. In G.R. Joubert, D. Trystram, and F.J. Peters, editors, *ParCo '93: Conference on Parallel Computing, Proc. of the Int'l Conference, Grenoble, France, 7–10 September 1993.* Advances in Parallel Computing, North–Holland, 1993.

15. R. Sahner. *A Hybrid, Combinatorial Method of Solving Performance and Reliability Models.* PhD thesis, Dep. Comput. Sci., Duke Univ., 1986.

16. F. Sötz. A Method for Performance Prediction of Parallel Programs. In H. Burkhart, editor, *CONPAR 90–VAPP IV, Joint International Conference on Vector and Parallel Processing. Proceedings*, pages 98–107, Zürich, Switzerland, September 1990. Springer–Verlag, Berlin, LNCS 457.

17. K.S. Trivedi and M. Malhotra. Reliability and Performability Techniques and Tools: A Survey. In B. Walke and O. Spaniol, editors, *Messung, Modellierung und Bewertung von Rechen- und Kommunikationssystemen*, pages 27–48, Aachen, September 1993. Springer.

18. U. Trottenberg. (ed). Special Issue on the 2nd International SUPRENUM Colloquium. *Parallel Computing*, 7, 1988.

19. H. Wabnig, G. Kotsis, and G. Haring. Performance Prediction of Parallel Programs. In B. Walke and O. Spaniol, editors, *Proceedings der 7. GI–ITG Fachtagung "Messung, Modellierung und Bewertung von Rechen- und Kommunikationssystemen", Aachen, 21.–23. Spetember 1993*, pages 64–76. Informatik Aktuell, Springer, 1993.

20. M. Wolfe. High Performance Compilers. Monograph, Oregon Graduate Institute, 1992.

21. N. Yazici-Pekergin and J.-M. Vincent. Stochastic Bounds on Execution Times of Parallel Programs. *IEEE Transactions on Software Engineering*, 17(10):1005–1012, October 1991.

SIMULATION OF
ATM COMPUTER NETWORKS
WITH CLASS*

M. Ajmone Marsan[1], R. Lo Cigno[1], M. Munafò[1] and A. Tonietti[2]

[1] Dipartimento di Elettronica, Politecnico di Torino,
Corso Duca degli Abruzzi 24, 10129 Torino – Italy
[2] CSELT, Via G. Reiss Romoli 274, 10148 Torino – Italy

Abstract. CLASS is a software package for the simulation of connectionless services in ATM networks, such as ATM LANs, or long distance ATM computer networks. CLASS was developed in standard C language in order to achieve good portability, and was implemented aiming at maximum CPU efficiency, in order to mitigate the impact of the CPU time requirements typical of the simulation of cell-based communication systems. The paper describes the main characteristics of the software tool and illustrates an application example.

1   Introduction

The Asynchronous Transfer Mode (ATM) was designed as the transport technology for the forthcoming Broadband Integrated Services Digital Network (B-ISDN), that will integrate a very wide variety of telecommunication services, ranging from the enhanced versions of traditional telephony, to the distribution of digital video signals on demand, and to the high-speed data transfer required for the interconnection of computers.

In essence, ATM networks [1, 2, 3] are based on the transfer of short constant-length packets from source to destination along pre-established connections, called virtual paths (VP) and virtual circuits (VC). The constant length packets are normally called *cells*, and they comprise 53 bytes, 5 of which are overhead for addressing and control, while 48 are payload at the ATM protocol layer. Of these 48 payload bytes, 4 are used as overhead at the layer above ATM (called AAL – ATM Adaptation Layer), so that 44 bytes only are available for user data. Thus, user messages must be segmented into 44-byte fragments that are combined with the protocol control information before accessing the ATM layer. The time required for the transmission of one 53-byte cell is normally called a *slot*.

ATM was conceived as a transport technique for the provision of connection-oriented services, but market forecasts indicate computer interconnects, or in

---

* This work was performed in the framework of a research contract between the Electronics Department of Politecnico di Torino, and CSELT.

general data services for the extension of Local and Metropolitan Area Network (LAN and MAN) services to wider areas, as the vastly predominant payload in the first wave of B-ISDN. Much attention is thus presently paid to the provision of connectionless services within ATM networks. The development of efficient solutions for computer communications within B-ISDN will make ATM the leading technology for the implementation of distributed computing applications over regional and national areas.

Actually, the availability of the first ATM interconnects, and the expectations for very high volumes (and hence low costs) in the production of B-ISDN user interfaces, has spurred a large interest in the field of ATM also in the local area, so that much research work is being devoted to ATM LANs, or local ATM networks, thus extending the impact of the ATM technology also to short distance computer interconnections for distributed applications.

This makes the analysis and design of ATM networks for connectionless services a very important problem, both in the short, and in the long distance scenarios. Since analytical models for ATM networks are not available, simulation appears to be the only viable approach for the quantitative assessment of the merits of different network configurations. However, the characteristics and requirements of ATM may make brute force simulation extremely resource consuming, specially in terms of CPU time. Much care must therefore be placed in the design of software tools for the simulation of ATM networks so as to try and obtain the best possible efficiencies.

In this paper we describe a simulation tool for the performance analysis of connectionless services in ATM networks. The tool is named Connection-Less ATM Services Simulator (CLASS). Section 2 presents an overview of the simulator structure, focusing on the models of the traffic sources, and the characteristics of the network nodes. Among the latter, a special attention is given to the so-called ConnectionLess Servers (CLS) that are entrusted with the management of connectionless services. Section 3 illustrates a representative example of the use of CLASS and of the results that can be obtained with it.

2   CLASS Overview

CLASS is a time-driven, slotted, synchronous simulator for the performance analysis of connectionless services in ATM networks.

CLASS is entirely written in standard C language, which makes it a highly portable simulation tool.

The choice of the slotted simulation approach is dictated by efficiency requirements, and is justified by the fact that the data units within the network (the cells) have constant size and by the interest in performance indices related to user messages and cells, not to smaller units of data.

The shortest time interval that is handled by the simulator, thus defining the simulation time unit, is the slot duration on the link at highest data rate, or, more precisely, the maximum common divisor of the slot durations on all the links in the network. The two definitions coincide in the normal case when the data rates on all links are either equal or multiples of one another, but differ otherwise. Whereas ATM LANs may be expected to employ links with identical data rates, the same assumption may not be reasonable for long distance networks.

CLASS is not efficient when simulating networks where the data rates on the different links are arbitrary. This is not a serious limitation in ATM networks, that are expected to employ digital links obtained from an SDH transmission facility, where data rates are multiples of one another.

The temporal resolution in CLASS requires an approximation on the propagation delays along point-to-point links, and on the processing delays within nodes, whose durations must correspond to an integer number of time units.

The time-driven simulation technique is best suited for the simulation of systems where each component is involved in some operation at any time instant, while systems whose components are idle most of the time are in general more efficiently simulated with the event-driven technique. ATM networks belong to the first category when heavily loaded, while they may belong to the second when lightly loaded. It must however be observed that the most critical situation for the analysis of systems through simulation is the heavy load, because of the greater stochastic variability of their behaviour that induces longer CPU times for a constant accuracy of the performance estimates. On the other hand, the quantitative estimation of the network performance to help in the choice among alternate configurations must consider reasonable operations scenarios, that normally assume a significant traffic load, so as to justify the necessary investments. In conclusion, since it is expected that most of the simulations concerning the characterization and performance analysis of connectionless services in ATM networks will be carried out under load conditions ranging from medium to high, a time-driven simulation technique was chosen, with the aim of minimizing the CPU requirement of the simulation runs. The result of this optimization effort is a tool that can simulate the generation, transport and delivery of a number of cells in the order of the few thousand per CPU second, depending on the complexity and load of the network.

The simulator is divided into five main functional blocks, or *entities*, whose relationships are shown in Fig. 1. Three of the five functional blocks sketched in the figure, namely USER, NODE and CLS (ConnectionLess Server) are described in some detail in Subsections 2.1, 2.2, and 2.3, respectively. The other two are shortly described below, together with the performance indices that are measured by the simulator, and the input and output sections. In the rest of the paper, when referring to the *entities* of CLASS small capitals will be used (e.g. USER, NODE, MANAGER, etc.) to distinguish these entities from the generic objects with the same name.

INPUT and OUTPUT — The output section requires no special comments, since this part of the simulator just writes the results collected during the simulation on standard ASCII files.

The input section instead is more complex, since this is the section through which the user must characterize the topological properties of the network to be simulated, and possibly control the consistency of the network description.

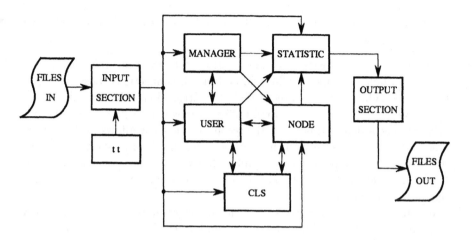

Fig. 1. The functional blocks of CLASS and their mutual relationships

To this purpose a formal grammar called NICE (Network Input for CLASS Experiments) was defined, and its parser was generated with the help of the YACC parser generator. This grammar allows a fairly easy description of the network to be simulated by defining each NODE, its connections to adjacent NODES and the USERS connected to it. With the help of the parser, it is possible to check the consistency of the described network, and trap most of the conditions that would lead to errors during the simulation (unreachable NODES or USERS, repeated identifiers, etc.).

MANAGER — The MANAGER entity provides some of the functions pertaining to network administration and management. Typical functions are the definition of the connection routing within the network, and the bandwidth allocation to the connections. Connections are opened at the beginning of the simulation, and remain open until the end of the simulation experiment.

STATISTIC — The CLASS simulation experiments are automatically stopped when the desired user-specified accuracy is reached on a selected performance parameter. The entity STATISTIC automatically determines the duration of the initial simulation transient period, collects the measured data, computes aggregate performance figures, and provides an estimation of the accuracy reached by the results, using the "batch means" technique. The implementation of the batch means technique follows the guidelines in [4], while details about the used parameters can be found in [5].

Performance Measures — CLASS aims at the estimation of the performances of connectionless services in ATM networks, i.e., services that are based on the exchange of *messages*. Messages are segmented into cells by the ATM Adaptation Layer (AAL) before transmission, and they are reconstructed before the information is delivered at the destination. Thus, the estimation of the performances of connectionless services requires the computation of performance parameters both at the cell and at the message levels.

The performance parameters computed with CLASS can be divided into two categories, as follows.

Cell and message loss rates — These parameters are measured considering both the whole network, and individual links or connections. Loss rates can be further subdivided in three different contributions: the first is due to overflows in the USER transmission buffers (the whole message, i.e., all of its cells, are discarded in this case); the second is due to overflows in the buffers associated with links between network NODES; the third is due to overflows in the buffers associated with the links that connect the destination NODE to the final destination (a USER). A message is declared lost when at least one of its cells is lost.

Cell and message delay jitters — Cell and message delays are made of a constant part due to propagation delays and processing delays within nodes, and of a variable part, due to the waiting time in the USER transmission buffer and to queuing in the buffers associated with links between nodes. Only the variable part is considered and measured. For each delay, the histogram, the average value, and the variance are computed. Also these performance parameters are computed considering both the whole network, and individual links or connections.

As pointed out above, results are collected separately for each component of the network. There are two different reasons for keeping separate the contributions from different components of the network. The first is due to the need of identifying bottlenecks and pathological conditions within the network. The second, instead, concerns the difference between the internal components of the network and the components at the border of the network. The former are for example the links connecting two NODES, and the latter are for example the links connecting a NODE and a USER. This separation is useful because the phenomena observed within the network and at its interface with the external world quite often have a different origin, and may require different actions. For instance, overflows within a user transmission buffer may be solved with a better shaping or policing of the input traffic, while a congestion on a node-to-node link probably calls for a different network topology, or a faster link.

2.1 USER Entity

In the simulation of computer networks, and of many other types of systems, two different modeling aspects must be considered: the first concerns the development of functional models of the internal operations of the network itself, while the second concerns the development of models of the workload, i.e., of the traffic that is handled by the network. The USER entities in CLASS serve the purpose of generating a representative workload for the functional model of the ATM network operations. Therefore, the USER entities generate and consume the traffic that is transported by the network. After a cell has been received by the destination USER and its contribution to the performance statistics has been recorded, the cell is discarded by releasing the memory it occupies.

The wide variety of possible application scenarios in B-ISDN translates in the need for USER models of different nature, so as to adequately describe different characteristic traffic patterns. The USER models available in CLASS can be grouped into two categories: connection-oriented generators and connectionless generators.

Connection-oriented Generators — These generators model two possible traffic patterns: Constant Bit Rate (CBR) traffic, or highly bursty (On-Off) traffic. The latter model the generation of cells at fixed intervals during the On period only. The On and Off periods are geometrically distributed, but while the mean value of the On period is fixed by the user, the average duration of the Off period is automatically computed so as to meet the average traffic level requirements. Connection-oriented CBR USERS generates a continuous stream of cells (i.e., cells are not grouped into messages), with the same destination address (defined at the beginning of the simulation) throughout the whole simulation. The main goal of these source models is to create a background traffic for the connectionless services analysis.

Connectionless Generators — The generators belonging to this category produce user messages that are subsequently segmented by the AAL sublayer into cells that must be transmitted on the network. All the connectionless USERS generate traffic whose average level and whose destination are controlled by a traffic matrix that specifies each individual traffic relation. The destination is randomly selected on a message basis. Several simple generators belong in this category; they generate messages according to a Poisson arrival process. The length of the messages can be chosen among three possible distributions: constant, bimodal (i.e., two different message lenghts) or truncated geometric. These generators can be very helpful in understanding and interpreting the network operations, but may not provide a realistic description of the behaviour of the traffic arriving at an ATM network from a tributary network, such as a LAN or MAN. For the particular case of a DQDB tributary network [6] (either a LAN or a MAN), a more sophisticated traffic source model was implemented.

The DQDB USER — The traffic pattern that is generated at the interface between a DQDB LAN or MAN and an ATM network is fairly complex, as shown in [7], due to the interaction and multiplexing of the traffic from different stations on the transmission medium. On the other hand, the simulation of the DQDB MAC protocol [6] is computationally very heavy, making the simulation of the whole chain comprising the source user – the source DQDB LAN or MAN – the ATM network – the destination LAN or MAN – the destination user, absolutely not feasible. Thus, a simple and fast model is needed if a DQDB-like generator is to be integrated into an ATM simulator.

The DQDB MAC protocol implements a round-robin service discipline, whose behavior is not ideal due to the network latency. This phenomenon leads to unfairness in the bandwidth sharing among stations, whose consequences were investigated by many authors and mitigated by the introduction of the bandwidth

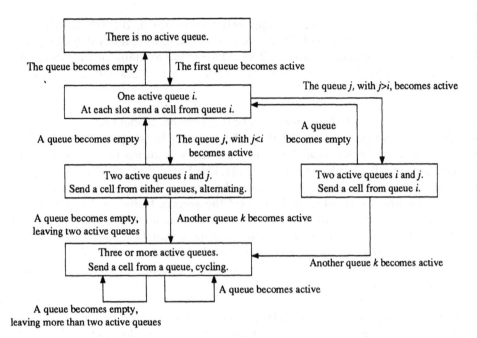

Fig. 2. Flow chart of the "scheduler" in the DQDB user

balancing algorithm in the last version of the DQDB standard (see for example [8]). The same phenomenon also leads to the particular shaping of the traffic described in [7]. The DQDB USER in CLASS implements a simplified functional model of the DQDB MAC protocol behaviour based on a simple queuing system. The rationale of this model and its detailed description can be found in [9]. The basic idea is that under certain conditions a transmitting station can block the access of downstream stations to the transmission medium. The generator is thus based upon a set of queues, called *miniusers*, and a scheduler that defines which of the queues must be served in the present slot. The *miniusers* generate messages that can be labeled either "external" or "internal". External messages represent the traffic effectively offered to the ATM network, while internal messages represent the messages between DQDB stations that are not injected onto the ATM network, and that in the model do not leave the DQDB USER.

Fig. 2 sinthetically describes the policy followed by the scheduler of the DQDB USER, with the assumption that a queue with higher order corresponds to a downstream station; for the sake of simplicity it is supposed that the gateway is located at one of the head-ends. In practice, the service policy always follows a round robin schedule, except in the case when queue i is being served alone and a message arrives at a queue $j > i$ (corresponding to a downstream station); in this case queue i is served exhaustively, and queue j is served only when queue i is empty. If a message arrives at a third queue, the round robin discipline is immediately resumed.

The message interarrival times at the *miniusers* can be governed by either a

Poisson process or an On-Off process. If the On-Off process is selected, messages are generated with fixed spacing during the On phase, and the On and Off phases are geometrically distributed. All messages generated during an On phase have the same destination, because it is assumed that they are generated by the same application process, for example a file transfer. The average duration of the On phase is equal for all the *miniusers* and is defined at the beginning of the simulation. The mean duration of the Off phase is automatically computed in order to meet the required average traffic level.

The length of the messages can either be a mix of constant length messages with up to three different message lenghts and arbitrary proportions, or have truncated geometric distribution, or both.

2.2 NODE Entity

The NODE entities perform the key internal operations characterizing the behaviour of an ATM network. The NODES route and switch cells from one link to the next; in other words they "move" the cells through the network, following, for each connection, the path assigned by the MANAGER during the setup of the simulation.

The NODES execute a fairly simple algorithm: they inspect input cells, look up a routing table in which, for each VC, the output link where the cell must be routed was written by the MANAGER, and queue cells for transmission on the output link.

The simplicity and efficiency of the simulation of this task is of paramount importance, because this is the most frequently performed operation during the execution of a simulation run: the number of times this operation is executed is roughly equal to the number of simulated cells multiplied by the mean number of nodes crossed by each connection.

The switching operation itself consist in moving the pointer to the cell from the linked list representing the input transmission link to the linked list representing the output buffer and link. When links with different transmission speeds exist, this operation can be performed only for few links during each simulation step: testing all the links of each node during each simulation step leads to a significant waste of resources.

To avoid this waste, CLASS builds a *minimum scheduling sequence* that represent the shortest periodical sequence of NODE invocations, so that the only test that must be performed when a NODE is activated is the presence of cells at the input ports. Fig. 3 shows an example of minimum scheduling sequence in a network with links with three different speeds. For the sake of simplicity, the network comprises only four NODES , and no USERS are shown. The data rate on two links is taken to be 200 Mbit/s, while on the two other links it is assumed to be 300 Mbit/s, and 600 Mbit/s, respectively. For each link a parameter is defined, called `timescale`. The value of the parameter `timescale` for the fastest link is by definition set equal to 1. The value of the parameter `timescale` for the remaining links is computed by dividing the highest data rate in the network by the link data rate; thus, the link with data rate 300 Mbit/s has `timescale = 2`, and the two links with data rate 200 Mbit/s have `timescale = 3`.

The periodicity of the minimum scheduling sequence equals the minimum common multiplier of the link `timescales`; each NODE appears in the minimum scheduling sequence each time the `timescale` of one of its links divides the index of the slot within the sequence. The first slot within the minimum scheduling sequence is labeled "0" and every node is invoked in this slot.

The slots of the sequence are separated inserting a *timing event*, that represent the end of a simulation step.

It is easy to verify in the example that starting from slot number 6 (the number of the slot is reported below the minimum scheduling sequence) the scheduling of entities is repeated periodically.

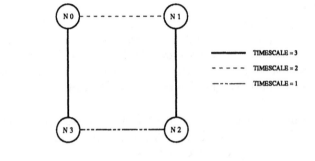

0,1,2,3,T - 2,3,T - 0,1,2,3,T - 0,1,2,3,T - 0,1,2,3,T - 2,3,T

  0   1   2   3   4   5

Fig. 3. Minimum scheduling sequence for a network with different link data rates; T is the timing event

2.3 CLS Entity

One of the possibilities being considered for the management of connectionless traffic in ATM networks is based on the use of specialized functions located in special network objects, called connectionless servers (CLS). The CLS operate on messages generated by the users of connectionless services. The main reason for the use of CLS is the reduction of the overhead due to the allocation of connections. The main functions of a CLS include reading the destination address of the message, making a routing decision, (this may require a considerable amount of time), and finally queuing all the cells of the message to the appropriate output queue. These functions are typical of a network layer protocol, thus the CLS functions cannot be implemented within a standard ATM node that operates at the ATM layer.

A major issue in the definition of the characteristics of the CLS concerns the level of integration of the CLS with the ATM node to which it is connected.

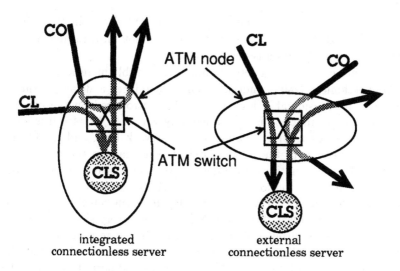

Fig. 4. Graphical representation of the possible levels of integration between connectionless servers and nodes in ATM

Two possible solutions are sketched in Fig. 4. The left one assumes a high level of integration between the node and the CLS, while the right one assumes that the CLS is a standalone network object completely separated from the node. The main difference lies in the techniques used for the data transfer between the ATM switch and the CLS. In the first case the transmission is not constrained by the ATM technology and may be implemented on a dedicated bus, and the bus throughput can easily be chosen in such a way that it will never become a bottleneck. In the second case, the transfer of data is implemented with a standard ATM link, that may become a bottleneck if the data transfer between the switch and the CLS is very heavy.

CLS in CLASS provide a model for both cases of tight and loose integration between the node and the CLS. Apart from this difference, the two CLS types implement the same algorithm. When the first cell of a new message (BOM – Begin Of Message) is received, a FIFO queue is created for the new message, and all the incoming cells of that message are stored in such queue. The CLS first computes the routing of the message and then begins to forward the message cells towards their destination, that can be either the final user or another CLS in the network. The computation of the routing requires a fixed amount of time, of the order of a few hundred slots, that can be chosen by the user. This amount of time is made constant in real CLS (as stated in [10]), in order to upper bound the real time taken for the routing decision — the reason is to avoid that messages between the same source and destination arrive in the wrong order, a condition that may lead to undesired retransmission requests by higher level protocols. Another reason for setting a fixed delay in CLSs is due to the SMDS (Switched Multi-megabit Data Service [11, 12, 13]) requirements for which a maximum delay of 7 ms is allowed: if a message was not routed within that time, the message is discarded. This feature, however, was not implemented in CLASS since the routing information is always available.

The presence of CLS into a network may considerably increase the burstiness of the connectionless traffic. Indeed, a CLS introduces a delay on the BOM cell that can be considerably higher than that introduced on the other cells, because the routing function is performed only when the BOM cell is received. If the CLS processes a large amount of the network load, and is linked with a very fast connection to the ATM switch, the increase in the burstiness can lead to pathological situations. In order to avoid such situations some kind of traffic shaping must be performed at the CLS output. One possible solution could be the introduction of a constant delay for each cell, but this entails a waste of resources, and was thus discarded. A second, very simple solution is to constrain the CLS to output cells from the same message at a rate not higher than the one of the links outgoing from the ATM node to which the CLS is connected. This latter solution wastes less resources, but permits a certain degree of burstiness increase. This solution was implemented in CLASS. A study of more sophisticated algorithms can improve the understanding of the impact of CLS on the network performance.

3 An Application Example

Fig. 5 shows the topology of the simulated ATM network, whose layout was selected as a possible candidate to cover the Italian territory with a high-speed network for connectionless services. Also the traffic patterns that define the network workload were devised imagining a realistic scenario within Italy. For the sake of simplicity, the simulations were run assuming that only connectionless traffic exists in the network; this may be a reasonable scenario for an ATM crossconnect operating as a backbone among MANs. Table 1 gives the relative traffic values used in the simulation runs. The traffic matrix defines the traffic load on a NODE-by-NODE basis; the NODE traffic is then equally distributed among the USERS connected to the NODE. Nevertheless, the overall traffic is not uniformly distributed; this justifies the uneven distribution of the link data rates in the network of Fig. 5. The last row of the traffic matrix gives the total traffic generated by the USERS connected to the NODE.

The network of Fig. 5 was simulated under different configurations, as regards the buffering capacity at the nodes, the types of USERS, and the presence of CLS. The buffering capacity at the nodes was varied between 50 and 200 cells, while the USER transmission buffering capacity is equal to 1000 cells. The two types of USERS that were selected are the DQDB USER and a USER with Poisson generation. Results were obtained both with and without CLS.

The first set of numerical results refers to the total cell loss probability in the network, for a global load equal to 800 Mbit/s, as a function of the size of the buffer associated with the links between the network nodes. Fig. 6 shows the curves of the total *cell* loss rate when no CLS is present in the network,

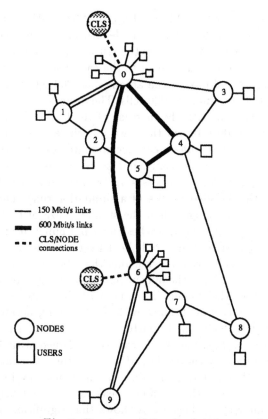

Fig. 5. Topology of the simulated network

Node	0	1	2	3	4	5	6	7	8	9
0	0	52	20	2	18	3	206	17	2	17
1	52	0	3	0	3	5	32	3	0	3
2	20	3	0	0	2	2	12	1	0	1
3	2	0	0	0	0	0	1	0	0	0
4	18	3	2	0	0	2	11	1	0	1
5	35	5	2	0	2	0	22	2	0	2
6	206	32	12	1	11	22	0	10	1	10
7	17	3	1	0	1	2	10	0	0	1
8	2	0	0	0	0	0	1	0	0	0
9	17	3	1	0	1	2	10	1	0	0
Total	369	101	41	3	38	70	305	35	3	35

Table 1. Traffic matrix used in the simulation runs; the traffic is generated by the node in the column and goes to the node in the row; the relations are expressed in thousandths of the global generated traffic

and for the two cases of Poisson USERS or DQDB USERS with Poisson arrivals at the *miniusers*. In both cases, the message length has a trucated geometric distribution with average equal to 20 cells; the trucation value is 200 cells. Two pairs of curves are reported, the first refers to the case when the *output buffers*, i.e., those between NODES and USERS, can accomodate 200 cells, and the second with the size of these buffers equal to 1000 cells. The decrease in the cell loss rate has an exponential behaviour with the increase of the buffer size if the output buffers can accomodate 1000 cells (in this case no cells are lost in the output buffers). Instead, if the output buffers can accomodate only 200 cells, the cell loss rate does not decrease exponentially, but shows an asymptotic behavior, indicating that the majority of the cell losses occur in the output buffers (as can be seen by analysing the detailed results of a simulation run). The reason for letting the output buffers be longer than the standard size of buffers between nodes is thus twofold: first, when a cell reaches an output buffer, it has practically been brought to the destination, it must only be delivered to the end user, thus its loss in these buffers has a global cost much higher than its loss into an inter-node buffer; second, as shown by Fig. 6, the output buffers have a greater influence on the cell loss rate than the other buffers in the network. It is interesting to notice the difference between the curves obtained with the Poisson generators and with the more realistic DQDB USER. In the latter case, the cell loss rate is always lower than in the former. This behaviour can be explained with the intrinsic traffic "shaping" performed by the DQDB MAC protocol if the gateway routes the cells "on the fly", thus not reassembling the messages: the interleaving of cells on the DQDB bus reduces the sustained peak rate of the generated traffic. This result is fairly interesting, because it shows that significant performance improvements can be obtained with an adequate shaping of the input traffic.

Fig. 7 shows curves of the *message* loss rate within the network, for the same simulation experiments considered in Fig. 6. It must be noted that no messages are lost within the USERS transmission buffers. The shape of these curves is very similar to that of the curves of the cell loss rate, although the absolute value is higher. This difference in the loss rate is due to the fact that a message is lost even if only one of its cells is lost, thus the message loss rate may be significantly larger than the cell loss rate (on the average a message contains 20 cells). This fact is very important, since higher level protocols will re-transmit whole messages, not cells.

The high influence of the output buffer dimension on the cell and message loss rates is enhanced by the fact that the links between NODES and USERS are always the slowest in the network, while they may receive cells from high speed links as in nodes 0-4-5-6. Increasing the buffering capacity to 1000 slots ensures that up to 5 messages of the maximum length can be buffered at the output, so that only collisions among a greater number of messages (a rare event!) lead to cell and message loss.

Figs. 8 and 9 report the network cell and message loss rates, as a function of the network load, for a network buffer size of 100 slots. As expected, the loss rate increases exponentially with the network load.

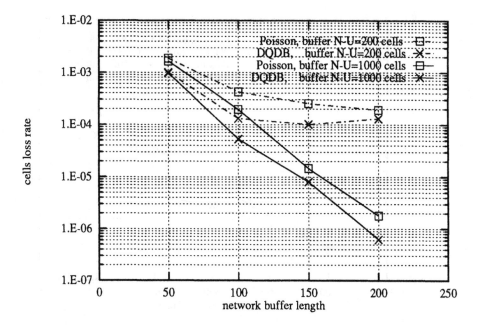

Fig. 6. Network cell loss rate as a function of the size of the buffers between the nodes; the total traffic offered to the network is 800 Mbit/s

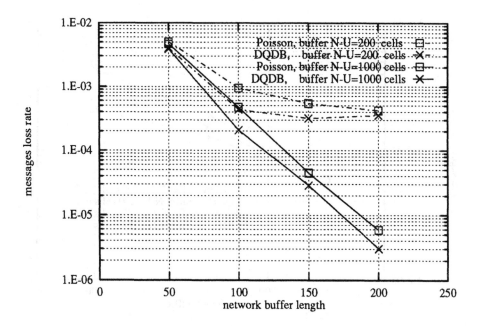

Fig. 7. Network message loss rate as a function of the size of the buffers between the nodes; the total traffic offered to the network is 800 Mbit/s

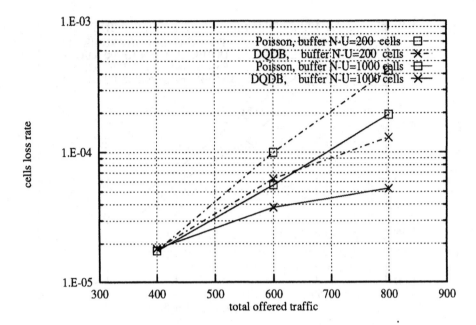

Fig. 8. Network cell loss rate as a function of the network load; the network buffer size is 100 cells

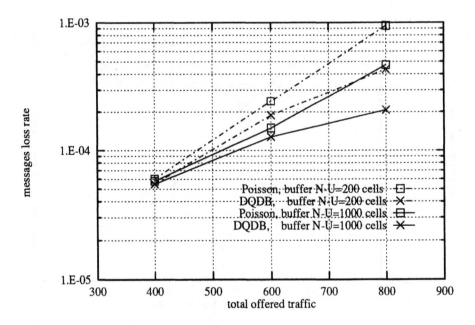

Fig. 9. Network message loss rate as a function of the network load; the network buffer size is 100 cells

Buffer	Network load (Mbit/s) and user type					
	400		600		800	
Size	Poisson	DQDB	Poisson	DQDB	Poisson	DQDB
50	4.82	5.40	8.20	7.70	12.87	10.00
100	4.85	5.42	8.36	7.71	13.09	10.22
200	4.84	5.42	8.33	7.73	13.16	10.42

Table 2. Average cell network delay as a function of the network load, user type, and network buffer size; the output buffer is 1000 cells

Buffer	Network load (Mbit/s) and user type					
	400		600		800	
Size	Poisson	DQDB	Poisson	DQDB	Poisson	DQDB
50	28.44	32.99	35.27	43.40	44.68	60.22
100	28.54	33.11	35.50	43.55	44.92	61.80
200	28.52	33.05	35.46	43.46	45.01	61.99

Table 3. Average cell user delay as a function of the network load, user type, and network buffer size; the output buffer is 1000 cells

Tables 2 and 3 present the average values of the cell network and user delays, expressed in number of slots on the 150 Mbit/s links, as a function of the network load and of the network buffer size; the output buffer dimension is set to 1000 cells. As already stated, the *network* delay is the variable part of the delay suffered by cells, not considering the waiting time in the USER transmission buffer, while the *user* delay takes into account also this contribution. Notice that both delays are practically independent from the network buffer dimension, while they increases linearly with the network load. Notice that the contribution from the USER transmission buffer is dominant. This is due to the fact that the cells of a message arrive together to this buffer, but are transmitted sequentially, thus there is a contribution that cannot be avoided. The infuence of the user type on the mean delay is also interesting.

- The mean of the network delay is almost independent from the generator type: the DQDB USER produces lower delays only at high load; this is due to the interleaving within the DQDB USER that "shapes" the traffic offered to the network, reducing its burstiness.
- The mean of the user delay is always larger for the DQDB USER. The reason is again the interleaving within the DQDB USER: if an *external* message is enqueued after an *internal* message, the former suffers a greated delay.

Figs. 10 and 11 report the standard deviation of the network and user delays as a function of the network buffer size, with the load set to 800 Mbit/s; and of

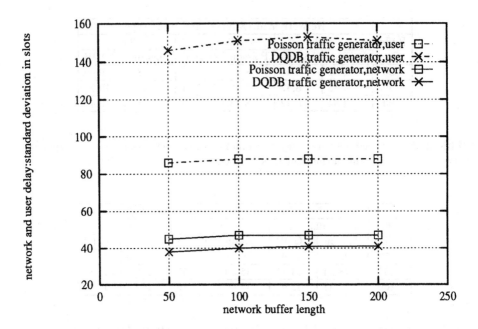

Fig. 10. Standard deviation of the network cell delay as a function of the size of the buffers between the nodes; the total traffic offered to the network is 800 Mbit/s

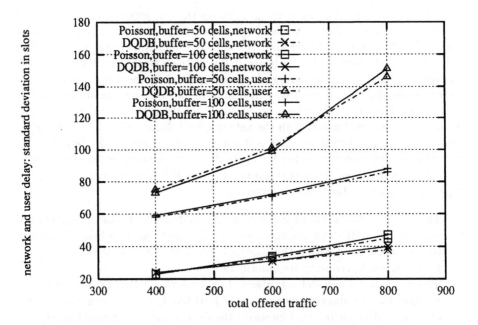

Fig. 11. Standard deviation of the user cell delay as a function of the network load; the network buffer size is either 50 or 100 cells

the network load, with the network buffer size set to 50 or 100 cells. As expected, the standard deviations increase with the load, and the standard deviation of the user delay is always greater. It is instead interesting to notice that also the standard deviations, like the means, of both the user and network delays are practically independent from the network buffer size. This result is not obviuos, and contrasts with the general opinion that increasing the network buffer size heavily deteriorates the quality of service in terms of delay jitter. The reason for this behaviour is probably due to the highly asimmetric network assumptions, that induce a clear bottleneck effect. The delay is not randomly accumulated in many different buffers, but is generally due to a single congested buffer, while all the other buffers are empty or nearly empty and do not contribute to the delay.

The last results refer to a network configuration with two CLS (see Fig. 5). These results were obtained in the case of Poisson and DQDB USERS, with connectionless traffic only. All the traffic is handled by the CLS. The CLS are integrated within the node, thus the communication between the NODE and the CLS is not constrained by the ATM technology. As expected, the network performs poorly under these conditions. The nodes where the CLS are located represent two clear bottlenecks, increasing drastically the cell and message loss rates, as shown in Fig. 12. Comparing these results with those in Figs. 6 and 7 it is clear that, given a target cell loss rate, the throughput reachable in a network with CLS is less than half of that without CLS. A peculiarity of these results is that the network with DQDB USERS now suffers higher cell and message loss rates than the network with Poisson generators, which is exactly the opposite of the case without CLS. This phenomenon requires further investigation, and may be due to the reassembly of messages that is performed within the CLS, so that the advantage of the intrinsic shaping of the traffic within the DQDB MAN is lost.

Fig. 13 reports the standard deviation of the network and user delays in the same scenario. Also these results are rather poor if compared with those without CLS: the CLS themselves introduce a very long delay on the BOM cells, and this greatly affects the overall performance.

It must be noted that the use of CLS was proposed for B-ISDN with a traffic mix in which connection-oriented services produce the dominant traffic load, and connectionless services must be handled in such a way to obtain good performance with minimum impact on the quality of connection-oriented services. In such an environment, the advantages of the CLS approach stem from the easier network management and the reduction in the overhead due to the opening and closing of the connections for the transfer of connectionless services data. The environment of our simulations is very far from the intended application of CLS: all the simulated traffic derives from connectionless services, and must go through the CLS; furthermore, the connection dynamics (opening and closing) are not simulated. Our results indicate that the disadvantages of CLS in the simulated environment are significant, but a deeper study should be performed to investigate operating conditions that take into account also the presence of connection-oriented traffic.

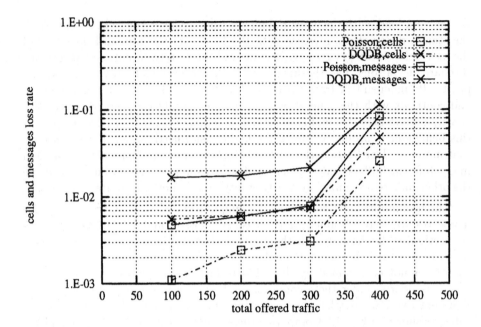

Fig. 12. Cell and message loss rates versus the network load when two cls are present; the network buffers size is 50 cells and the output buffer size is 200 cells

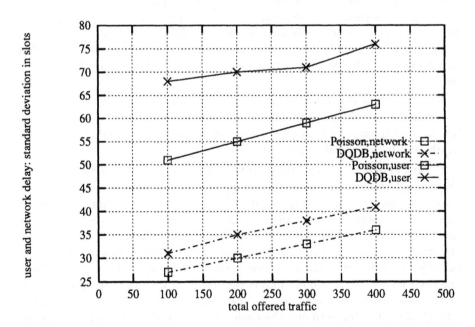

Fig. 13. Cell network and user delays: standard deviation versus the network load when two cls are present; the network buffers size is 50 cells and the output buffer size is 200 cells

The numerical results presented in this section were obtained with 36 different simulation runs. The CPU requirements of the different simulation runs are determined by the stopping rule of the simulation experiment. In order to obtain accurate estimates of the cell loss probability, simulation runs were stopped when a predefined accuracy on the average cell delay was reached, *and* at least a given number of cell loss events had been observed. This greatly increases the duration of low traffic simulation runs, where cell loss events are extremely rare.

The longest simulation run refers to the 800 Mbit/s network load, with node buffer size equal to 200 cells, and no CLS. The simulation run stopped after the delivery of 323,342,815 cells, after 128,280 seconds of CPU time, producing 202 cell loss events. Thus, on the average, the simulator was able to simulate the transfer of about 2,500 cells per second through the ATM network.

Other simulation runs are much shorter. For example, in the same situation as above, but with CLS, the simulation run stopped after the delivery of 2,431,954 cells, after 1,560 seconds of CPU time, producing 99,208 cell loss events. Thus, on the average, the simulator was able to simulate the transfer of about 1,500 cells per second through the ATM network. In this case it must be noted that the simulator had to perform a considerable amount of I/O due to the fact that each cell loss event is recorded into a file, and this reduces the efficiency of the simulation.

These results were obtained with a DEC VAX 6000-500 under the VMS operating sistem, whose speed is similar to that of state-of-the-art Unix workstations.

4 Conclusions

The simulation of ATM networks requires a great deal of attention in order to keep the CPU time requirements at reasonable levels, so as to be able to examine the merits of alternate configurations with acceptable cost.

In this paper we presented the main characteristics of CLASS, a software tool for the simulation of connectionless services in ATM networks, that was developed trying to achieve a good compromise between efficiency and flexibility.

The resulting software is such that the simulation of the transfer of a quarter million cells in a network of medium size can be completed with about one minute of CPU time on state-of-the-art workstations.

The future evolution of CLASS will follow the progress of the standardization in the field of ATM LANs, and B-ISDN. The next step in the development of the simulator capabilities will consist in the inclusion of several traffic control algorithms both within the source models and within the ATM switches.

5 Acknowledgments

The authors wish to acknowledge the contribution of M. Molina in the development of portions of code of CLASS and in the derivation of some of the numerical results, and of R. Pasquali in the derivation of some numerical results.

References

1. The ATM Forum, ATM Data Exchange Interface (DXI) Specification, Version 1.0, August 4, 1993
2. The ATM Forum, BISDN Inter Carrier Interface (B-ICI) Specification, Version 1.0, August, 1993
3. The ATM Forum, ATM User-Network Interface Specification, Version 3.0, September 10, 1993
4. G. Balbo, "Appunti per il corso di simulazione", Dipartimento di Informatica, Università di Torino (in Italian)
5. M.Ajmone Marsan, R.Lo Cigno, M.Molina, M.Munafò, *CLASS: Un Simulatore di Servizi Connectionless in B-ISDN*, Final Report of the contract between CSELT and Politecnico di Torino "*Valutazione di traffico nell'interconnessione MAN/ATM*", Politecnico di Torino, August 1993 (in Italian)
6. IEEE P802.6 - Metropolitan Area Networks, *Distributed Queue Dual Bus (DQDB) Subnetwork of a Metropolitan Area Networks (MAN)*, Draft Standard - Version D15, October 1990
7. M.Ajmone Marsan, G. Albertengo, R.Lo Cigno, M.Munafò, F.Neri, A.Tonietti, *Characterization of the traffic at the output of a DQDB MAN*, Local Area Network Interconnection, Edited by R.O. Onvural and A.A. Nilsson, Plenum Press New York, 1993, pp. 251–267
8. E.L.Hahne, A.K.Choudhury, N.F.Maxemchuk, *Improving the Fairness of Distributed-Queue-Dual-Bus Networks*, Proc., IEEE INFOCOM'90, San Francisco, CA, pp. 175-184, June 1990
9. M.Ajmone Marsan, R.Lo Cigno, M.Munafò, A.Tonietti, *A Source Model for Connectionless Traffic in B-ISDN*, submitted for publication
10. Various Authors, "*Models for Heterogeneous Networks and Their Elements*", COMBINE (Composite Broadband Internetworking and End-to-end Models) Deliverable D3, Race Project 2032, 31/12/1992,
   Reference: R2032/KTA/WPI/DS/P/003/b1
11. G. Clapp, "*LAN Interconnection Across SMDS*", IEEE Network Magazine, September 1991, pp. 25–32
12. Bellcore Technical Reference TR-TVS-000772, "*Generic System Requirements in Support of Switched Multi-Megabit Data Service*", Issue 1, May 1991
13. Bellcore Technical Reference SR-NWT-002076, "*Report on the Broadband ISDN Protocols for Providing SMDS and Exchange Access SMDS*", Issue 1, September 1991

NetSim – A Tool for Modeling the Performance of Circuit Switched Multicomputer Networks

Vipul Gupta[1]* and Eugen Schenfeld[2]

[1] Dept. of Computer Science, Rutgers University, Busch Campus,
Piscataway NJ 08855, USA. E-mail: vgupta@cs.rutgers.edu
[2] NEC Research Institute Inc., 4 Independence Way,
Princeton NJ 08540, USA. E-mail: eugen@research.nj.nec.com

Abstract. Mathematical analysis is a popular approach for modeling network performance. However, a tractable analytic model often restricts the range of system characteristics that can be explicitly considered. Simulation provides an alternative approach when such restrictions can not be tolerated. This paper describes NetSim, a versatile simulation tool for modeling the performance of circuit switched multicomputer networks. NetSim is programmable as to network topology, communication pattern, routing algorithm, mapping strategy and network operating assumptions. Additionally, its modular design makes it easy for the user to extend NetSim's functionality when required. The tool provides a means for exploring the interplay of application and architecture characteristics in a parallel system.

1 Introduction

Parallel processing is an attractive means for meeting the ever increasing demand for greater computing power. A *multicomputer* is comprised of many processing elements (PEs) interconnected through a network. Processes in a parallel application are assigned to these PEs by a mapping function. Concurrently executing PEs use the interconnection network to exchange data and control information in the form of messages. The efficiency with which messages can be exchanged is a crucial factor in the performance of a parallel system. Two important criteria characterizing network efficiency are *throughput* and *latency*. Throughput measures the number of messages, per PE, exchanged across the network in unit time. Latency refers to the time elapsed between the injection of a message in the network and its delivery at the destination. Another important, but often overlooked, network characteristic is *scalability*. A parallel application is said to scale well on a network when increasing the system size (number of processes and PEs) results in only a small (if any) loss in efficiency.

These performance metrics are affected by a complex interplay of many factors. These include the network topology, the message routing algorithm, the communication pattern of the application and the mapping function. Two approaches

---

* This work was done while the author was a Ph.D. student supported by NECI.

can be used to model the interaction of these system parameters; purely analytical and simulative. The main advantage of the first approach is that closed form solutions provide explicit relationships between input parameters and performance indices. These are very helpful in understanding the behavior of the system. In most cases, however, closed form solutions can only be derived for extremely simple models. Simulation can be used to study a system's behavior under more general conditions. The down side of simulation is higher cost in terms of time complexity. In the space defined by all possible combinations of system parameters, each simulation run investigates only a single point. For complex systems, the use of simplifying assumptions can introduce significant discrepancies between analytically predicted and empirically observed behavior. Simulation is the preferred approach for modeling such systems.

A number of very powerful simulation languages such as GPSS [24], SLAM [21], and SIMSCRIPT [4] are commercially available. However, there is a considerable start up time associated with using them, since they require learning a new language. Libraries and toolkits based on popular programming languages (e.g. CSIM [25] and SIMPACK [9] based on C) do not suffer from this drawback, but they too require a computer architect to work with unfamiliar abstractions. NetSim is a simulation tool, designed specifically for modeling the performance of circuit switched multicomputer networks. For a multicomputer network analyst, the generality of the aforementioned languages and libraries is unnecessary; especially when it comes at an added cost — speed. Our experience indicates that the special purpose features of NetSim let us simulate anywhere between five to ten times more steps than a program using SIMPACK in the same amount of time. Tools for simulating computer networks have also been reported in [5, 7, 15, 26]. However, those tools are primarily designed to validate their accompanying mathematical analyses and as such lack the versatility of NetSim.

The rest of the paper is divided into five sections. In Sect. 2 we describe the model of a parallel system used by NetSim. Network operating assumptions are listed in Sect. 3. An outline of NetSim's design is presented in Sect. 4. Section 5 provides examples illustrating the utility of this tool and Sect. 6 summarizes our conclusions.

2   Model of a Parallel System

NetSim views a parallel system as the combination of a parallel architecture, a parallel application and a mapping function. In what follows, we first describe how NetSim models each of these components; and then, the mechanism used for their specification.

The parallel machine architecture is modeled by a *network graph* $G_N(V_N, E_N)$ in which the vertices are either PEs or switching elements (routers) and the edges are network links. Each PE is assumed to have one incoming and one outgoing link. Routers are crossbar switches; a switch of size $a \times b$ can have at most a incoming links and b outgoing links. Within these restrictions, any directed graph is allowed. This model is general enough to represent many network topologies including the three shown in Fig. 1. Figure 1a shows the GF11 network [1] consisting of three stages of twenty-four switching elements each. Consecutive stages are connected by a shuffle interconnection. Output links in the last stage wrap around to become input links for the PEs. Figure 1b shows a possible configuration of iWarp [3] chips. Each switching element is connected to one PE and four other switching elements in a 2-Dimensional mesh. Figure 1c shows a hypercube network topology of degree three.

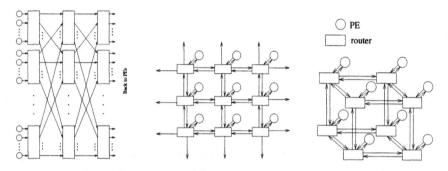

Fig. 1. Network topology of (a) GF11 (b) an iWarp configuration, and (c) a hypercube network

A parallel application is modeled by a *communication graph* $G_C(V_C, E_C)$ in which vertices represent processes and edges denote interprocess communication. This allows the communication pattern of an application to be considered explicitly. Some previous studies on network performance have assumed uniform, randomly distributed, communication traffic between computation entities [5, 7, 15, 20]. There is enough evidence [12, 18] to indicate that many real applications do not fit this assumption. This has prompted the use of more sophisticated models of network traffic, e.g. spherical distribution [14] or hot-spot traffic [17]. However, the model supported by NetSim is more general.

We assume that only one process is mapped to each PE. When the number of processes in the communication graph exceeds the number of available PEs, techniques similar to those described in [2] can be used to preprocess the communication graph.

The network topology along with routing information, the structure of the communication graph, and the mapping function are specified by the user in the form of three ASCII files. The format of these input files is explained in Fig. 2 and

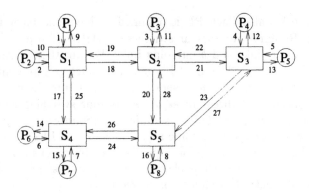

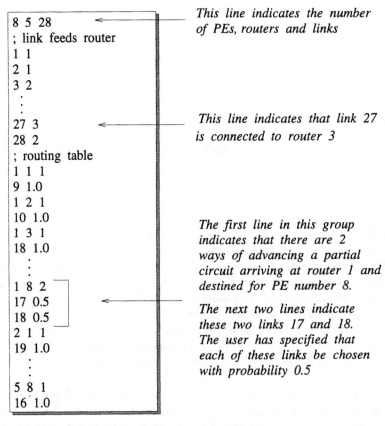

Fig. 2. A sample network and its specification. NetSim uses the convention that the ith PE has its output link numbered i and its input link numbered $N + i$ (N is the total number of PEs).

Fig. 3. For large systems, it is impractical to generate these files by hand. Towards this end, we have developed programs capable of generating such files. Currently, network description files for Delta [20], 2-D mesh and hypercube networks can be automatically generated. Similar programs exist to generate communication graphs in the shape of k-ary trees, and d-dimensional meshes and tori (hypercubes are a special case in which each dimension has only two nodes). Programs for generating linear mappings, in which process i is mapped to the ith PE; and random mappings have also been developed. The simple format of these files makes it easy to add to the suite of such programs.

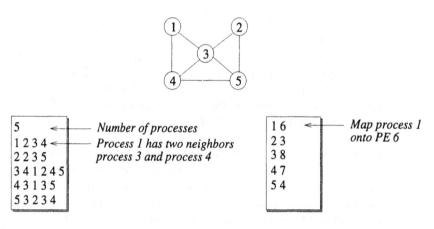

Fig. 3. A sample communication graph, its corresponding specification and a possible mapping of this graph in the sample network shown earlier.

3 Network Operating Assumptions

NetSim makes the following assumptions regarding network operation.

1. Circuit switching is used to transfer messages. A circuit is established between communicating PEs before data transfer begins. Partially set up circuits propagate through successive switching elements in steps of time t_c. These steps are called *cycles*.
2. A circuit that has reached its destination is said to be *complete*. Data transfer over a completed circuit takes d cycles. The value of d is determined by calling a function *genLength()* for each message.
3. Routing in NetSim is based on the use of lookup tables [22] maintained at every switching element. For each destination, the table lists all possible outgoing links on which a partial circuit may be advanced. In the case of multiple choices, the tie-breaking strategy is built into a function *getRouterLink()*.

4. A PE can not inject a new message into the network until its previous message has been delivered. During every cycle, eligible PEs inject new messages with probability m ($0 < m \leq 1$). The value of m is determined by a function *genProb()*. Using this function, it is possible to simulate different message generation rates at different PEs. This feature is potentially useful in modeling situations where the parallel machine is space-shared by applications having different communication characteristics.

5. A process only generates messages for its neighbors in G_C. For each message, a neighbor is picked by calling *getAneighbor()*. The PE to which this neighboring process is mapped becomes the destination PE. The neighboring process need not always be mapped to a neighboring PE. For this reason, communication edges in G_C are implemented as paths, consisting of one or more network links, in G_N.

6. Unless circuits progress along edge disjoint paths in G_N, they must compete to acquire network links. If a partial circuit requests an outgoing link, which has already been acquired, the circuit is *blocked*. When multiple partial circuits arriving at a switch request the same outgoing link, a conflict is said to occur. In this case, a circuit selected by *pickCir()* is advanced and others are blocked.

7. NetSim allows two strategies for handling blocked circuits. Following the terminology in [26], we refer to these as the *drop* and the *hold* strategies. Under the first, a blocked circuit releases all its links at the end of the cycle and starts all over again. Under the second, a blocked circuit retains possession of previously acquired links as it waits for the next required link. The user can specify which of these two strategies must be simulated by changing a flag.

8. After a circuit finishes its data transfer, it releases all acquired links.

In the past, some network analysts have assumed that dropped circuits can be ignored [20]. This is called the independent request model (IRM). While this assumption is useful in keeping mathematical analysis tractable, it is clearly unrealistic. As shown in [26], the use of this assumption can cause significant discrepancies between actual behavior and that predicted analytically.

Cycle time t_c represents the delay which a partial circuit suffers at a switch. This does not include any time spent waiting for a link to become free. It reflects the time needed to make routing decisions and to establish a direct connection between the input and output links. Different factors such as switch size, switch manufacturing technology, and network structure affect t_c in a network. In general, bigger switches have longer cycle times; and optical switches have longer cycle times than electronic switches [23]. Similarly, networks in which routing decisions are made centrally (e.g. Clos networks) have longer cycle times than others with distributed routing (e.g. multi-stage interconnection networks). The value of d depends on message length, the bandwidth of network links and t_c. If L is the message length in bits, B is the link bandwidth in bits per second and t_c is the cycle time in seconds, then $d = B/(L \cdot t_c)$.

The choice of only two parameters m and d reflects our attempt to balance simplicity against accuracy and broad applicability. Each of these parameters can model multiple system characteristics. The effect of longer messages as well as lower link bandwidth can be captured by increasing d. The effect of faster processors can be modeled by increasing m, since faster processors can be expected to generate more messages per unit time. Similarly, applications with coarser grain can be modeled by lower values of m compared to those with finer computation grain.

4 The structure of NetSim

Simulation of networks in NetSim is clock-driven. Each clock tick corresponds to one network cycle. The structure of the main loop is outlined in Fig. 4. An initialization routine reads the specifications of the parallel system from input files. It also initializes various data structures and a random number generator used by the simulator. The program variables *simtime* and *simlen* represent the number of cycles simulated and the total number of cycles to be simulated, respectively. For the purpose of describing the various activities that take place during one simulation step, it is convenient to divide each cycle into five subcycles. The activities associated with these subcycles are handled by five separate routines outlined in Fig. 5.

```
main()
{
    Initialize();
    Generate_messages();
    while (simtime++ < simlen) {
        Compete_for_links();
        Try_to_advance();
        Handle_complete_circuits();
        Tear_down_circuits();
        Generate_messages();
    }
    Print_statistics();
}
```

Fig. 4. Structure of the main loop in NetSim.

In the first subcycle, partial circuits determine the links they require next and compete for it. In the second, some of these circuits are selected to advance and others are blocked. Circuits that reach their destination are marked as complete.

```
Compete_for_links() {
    for (all circuits that are not complete) do {
        /* Use knowledge of current position and destination to determine
        the next required link */
        next_link = getRouterLink(routerId, destination);
        Register a request for next_link;
    }
}


Try_to_advance() {
    for (all links requested in the previous subcycle) do {
        Select one of the requesting circuits by calling pickCir();
        Update its position and if it reaches destination, mark as complete;
        if (drop assumption is in effect) then
            Mark all other circuits for tearing down;
    }
}


Handle_complete_circuits() {
    for (all circuits that are complete) do
        if (data transfer is over) then
            Update statistics and mark this circuit for tearing down;
        else
            Transfer data;
}


Tear_down_circuits() {
    for (all circuits marked for tearing down) do
        if (circuit had finished data transfer) then
            Release all its links and mark the circuit as inactive;
        else /* control reaches this point only with the drop assumption */
            Release all links except the output link of source PE;
}


Generate_messages() {
    for (all processes with inactive circuits) do
        if (gen_random() < genProb()) then {
            Determine destination process using getAneighbor();
            Determine d using genLength();
            Mark the new partial circuit as not complete;
            Position the partial circuit at the output link of source PE;
            Update statistics;
        }
}
```

Fig. 5. Routines implementing the five subcycles in each simulation step.

If the drop assumption is in effect, blocked circuits are marked to be torn down. The third subcycle deals with complete circuits. Those that have finished data transfer are marked to be torn down and others continue their data transfer. In the fourth subcycle, appropriately marked circuits are torn down and their links released for use by other circuits. New messages are generated in the fifth subcycle. Since PEs only have one outstanding message at any time, each circuit can be identified uniquely by the number of its source process.

In Sect. 2, we have already described NetSim's ability to handle a variety of communication patterns, network topologies, routing algorithms and mapping functions. The use of functions such as *getAneighbor()*, *genLength()*, *getRouterLink()*, *genProb()* and *pickCir()* provides additional flexibility. NetSim already includes multiple versions of these functions to simulate different situations, e.g. one one version of *getRouterLink()* tries to pick the first free link amongst multiple choices while another makes this decision in probabilistic manner. By rewriting one or more of these routines to simulate other assumptions, NetSim's functionality can easily be enhanced. Currently, *pickCir()* makes its selection equiprobably but it is fairly straightforward to implement some kind of a priority mechanism, e.g. one based on the waiting time of a partial circuit. Similarly, *genLength()* can be modified to implement whichever message length distribution the user requires. Some choices that have been suggested for network evaluation include exponential [14], fixed [5] (all messages have same length), and distributions based on multi-stage probability density functions [12].

The hold assumption can result in a deadlock in the network. A deadlock occurs when there is a cyclic wait situation between multiple partial circuits. If each circuit is waiting for a link that has already been acquired by another waiting circuit, none of these circuits can advance until exceptional action is taken to break this cycle. Networks with acyclic topologies (e.g. Delta) are deadlock free. For others, deadlock can be avoided by placing appropriate restrictions on the routing algorithm (see [19]). It is the user's responsibility to ensure that the particular combination of network topology, communication pattern, and routing algorithm simulated by NetSim is free of deadlocks and/or livelocks. Our programs for the automatic generation of network specification files use row-column routing for meshes, and e-cube routing for hypercubes.

5 Experimental Results

NetSim has been implemented in C++ and validated by comparison against other simulations reported previously (e.g. [26]). Below we describe some experiments involving NetSim.

5.1 Effect of Mapping on Network Latency

The use of techniques such as circuit switching and wormhole routing, rather than store-and-forward, makes network latency only weakly dependent on the distance between source and destination PEs [8]. This has undermined the importance of trying to map application tasks to PEs in a way that minimizes non-neighbor communication [10]. However, when a substantial portion of the communication involves distant PEs, more links are kept busy during each data transfer operation. This increases the chances that a partial circuit will find the link it needs busy. We describe an experiment using NetSim which illustrates the effect of sub-optimal mappings on network efficiency. We run each simulation for 105,000 cycles. Data collection is suppressed during the first 5,000 cycles to avoid transient effects.

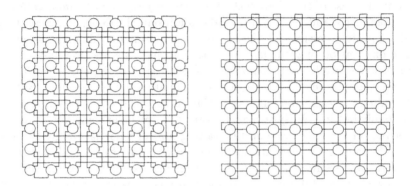

Fig. 6. Two different ways of mapping a 2-D torus pattern in a mesh network of the same size. Network links in the underlying mesh are not shown.

We choose a network of 64 PEs arranged as a 2-D mesh (similar to one shown in Fig. 1) of size 8×8. The communication graph is a torus of the same size. We consider three mappings of this torus in the mesh. Two of these mappings are shown in Fig. 6: circles denote PEs in the mesh and lines denote communication edges in the torus. The third mapping assigns a randomly chosen process to each PE. We refer to these as mappings A, B and C, respectively. To study scalability issues, we also consider a larger system of size 16×16. Simulation results from this experiment are tabulated in Table 1. The meshes use row-column routing, blocked circuits are held and d is chosen to be 128.

When m is very small, there is little contention in the network and average message latency is close to ideal: the sum of d and the average number of routers a message must pass through. Since d is much larger than the path lengths, the performance difference between the various mappings is not very perceptible. This difference becomes more prominent for larger values of m. In general, mapping A is better than mapping B which is better than mapping C. While this ranking holds for both system sizes, the performance gap between mappings is larger for the bigger system. This indicates that not only is mapping A the best

Table 1. Variation of average message latency with network load under three different mappings.

LOAD	LATENCY (IN CYCLES)					
m	8 × 8			16 × 16		
	A	B	C	A	B	C
0.0001	132.3	131.0	137.5	131.8	131.9	148.7
0.001	139.4	139.6	180.1	140.7	140.9	387.4
0.01	184.8	184.6	469.3	189.6	198.3	1097.2
0.1	216.1	220.1	544.5	222.9	235.1	1199.8
0.5	220.4	223.4	569.3	228.3	243.5	1203.7
1.0	224.9	227.1	588.6	232.3	246.4	1176.2

in terms of performance, it is also the best in terms of scalability. Mapping C is the worst on both counts.

These results can be explained by the following observations. Amongst the three mappings, C results in the longest communciation paths. Even worse, as system size increases, communication path lengths increase most rapidly for C. Irrespective of system size, communication paths for mapping A never go through more than three routers. For mapping B, most communication paths only go through two routers. The wrap around edges, however, travel through n routers in a system of size $n \times n$. This results in an average path length asymptotically approaching three from below. The main reason behind the poorer performance of B is the long paths implementing wrap around edges. When data transfer occurs over these long paths, a large number of other circuits get blocked. Consider the data transfer operation between the first and the last process in a row. While it proceeds, no other process within that row can communicate with any process to its right.

5.2 Evaluation of Reconfigurable Architectures

It is well known that an application's performance can often benefit from a close match between its communication pattern and the network topology [6, 13]. Parallel applications show wide variations in their communication patterns [16]. This makes reconfigurable networks, whose topology can be altered to suit individual applications, particularly appealing. Non-uniformity of communication patterns and the flexibility offered by reconfiguration, make it hard to model the performance of such systems analytically. In this situation, a simulation tool like NetSim is invaluable.

We have used NetSim to study the performance of a reconfigurable multicomputer network called the Interconnection Cached Network (ICN). Results of this study are reported in [11].

6 Conclusion

In this paper, we presented NetSim which is a versatile tool for studying the performance of circuit switched multicomputers. The tool derives its flexibility from choosing a small, but powerful, set of "core" functions. NetSim is able to handle a large variety of assumptions regarding communication patterns, network topologies, routing algorithms, mapping functions etc. It is useful for studying the complex interplay of many different application and architecture characteristics in a parallel system. Mathematical analysis cannot effectively deal with the intricacies and non-uniformity of such systems.

There are a number of directions in which our work might be extended. Wormhole routing has emerged as a popular network routing scheme. We believe that a tool, similar to NetSim, which employs wormhole routing will be widely useful. Other possible directions include improving upon the primitive user interface of NetSim and developing a parallel version of it. It is important to realize that a simulation can only be as accurate as the assumptions it is based on. While NetSim has the capability of handling many different assumptions, it is still up to the user to determine which assumptions are most realistic. Some work has been reported on characterizing applications empirically [12]. However, it is not clear which of these characteristics are inherent in the application, and which reflect peculiarities of the particular architecture used in data collection. A better understanding of these issues is crucial to the design of improved parallel systems.

The authors gratefully acknowledge the anonymous referees whose comments improved the organization and readability of this paper. NetSim is available by sending e-mail to the authors.

References

1. J. Beetem, M. Denneau, and D. Weingarten. The GF11 Supercomputer. In *Proceedings of the 12th International Symposium on Computer Architecture*, pages 108–115, Jun 1985.
2. F. Berman. *The Characteristics of Parallel Algorithms*, chapter Experience with an Automatic Solution to the Mapping Problem, pages 307–334. The MIT Press, 1987.
3. S. Borkar et al. iWarp: An integrated solution to high-speed parallel computing. In *Supercomputing'88*, Kissimmee, FL, Nov 1988.
4. CACI. *Simscript II.5 Programming Language*. CACI Products Company, La Jolla, CA, 1988.
5. W. J. Dally. Performance Analysis of k-ary n-cube Interconnection Networks. *IEEE Transactions on Computers*, 39(6):775–785, Jun 1990.
6. J. Deminet. Experience with Multiprocessor Algorithms. *IEEE Transactions on Computers*, C-31(4), Apr 1982.
7. D. M. Dias and J. R. Jump. Analysis and Simulation of Buffered Delta Networks. *IEEE Transactions on Computers*, C-30(4):273–282, Apr 1981.

8. T. H. Dunigan. Performance of the Intel iPSC/860 hypercube. Technical Report ORNL/TM-11491, Oak Ridge National Laboratory, Oak Ridge, Tennessee 37831, Jun 1990.

9. P. A. Fishwick. SimPack: Getting started with simulation programming in C and C++. Technical Report TR92-022, Dept. of Computer and Information Science, Univ. of Florida, Gainesville, FL, 1992.

10. J. M. Garcia and J. Duato. An Algorithm for Dynamic Reconfiguration of a Multicomputer Network. In *Proceedings of the 3rd IEEE Symposium on Parallel and Distributed Processing*, pages 848–855, Dec 1991.

11. V. Gupta and E. Schenfeld. A Comparative Performance Study of an Interconnection Cached Network. Submitted to the International Conference on Parallel Processing 1994.

12. J.-M. Hsu and P. Banerjee. Performance Measurement and Trace Driven Simulation of Parallel CAD and Numeric Applications on a Hypercube Multicomputer. In *Proceedings 17th International Symposium on Computer Architecture*, pages 260–269, May 1990.

13. L. H. Jamieson. *The Chracteristics of Parallel Algorithms*, chapter Characterizing Parallel Algorithms, pages 65–100. MIT Press, 1987.

14. J. Kim and C. R. Das. Modeling Wormhole Routing in a Hypercube. In *11th International Conference on Distributed Computing Systems*, pages 386–393, May 1991.

15. C. P. Kruskal and M. Snir. The Performace of Multistage Interconnection Networks for Multiprocessors. *IEEE Transactions on Computers*, C-32(12):1091–1098, Dec 1983.

16. H. T. Kung and C. E. Leiserson. *Introduction to VLSI Systems*, chapter Algorithms for VLSI Processor Arrays, pages 271–292. Addison-Wesley, 1980.

17. G. Lee. A Performance Bound of Multistage Combining Networks. *IEEE Transactions on Computers*, 38(10):1387–1395, Oct 1989.

18. J. C. Mogul. Network locality at the scale of processes. *ACM Transactions on Computer Systems*, 10(2):81–109, May 1992.

19. L. M. Ni and P. K. McKinley. A Survey of Wormhole Routing Techniques in Direct Networks. *Computer*, pages 62–76, Feb 1993.

20. J. H. Patel. Performance of Processor-Memory Interonnections for Multiprocessors. *IEEE Transactions on Computers*, C-30(10):771–780, Oct 1981.

21. A. A. B. Pritsker. *Introduction to Simulation and SLAM II*. Halsted Press, 3rd edition, 1986.

22. D. A. Reed and R. M. Fujimoto. *Multicomputer Networks Message-Based Parallel Processing*. The MIT Press, 1987.

23. A. A. Sawchuck, B. K. Jenkins, C. S. Raghavendra, and A. Varma. Optical crossbar networks. *Computer*, 20(6):50–60, Jun 1987.

24. T. J. Schriber. *An Introduction to Simulation using GPSS/H*. John Wiley, 1991.

25. H. D. Schwetman. CSIM: A C-based Process-oriented Simulation Language. In *Proceedings of the 1986 Winter Simulation Conference*, Dec 1986.

26. C.-L. Wu and M. Lee. Performance Analysis of Multistage Interconnection Network Configurations and Operations. *IEEE Transactions on Computers*, 41(1):18–27, Jan 1992.

ATMSWSIM An Efficient, Portable and Expandable ATM SWitch SIMulator Tool

Joan García-Haro, Rocío Marín-Sillué and José Luis Melús-Moreno
Department of Applied Mathematics and Telematics, Polytechnic
University of Catalonia, C/ Gran Capitán s/n, Mòdul C3, Campus Nord,
08034 Barcelona, Spain.
e_mail: teljgh@mat.upc.es

Abstract. The Asynchronous Transfer Mode (ATM) has been adopted by CCITT as a world-wide standard for the Broadband Integrated Services Digital Network (B-ISDN). The Switching architectures developed for conventional circuit and packet switching cannot be directly applied to ATM. In spite of the large number of ATM switching architecture proposals, to the knowledge of the authors, there are few or none description in the literature of the simulation model and tools adopted to analyze them. Therefore the main contribution of this paper is to present and describe, with a practical point of view, a discrete event ATM SWitch SIMulator (ATMSWSIM) designed to evaluate the performance of ATM switching fabrics under different traffic and architecture parameters.

1 Introduction

The Asynchronous Transfer Mode (ATM) has been standardized by the CCITT as the multiplexing and switching principle of the Broadband Integrated Services Digital Network (B-ISDN). ATM is a packet and connection-oriented transfer mode based on statistical time-division multiplexing techniques. The information flow is organized in fixed-size packets called cells, consisting of a user information field (48 octets) and a header (5 octets). The primary use of the header tag is to identify cells belonging to the same virtual channel and to make routing possible. The lines speeds are also specified with nominal rates of 150 Mbps and 600 Mbps [1].

The switching architectures developed for conventional circuit switching or the architectures developed for pure packet switching cannot be directly applied to ATM. The major and different requirements imposed by the ATM involve the high speeds at which the switch has to operate as well as the statistical behavior of the ATM streams passing through the ATM switching nodes. Performance evaluation of ATM switches in terms of the input traffic characteristics and concrete switch architecture is, thus, critical and useful for comparative performance testing among different switch design alternatives.

A variety of ATM switching architectures has been described and their performance reported in the literature [2]. The analytical models developed to characterize such systems are not yet well established. The approximations and constraints that they impose often restricts the range of system characteristics to evaluate, regarding input traffic as well as internal architecture. In the majority of the cases, the complexity of the system make it analytically not tractable and the unique manner to predict the performance of various design options is through computer simulation techniques [3].

Despite the large number of ATM switching architecture proposals and related performance evaluation studies, there are few or none description about the simulation models and tools adopted. Therefore, the main contribution of this paper is to present and describe, with a practical point of view, a discrete event switch simulation environment designed to evaluate the performance of ATM switching fabrics under different traffic and architecture parameters.

The ATM SWitch SIMulator (ATMSWSIM) is a modular and hierarchical software system, its design is aimed to be efficient, portable onto several hardware platforms, easy to extend to incorporate new features and easy to use. ATMSWSIM allows researchers to evaluate and study different buffering strategies, several selection policies and backpressure mechanisms for internally nonblocking and blocking ATM switches under various input traffic models.

The rest of the paper is organized as follows. In section 2 the ATM switch simulator model is introduced and all its functional blocks carefully explained and justified. Section 3 describes the simulation algorithm. A performance evaluation example is outlined in section 4. Finally, section 5 concludes this paper.

2 ATM Switch Simulation Model

Mainly two factors have a significant impact on the performance of an ATM switch system. The first of them is the incoming traffic that feeds the inputs of the switch, therefore the first step to implement an ATM switch simulator is to characterize the statistical properties of the traffic that the switch have to support.

The second factor is the concrete architecture of the switching network. From a general point of view, switch architectures can be divided into internally nonblocking and blocking ones [4] [5]. In a nonblocking packet switch every input has a nonoverlaping direct path to every output so that, no blocking or contention can occur internally. However, due to the statistical nature of the input traffic, ATM switches require buffering to avoid packet loss whenever there are packets arriving at different input ports that attempt to reach the same output port simultaneously. In this case, only one packet at a time can be transmitted over an output link, the rest must be temporarily stored in a buffer for later transmission. The placement and arrangement of the buffering system have notable effects on the switch performance as well as on its implementation complexity [6].

ATMSWSIM is an event driven simulation tool written in C language and designed to be efficient, portable and easy to use. Its modularity and hierarchical design allows it to be easily extended or tailored for a specific problem. The program runs on any platform that supports ANSI C. For example, it has been successfully run on SUN Microsystems as well as on HP apollo workstations.

ATMSWSIM takes into consideration the above mentioned factors that affect the system performance and in accordance to them it can be functionally divided into three main logical modules, namely the traffic generation module, the switching architecture module and the output module that is responsible for gathering statistics. These modules work together to form a single simulation facility.

The overall ATMSWSIM simulation environment also provides a user interface (UI), a debugging module, a monitoring module and a statistical module for processing the output results that are recorded in an output file when the program reaches the end of simulation condition.

Figure 1 shows the relation among all the functional blocks.

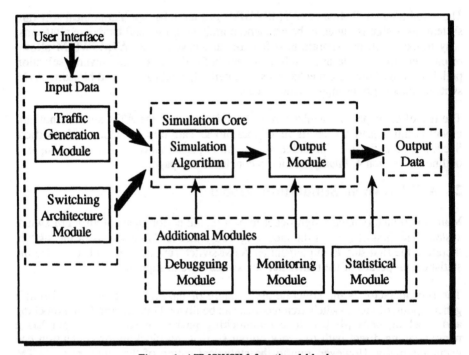

Figure 1. ATMSWSIM functional blocks

Before the discussion of every module and submodule as well as the overall switch operation, the following general assumptions have to be stated:

-   The number of switch input ports N is equal to the number of switch output ports. Therefore only symmetric $N \times N$ switching systems are simulated.

-   The system operates synchronously, time is divided into units (cycles or slots) of cell transmission time duration. This value is taken as the unit of time in the simulation.

-   The speed of the input and the outputs links is the same. That is, cells can arrive to the input ports to a maximum rate of one cell per unit of time. Cells depart from the output ports at a maximum rate of one cell per unit of time.

-   Packet arrivals are assumed to occur at the beginning of a time slot.

-   Packet departures are assumed to occur at the end of a time slot.

Let us follow with the three main modules and its design and implementation issues.

2.1 Traffic Generation Module

Prior to explaining the traffic generation module, some comments about the pseudorandom number generator along with the probability distribution functions implemented are briefly explained.

ATMSWSIM uses an independent platform pseudorandom number generator. It is a multiplicative linear congruential generator based on the Lehmer's algorithm. It produces a sequence of pseudorandom numbers uniformly distributed in the range (0,1). The generation function is $f(z) = az \bmod m$ with multiplier $a = 7^5$ and prime modulus $m = 2^{31} - 1$.

In [7] it is shown that with this parameters $f(z)$ is a full period generating function, the output sequence is "random" and it can be implemented efficiently using 32-bit arithmetic. The efficient implementation in a high level programming language is also presented in [7].

The program also keeps a list of "good" initial seeds to be able to reproduce experiments under the same conditions.

From this pseudorandom number generator the following distributions are implemented:

- *Uniform distribution in the range [a,b].*

- *Negative exponential distribution.*

- *Geometric distribution.*

Modelling of B-ISDN traffic is an active research field. The exact B-ISDN traffic characteristics are still unknown, among other reasons because B-ISDN is not deployed yet. However, some integrated traffic models have been developed to specify the stochastic process describing cell traffic on the switch input lines [8].

When describing the input traffic of an ATM switch, two facts have to be taken into account. First, how cells are generated on each switch input port. That is, what is the probability law that governs the generation of packets at the inputs. Second, the addressing of incoming cells among the output ports. Addressing, thus, refers to the probability distribution that controls the output port destination of each packet.

At the present time, ATMSWSIM is able to simulate random traffic and two types of bursty traffic. Three different output port addressing distributions are also available. The traffic generation patterns can be described as follows.

Random Traffic. Cells arrivals to each input port are generated according to a Bernoulli process of parameter pr, $0 \le pr \le 1$, (pr is a user parameter). Figure 2 shows the flowchart describing the process to schedule and generate a cell with probability pr.

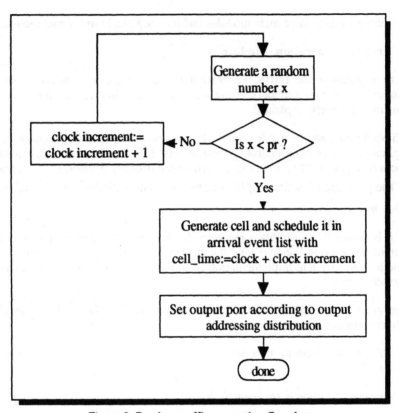

Figure 2. Random traffic generation flowchart

Bursty Traffic. In this case, the offered traffic on each input line arrives at batches and it is modelled by a two-state Markov chain which alternates between active and idle periods. The active an idle period lengths follow a random geometric distribution of a given mean. The minimum active and idle period length is of one cell-length. No cell generation occurs during the idle period. Cells generated during the same active period have a common destination. However, cells from different bursts may be going to different output port destinations.

Therefore, the probability that the active period lasts for i units of time is (analogous for the idle period)

$$Prob(i \ active \ slots) = p(1-p)^{i-1} \ , \qquad i \geq 1$$

and the average active period length $T_{ON} = \displaystyle\sum_{i=1}^{\infty} i \cdot Prob(i \ active \ slots) = \dfrac{1}{p}$

Depending on the cell generation process during the active period two types of bursty traffic can be characterized:

- *Bursty Traffic with Cell Random Generation.* Cells are generated during active periods according to Bernoulli trials of probability α, so that, the space between consecutive packets is not constant. This type of traffic is shown in figure 3.

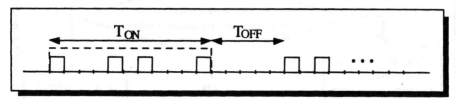

Figure 3. Bursty traffic with cell random generation

In order to vary the value of the average applied load on an input line ρ, users are allowed to select any of the parameters that define the traffic, namely the average active period length T_{ON}, the mean of the idle period length T_{OFF} and the probability of cell generation inside a burst α.

$$\rho = \frac{T_{ON}}{T_{ON} + T_{OFF}} \cdot \alpha , \qquad 0 \le \alpha \le 1$$

Once one of these parameters is selected as independent variable, the users have to fix the value of the remainder two parameters.

- *Bursty Traffic with Periodic Cell Generation.* During active periods cells are generated periodically. Therefore the space between successive packets is constant (see Fig. 4). This traffic pattern try to capture the inherent periodicity in most of the real traffic sources. Such periodicities usually arise due to the fact that it takes a constant amount of time to gather bits to form a cell to offer to the network [9].

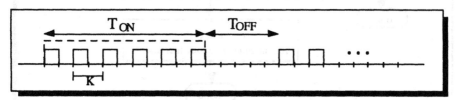

Figure 4. Bursty traffic with periodic cell generation

The user parameters are T_{ON}, T_{OFF} and the interarrival time k. The applied load can be denoted by

$$\rho = \frac{T_{ON}}{T_{ON} + T_{OFF}} \cdot \frac{1}{k}$$

The flowchart for the bursty traffic with cell random generation is presented in figure 5. The variable called remanent cells governs the active period length.

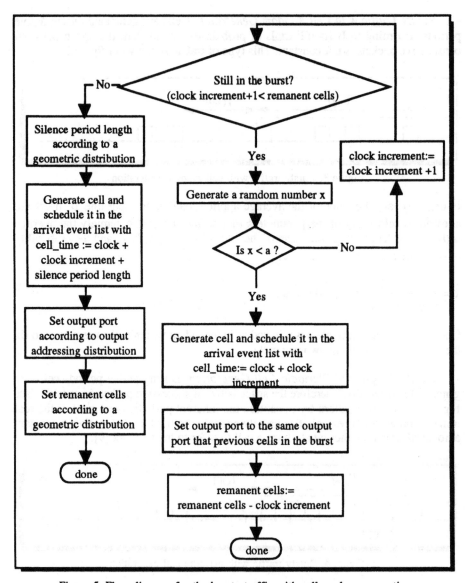

Figure 5. Flow diagram for the bursty traffic with cell random generation

Regarding the output addressing distributions the following three distributions are implemented in ATMSWSIM.

-   *Uniform Distribution*. Each incoming cell/burst has equal probability $(1 / N)$ of being addressed to any output port.

-   *Hot Spot Distribution*. One of the (N) output ports has a higher probability of being addressed by any incoming cell/burst at any input port than all other output ports.

- *Mixed Uniform and Point-to-Point Distribution.* All the traffic from one input port is always destined to a same output port. The traffic from other input ports is uniformly distributed among all the output ports.

Users can select the addressing distribution of interest as well as its parameters.

Note that in the case of random traffic, addressing is defined per incoming cell, however for bursty traffic addressing is best defined per burst, since all the cells belonging to an active period have the same output port destination.

2.2 Switching Architecture Module

As it was mentioned above, the overall operation of any packet switch under statistically varying input traffic can be broadly divided into a buffering function and a switching function. The buffering function is needed to store cells that cannot proceed due to contentions. The switching function refers to the effective transfer of input cells to their corresponding output port destinations. In addition, depending on how the switching function is done switching architectures can be further classified into internally nonblocking and blocking ones.

Since the concrete implementation of both operations greatly influences the system performance ATMSWSIM is, thus, able to simulate different buffering strategies as well as different switching architectures.

2.2.1 Buffering Strategies

The following buffering strategies are available:

Input Buffering

Two options are implemented. The individual FIFO input buffering and the shared-memory input buffering.

- *Individual FIFO Input Buffering.* It refers to a classical buffering strategy, which simulates dedicated first-in first-out buffers for each input port. Note that the input FIFO buffers are only required to work at the same speed as the I/O links. That is, they are able to write and to read one cell per time slot.

- *Shared-Memory Input Buffering.* It is aimed to simulate Random Access Memories (RAM). In this strategy, also referred as to virtual queuing, each switch input has a buffer that is physically shared by all the waiting cells, but logically divided into N FIFO queues, one for each output port destination [10]. Such buffer arrangement has associated a control function that is responsible to simulate the way cells are read out from each one of the N logical FIFO queue.

Another feature of this strategy is the input grouping by which several input ports share a common-memory input module. If the input grouping factor (G_i) is higher than one, the access time of he simulated RAM must fulfil G_i write and G_i read operations in a time slot.

Output Buffering

Only an option is already implemented.

- *Individual FIFO Output Buffering.* Each switch output port is provided of a FIFO output queue. It must be able to receive up to N cells and to release one cell at a time. Therefore the switch does not suffer from the Head-Of-Line (HOL) blocking effect as they do switches with individual FIFO input queues [6].

- *Shared-Memory Output Buffering* with output grouping capabilities is under development.

Input/Output Buffering

In this case, input and output buffering techniques are combined in such a way that the speed of operation of the switching fabric as well as the number of simultaneously accepted cells by the output buffers can be lower than the switch size (N) [11]. An input/output buffered nonblocking ATM switch is shown in figure 6.

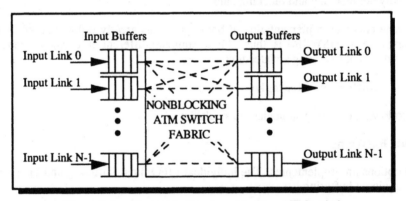

Figure 6. Input/output buffered nonblocking ATM switch

Related to this buffering strategy, ATMSWSIM also allows users to incorporate into the simulated switch design a backpressure mechanism to avoid cell losses at the output queues. One of such backpressure mechanisms can be activated and it operates as follows, when an output queue reaches a user specified level (e.g. the maximum queue capacity) a feedback signal from the output queues is transmitted to the input queues so that cell arrivals to that output queue are stopped. Incoming cells for that destination have to wait at the corresponding input queue until room at the output queue becomes available.

It should be noted that if input/output buffering is used a speed-up factor of the switching architecture has to be defined by the user that is, the operation of the switch must be higher than the I/O link speed.

Central Buffering

All input and output ports have access to a shared-memory module able to write up to N incoming cells and to read out N outgoing cells in a switch time cycle. In fact, this is another possibility to implement output buffering and therefore, to attain the best throughput/delay/cell loss performance [6]. In addition, considering a shared-memory module of size $N \times N$ is an efficient way to construct a nonblocking switch. Note that, in this case, the two required functions of any packet switch, queuing and switching are carried out via buffering. Despite some limitations, this along with other reasons make this design very attractive for the industry and the majority of the implemented prototypes today use it, alone or as the building blocks of larger switching fabrics.

For all the buffering strategies, the buffer size is selected by users. The program allocates memory dynamically as the simulation evolves. The maximum value permitted is the user entered buffer size and it is only limited by the physical memory of the workstation.

2.2.2 Selection Policies

When multiple cells at the head of the input queues have the same destination address, only some of them (depending on the speed-up of the architecture) will be delivered simultaneously. The order in which cells are served to avoid output port contention is known as the selection or arbitration policy of the switch.

The ATMSWSIM simulation environment provides several selection policies for the case that individual FIFO input buffering and a nonblocking switch architecture options are selected. Specifically, eight different selection policies were tested to ensure a certain degree of fairness in the service of input queues [12]. Here, we only discuss the well known Modified Cyclic mechanism in which the input queues are visited following a round-robin algorithm, but in each switching cycle the starting port is incremented according to the formula: $starting\_port = i \, mod(N)$, where i is the current switch cycle, N the switch size and the input ports are labelled as $0, 1, 2, ..., N - 1$. For this selection criterion, the scanning order of the input queues, itself, establishes the priorities to resolve any possible output port contention.

2.2.3 Switching Architecture

Regarding the switch operation and at the current version of the simulation tool, ATMSWSIM implements two switching architectures. An internally nonblocking switch architecture and a blocking one are, thus, available.

As it was stated before, time is slotted and the switch is assumed to operate on per unit of time. The switch function is assumed to take one time slot.

Nonblocking Switch Architecture

It is simulated at a high level of abstraction without considering the internal structure details of the switch. It operates as follows, according to the buffering strategy, speed-up factor and the selection policy (if required) used, cells at the inputs are moved to their corresponding output port destinations. The determination of the input cells to be switched in accordance to the selection policy and criterion imposed along with the effective transfer of the input cells to the output ports is done in a unit of simulation time. The pseudo-code of the nonblocking switch operation is shown below. Figure 6 also shows a possible configuration of such architecture.

```
FOR (each iteration due to speed-up)
        Set reservation flags for output ports to free
        FOR (all input ports)
                Select input port  and queue to serve according to selection policy
                and input buffering strategy
                IF (cell can be switched)  /*if reservation flag indicates output port
                        is free and there is room at output queue (if feedback
                        is on)*/
                        Remove cell from input queue
                        Update input queue length counters
                        IF (output buffer is full)
                                Increment output loss counter
                        ELSE
                                Store cell in output queue
                                Update output queue counter
                        END {IF}
                        Accumulated_Input_Delay :=
                                Accumulated_Input_Delay + (clock-cell_time)
                        Set reservation flag  to busy
                END {IF}
        END {FOR}
END {FOR}
```

Internally Blocking Architecture

A generalized cube network (banyan) is modelled. It is a unique path multistage interconnection network. An $N \times N$ banyan network is implemented in ATMSWSIM as an interconnection network made of $n = \log_2 N$ stages. Each stage consisting of $N / 2$ 2×2 switching elements [5]. The switching elements are considered to be internally nonblocking and memoryless. Link connections between stages as well as the link topology at the input ports and at the output ports take the pattern of a generalized cube network as it is shown in figure 7.

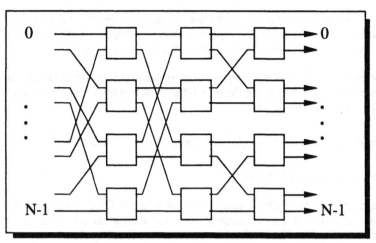

Figure 7. Generalized cube network (N = 8)

The network is said to be selfrouting that is, routing is carried out using an address tag of length $\log_2 N$ digits that is appended to the header of the transmitted cell. Each stage in the network is associated with the position and value of one bit of the destination tag. If such bit is '0' the cell is routed to the upper output link of the corresponding switching element, however if the bit is '1' the cell go to the lower output link and it is finally delivered to the next stage where the operation is repeated again until the output port destination is reached if no internal contention arises.

ATMSWSIM simulates this switching architecture at a low level of abstraction. The switching elements are defined as well as the interconnection topology that is determined using recursive functions controlled by the required number of stages depending on the switch size N selected by the user. Once the architecture is fully configured according to the generalized cube network pattern and the buffering strategy along with the speed-up selected, the routing of cells through the switching network is completely traced from the input ports to the output lines. It has to be mentioned that in case of collisions for an internal link, one of the cells involved is lost, and this event accounted for the subsequent calculus of the blocking probability.

2.3   Output Module

This module contains a set of functions that instruct the accurate statistical data gathering which will be further used to compute the entire switch performance. These functions are called while the simulation experiment progress at specific points when some changes of interest occur in the system status (e.g. changes in the queue length, delay updates, cell loss, etc.). The output module is also in charge of reporting the simulation results that are written, when the simulation time expires and several independent runs are already executed, into an output file. The name of the output file is introduced by the user before the simulation begins. The output file first echoes all the input parameters previously chosen by the user continuing with the simulation results and the performance measures of interest. The output file is arranged in such a way that it is possible to plot some results using a highly available graphical tool (e.g. Xvgr) not integrated in the ATMSWSIM simulation package.

The following performance measures are printed into the output file, for several independent executions, and their respective average computed. Concretely we usually employ five independent runs, however this value is a user parameter.

- The *average length of each input buffer* is calculated by dividing the sum of all products of the queue length and the time interval of that queue having such length by the total simulation time interval.

$$(\sum(queue\_length \times interval\_of\_such\_length)) / total\_interval$$

- The *average output buffer length.*

- The *average input buffer length distributions* are defined as the probability that the input buffer length (L) exceeds or be equal than a certain value (k) in cells ($F(k) = Prob(L \ge k)$). The levels of interest are introduced by the user.

- *Cell loss probability* measured over the entire switch and computed as (# means number)

$$Cell\_loss\_prob = \frac{\# in\_loss + \# out\_loss + \# int\_blockings}{\# succes\_switched\_cells + \# in\_loss + \# out\_loss + \# int\_blockings}$$

- *Probability of internal blocking.* Only measured in the banyan network.

- *Average switching delay* (in cell units). It measures the average total delay from the time a cell enters an input to the time that it leaves its corresponding output queue.

- *Average input delay.* It refers to the average delay cells encounter waiting at their input queues before being switched to their destination output ports.

- *Average output delay.* It is the average delay that cells have to wait at their corresponding output queues before they are forwarded to the output links.

- *Utilization.* It measures the average utilization of the switch.

- *Throughput.* It is computed as the ratio of the total number of cells that successfully exit the system to the total number of cells that enter it.

All these values are tabulated at the output file being the first column the independent variable, the second the average of five independent simulation runs and the following columns contain the results obtained by using each one of five different initial seeds.

2.4 Additional Modules

In this subsection, some additional modules are outlined.

The **User Interface Module**, at the current version it is textual due to the portability requirements. It is an independent piece of software that allows users to interactively define all the system parameters needed to configure the switch architecture and traffic pattern to study. The introduced values are stored in an input file readable for the simulation program that can, thus, initiate the simulation experiment.

The **Debugging Module** contains several functions useful to trace and debug the program. If the debugging mode is turned on (switching a global variable) the contents of critical data structures or the steps of some switching actions will be printed to help designers to trace the program or to fix some bug.

The **Monitoring Module** is in charge of generating a log file that records and continuously updates the status of all simulation runs during the execution time.

The **Statistical Module** processes the statistical simulation results. It basically computes confidence intervals.

The **Launching Module**. ATMSWSIM currently runs on a single processor. The launching module is able to launch several independent copies of the simulation tool, with different input parameters, over different workstations connected to a local area network. This is a very easy technique to minimize the time needed to simulate a switch model. In addition, it does not require synchronization among processes. In [3] it is shown that if the run length is long or the initial transient period is short, replicating a simulation is statistically more efficient than distributing it that is, partition the simulation into numerous logical processes (parallelizing the simulation program) and execute them in a multiprocessor environment.

3 Simulation Algorithm

All the above described modules work together to form a single simulation facility. Below, it is shown the pseudo-code representing the ATMSWSIM main program. It controls the correct sequencing and scheduling of events and activities in the simulation.

```
BEGIN
        Read input data.
        FOR (every simulation point selected by user)
                Initialize data structures
                FOR (first initial seed) TO (last different initial seed)
                        Arrival event list creation
                        Scheduling of the first arrival to each input port according
                                to the traffic generation patterns
                        Set clock to first cell arrival time
                        WHILE (NOT end of simulation)
                                WHILE (time of first cell in the
                                                arrival event list = clock)
                                        Store cell at input queue
                                        Generate new cell for the
                                                corresponding input port
                                        Update the arrival event list
                                END {WHILE present clock}
```

```
                              Increment clock value
                              IF (nonblocking switch)
                                      Cell transfer
                                      Store cell at corresponding output queue
                              ELSE {Banyan}
                                      Cell transfer
                                      Detailed routing operation
                                      Store cell at corresponding output queue
                              END {IF}
                              Remove head of line cells from output queues
                              Increment clock value
                      END {all cells done}
                      Statistical gathering
              END {FOR}
              Compute averages
      END {FOR}
      Write results
END
```

Once the file containing the input data (previously created interactively by the user) has been read, the program initializes all the variables to their appropriate values, creates the system configuration to be evaluated and generates arrivals according to the input traffic pattern selected.

ATMSWSIM is an event driven simulation program, however some of the activities associated to particular events only last for a unit of time (e.g. storage of cells at the input buffers, duration of the switching function, etc.).

At the start of the simulation experiment the arrival of cell events for each one of the N input ports are generated in accordance to the input traffic generation parameters. An arrival event list is created dynamically. The arrival event list contains a list of cell arrivals to occur in the future (one arrival event list entry for each input port), arranged in increasing occurrence time order. The simulation clock is set to the time associated to the first scheduled arrival in the arrival event list. Whenever an arrival event gets into the switching system process, a new arrival event for the same input port is generated and scheduled again, in such a way that the arrival event list is always full.

When a cell arrival is processed it progresses according to the following events:

(a) *Placement of arriving cells at the input queues* and generation of output port destination in accordance to the selected addressing distribution. This actions take one unit of simulated time.

(b) *Switching of cells at the head of the input queues*. It includes the contention resolution process using a selection policy.

(c) *Removal of cells out of the heads of output queues*.

The two last events also last for a unit of simulated time.

In this simulation model, a cell has to stay at its input queue for at least one time slot before being switched. The switching function also takes a time unit. Therefore the minimum amount of switching time for a cell to be switched is two units of time.

As long as there are cells in any input or output queue the clock is incremented according to the above mentioned rules. However when all queues are empty the clock is advanced to the time of the next arrival event in the arrival event list.

The condition that controls the end of the simulation is a user parameter that accounts for the total effective number of cells to be processed during the simulation. This number includes the total number of cells that are successfully switched as well as the cells that are lost anywhere in the system. For effective number we mean that the simulation results try only to account for the steady-state results.

In ATMSWSIM two factors contribute to minimize the effects of the initial or transient state of the simulation on the performance measures of interest. One is the discarding of the performance statistics collected during the transient portion of a simulation run. This is done through a user parameter that represents the number of initial transactions whose statistics will be discarded by the simulation. This number is not a well-specified one. It is found heuristically through previous simulation trials to gain insight in it. However the second factor is running the simulation for a large number of transactions so that the transient portion becomes negligible.

On the other hand in a simulation experiment it has to be some means to quantify the confidence in the performance measures obtained. In ATMSWSIM five independent runs (using five distinct initial seeds) are consecutively executed for the effective number of transactions and for each performance measure of interest. This quantification is expressed by confidence intervals. They are determined with the aid of the Student's t distribution table having $n-1$ degrees of freedom, where n is the number of independent observations. This is done by an external module.

Finally it has to be said that the model was verified and validated. Verification is the process through which the modeler establishes that the simulation program is a correct formulation of the model. On the other hand, validation refers to the process of demonstrating that the model is a realistic and satisfactory representation of the actual system.

ATMSWSIM was verified by using standard debugging tools as well as special purpose functions included into the Debug Module. This functions are used to test and print system status regarding input traffic and internal switch architecture. The validation of the model has been carried out by comparison to previous results presented in the related literature. Some simulation results have been also checked through the resolution of analytical models when a mathematical solution is possible. In this case, both type of results also matched.

4 Example

As an example the switch configuration shown in figure 8 was evaluated [10].

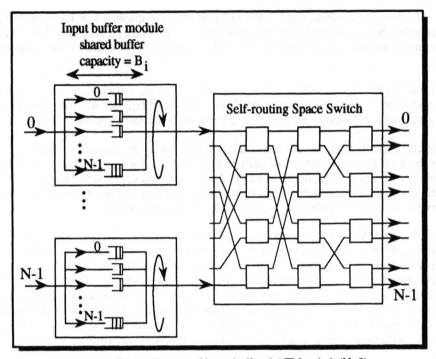

Figure 8. Block diagram of input-buffered ATM switch (N=8)

Basically, it consist of a generalized cube network with shared-memory input buffering. The logical input queues are scanned as follows, at the beginning of each time slot, an input pointer for each input module indicates the input virtual queue from which a cell will be extracted. If there is a cell to switch (at the head of the line) in the corresponding logical input queue, it will be forwarded to the switch. If the corresponding logical input queue is empty, no cell will be switched. The pointer for the input buffer module j (p_j) is incremented in each time slot and for a particular time slot or switch cycle (i) follows the relation: $p_j = (i + j) \bmod N$, where i is the current time slot number, j is the input buffer module label (varies from 0 to $N - 1$), and N is the switch size. There is a shift by one between two consecutive input modules and all the selected cells have a different output port destination and therefore, output port contention is not possible. In addition, the cells are ordered in such a way that the banyan network can route them without incurring internal blocking. Therefore, the entire system is internally and externally nonblocking with a very low hardware complexity.

Some performance results are also shown in figures 9 and 10 for the basic operation as well as for several input grouping factors (G) to improve performance results. From figure 9 it can be concluded that the switching bandwidth is evenly distributed among the different input ports. It can also be observed that as G grows there is a more efficient sharing of the memory resources. From figure 10 it is seen that the maximum throughput approaches 100%. In addition, if input grouping is used the delay degradation is minimized.

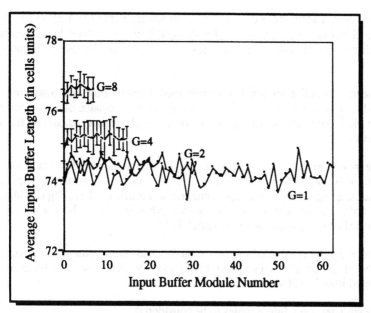

Figure 9. Average Input buffer length (N = 64, random and uniform traffic, p = 0.7)

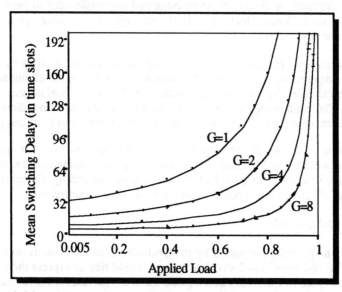

Figure 10. Mean switching delay vs. applied load performance (N = 64, random and uniform traffic)

5 Conclusions and Future Work

This paper has described an ATM SWitch SIMulator tool (ATMSWSIM). It is developed to provide a computer-aided engineering software environment that allows the emulation of different ATM switching techniques (classical and new ones) and their effect on the performance of the switching system.

The ATMSWSIM software is designed to be modular, hierarchical, efficient, portable onto several workstation platforms and with capabilities to be extended. It can be used for switching networks of practically any size, depending on the available system memory.

We proposed an ATM switch parametric model explaining and justifying all its functional blocks. It has to be specially highlighted the relevance of all the parameters related to the input traffic and switching modules for their direct impact on the system performance.

We have made extensive use of this tool in the research of several switching configurations regarding selection policies, buffering strategies, switching architectures, grouping and speed-up techniques and traffic models [10][12]. In addition to these observations, the simulation studies are expected to be useful for future mathematical analysis in this active research field.

Future work can be conducted on the expansion and improvement of the simulation tool as well as on the subsequent evaluation of the new architectures to be incorporated into the software tool.

The following are some future issues to be considered.

-    Due to the modular design, it can be estimated the possibilities to parallelize the code and run the different blocks (e.g. traffic generation, switching operation) on a multiprocessor environment.

-    Improvement of the user interface and the output presentation using standard windowing packages (e.g. X-windows) to maintain portability, but taking advantage of the graphical capabilities of current workstations. Even graphical animation can be considered. The main objective is to use this tool in the academia and allow students to test, experiment and fix their theoretical knowledge in broadband switching.

-    Extension of the input traffic and switching modules to integrate new traffic models and switching architectures. For example, detailed simulation of nonblocking switching architectures composed of shared-memory switching elements and the evaluation of different routing schemes are currently under investigation.

Acknowledgments

This work was partially supported by the Spanish Research Council under project CAYCIT TIC92-1306-C02-02-PB. The authors would like to express their gratitude to Fai Fan of Queen's University at Kingston, Ontario, Canada for his assistance with the simulation program, his helpful comments, discussions and sincere collaboration. Thanks are also due to Josep Paradells-Aspas, Armando Martín-López and Josep Maria Malgosa-Sanahuja of the Polytechnic University of Catalonia for his unreserved cooperation.

References

1. CCITT Recommendation I.121. "Broadband Aspects of ISDN", Blue Book, Fascicle III.7, Geneva 1989.

2. F. A. Tobagi, "Fast Packet Switch Architectures for Broadband Integrated Services Digital Networks", Proc. of the IEEE, Vol. 78, No. 1, Jan. 1990, pp. 133-167.

3. J. F. Kurose, H. T. Mouftah, "Computer-Aided Modeling, Analysis, and Design of Communication Networks", IEEE J. Sel. Areas on Comm., Vol. 6, No. 1, Jan. 1988, pp. 130-145.

4. A. Pattavina, "Nonblocking Architectures for ATM Switching", IEEE Comm. Magazine, Feb. 1993, pp. 38-48.

5. X. Chen, "A Survey of Multistage Interconnection Networks in Fast Packet Switches", International Journal of Digital and Analog Communication Systems, Vol. 4, 1991, pp. 33-59.

6. M. G. Hluchyj, M. J. Karol, "Queueing in High-Performance Packet Switching", IEEE J. Sel. Areas on Comm., Vol. 6, No. 9, Dec. 1988, pp. 1587-1597.

7. S. K. Park, K. W. Miller, "Random Number Generators: Good Ones are Hard to Find", Communications of the ACM, Vol. 31, No. 10, Oct. 1988, pp. 1192-1201.

8. I. W. Habib, T. N. Saadawi, "Multimedia Traffic Characteristics in Broadband Networks", IEEE Comm. Magazine, Jul. 1992, pp. 48-54.

9. T. E. Eliazov, V. Ramaswami, W. Willinger, G. Latouche, "Performance of an ATM Switch: Simulation Study", in Proc. of the INFOCOM'90 Conf., San Francisco, Jun. 1990, pp. 644-659.

10. J. García-Haro, F. Fan, A. Jajszczyk, "HITIBAS - A High-Throughput Input-Buffered ATM Switch", in Proc. of the SICON/ICIE'93 Conf., Singapore, Sep. 1993, pp. 359-363.

11. Y. Oie, M. Murata, K. Kubota, H. Miyahara, "Effect of Speedup in Nonblocking Packet Switch", in Proc. of ICC'89 Conf., Boston, MA, June 1989, pp. 410-414.

12. J. García-Haro, C. Cervelló-Pastor, J. Paradells-Aspas, H. T. Mouftah, "Evaluation Study of Several Head-Of-Line Selection Schemes for High Performance Non-blocking ATM Switches", in Proc. IEEE Pacific RIM'93 Conf., Victoria, B.C., Canada, May 1993, pp. 327-332.

A Model for Performance Estimation in a Multistreamed Superscalar Processor *

Mauricio J. Serrano, Wayne Yamamoto,
Roger C. Wood, and Mario Nemirovsky

Electrical and Computer Engineering Department
University of California, Santa Barbara, CA 93106-5130

Abstract. The current trend is integrating more hardware functional units within the superscalar processor. However, the functional units are not fully utilized due to the inherent limit of instruction-level parallelism in a single instruction stream. The use of simultaneous execution of instructions from multiple streams, referred to as multistreaming, can increase the number of instructions dispatched per cycle by providing more ready-to-issue instructions. We present an analytical modeling technique to evaluate the effect of dynamically interleaving additional instruction streams within superscalar architectures. Estimates of the instructions executed per cycle (IPC) are calculated given simple descriptions of the workload and hardware. To validate this technique, estimates obtained from the model for several benchmarks are compared against results from a hardware simulator.

1 Introduction

Superscalar processors employ multiple functional unit designs that can dispatch several instructions every cycle. Two factors limiting the number of instructions dispatched per cycle are: 1) the number of functional units available (hardware) and 2) the amount of parallelism in the workload (software). While the number of functional units determines the peak throughput of a processor, the instruction-level parallelism limits the actual performance obtained. Data dependencies and control breaks constrain instruction level parallelism resulting in a sustained system performance that is well below the peak.

Combining hardware and compiler techniques can increase the parallelism. Register renaming, out-of-order execution, branch prediction, and speculative execution are some of the hardware techniques. Loop unfolding, prefetching, and instruction reordering are some of compiler techniques. However, as the number of functional units increases, it is unlikely that a single stream can produce enough parallelism to effectively utilize the additional resources.

The ability to execute simultaneously multiple instruction streams, referred to as multistreaming, significantly increases the number of independent instructions that could be issued in a cycle. A multistreamed, superscalar processor

---

* This research was supported by the State of California and Apple Computer Inc. via MICRO grant #92-178. E-mail: mauricio@misd.ucsb.edu.

can dispatch instructions from multiple streams simultaneously (i.e., within the same cycle). If the active streams are independent then the total number of instructions that can be dispatched per cycle will increase as the number of active streams increases. By storing each stream context internally, the scheduler can select independent instructions from all active streams and dispatch the maximum number of instructions on every cycle. The processor adjusts the scheduling policy as the workload changes to maximize throughput.

A simulation study demonstrating performance benefits of multistreamed, superscalar processors was performed by emulating a multistreamed version of the IBM RS/6000 [10]. Due to the cost of the simulations in this study, both in complexity and execution time, we developed a simple analytical model to estimate the overall performance of these architectures. The model uses simple workload and architectural descriptions that are estimated or obtained using commonly available tools. The model produces instructions executed per cycle (IPC) estimates for an architecture as the number of streams is varied.

A purely analytic model is often too complex to be developed; often, many simplifying assumptions are needed for the model to be tractable. Instead, we propose a model which combines simulation and analysis to get performance measures faster than a purely simulation-driven method. Our approach views the instruction stream as a random source of instruction types. The characteristics of the instruction stream are extracted from simulations and used in the analytic model. Our results show that this interpretation gives a good approximation in the context of independent instruction streams. In addition, the model allows to analyze the behavior of programs for which we know the characteristics but do not have the actual code.

2 Background

Superscalar machines exploit the instruction-level parallelism of a program by issueing of several instructions in parallel. Data, control, and structural hazards limit the amount of parallelism [6]. Several compiler techniques increase the amount of parallelism, for example loop unfolding, software pipelining, and instruction reordering [6]. At the hardware level, techniques such as speculative execution and register renaming reduce but not eliminate the effect of the hazards. Also, the complexity of the hardware increases dramatically without a corresponding increase in performance. For example, studies have shown that very large window sizes and aggressive register renaming are needed to adequately exploit instruction parallelism [14]. However, the complexity of implementing a large window size is prohibitively expensive in both time and chip area.

Another technique that increases throughput is instruction interleaving. Interleaving is a way to share the processor resources between multiple streams. A pipeline is interleaved if an instruction from a different instruction stream enters the pipe at every cycle, and there are at least as many instruction streams as pipe stages. Instruction interleaving was introduced in the peripheral processor of the CDC6600 [15]. The main feature of an interleaving architecture is its ability to hide high latency operations. For example, the CDC6600 used interleaving because the memory latency was too large. Ten processes were interleaved in a cyclic way in the pipeline and the instruction cycle time was equal to the memory access time. Advantages of this scheme are the elimination of hazards from pipeline execution by interleaving independent instructions from different streams. More recently, the Tera Computer System [13] uses multistreamed processors in a massively parallel system. Each processor in a Tera computer can execute multiple instruction streams simultaneously. On every tick of the clock, a stream that is ready to execute is selected and allowed to issue an instruction. When an instruction finishes, the stream to which it belongs becomes ready to execute the next instruction. The processor is fully utilized if there are enough instructions streams in the processor so that the average instruction latency is filled with instructions from other streams.

Other examples of interleaved architectures are the multiple instruction stream processor of Kaminsky and Davidson [7], HEP [8], CCMP [12], DISC [9], and APRIL [1].

3 Multistreamed, Superscalar Processors

A multistreamed, superscalar processor is organized to support the simultaneous execution of several instruction streams. Each hardware stream is viewed as a logical superscalar processor. Each stream has an associated program counter, register set, and instruction buffer. The multistreamed fetch unit brings instructions from the instruction cache to the various buffers. Instructions are checked in the buffers for dependencies on previously issued instructions that have not yet completed. The scheduler dispatches instructions that are free of dependencies to the appropriate functional unit provided structural hazards do not exist. Functional units return results to the corresponding stream register file based upon a stream identification tag appended to each instruction. Figure 1 shows a functional block diagram of a multistreamed, superscalar processor.

Our model of a multistreamed, superscalar processor, is partitioned into three major parts: the instruction streams, the scheduler, and the functional units. An instruction stream consists of the context of a process and a buffer containing the next instructions to be executed. The section of the buffer considered by the scheduler for dispatch is called the stream issue window. Instructions within the stream issue window can be dispatched out-of-sequence provided no dependencies exist. The scheduler checks, in round-robin fashion, the instructions in each stream's issue window for data and control hazards and then moves the unblocked instructions into the global issue window. The global issue window contains all the instructions in the stream issue windows that are ready to execute, i.e., that have no data or control dependencies. From the global issue window, instructions are dispatched to the appropriate functional units provided no structural hazards are present. All functional units are assumed fully pipelined so each unit can accept a new instruction on every cycle. However, different instruction types have different execution times, i.e., an integer divide

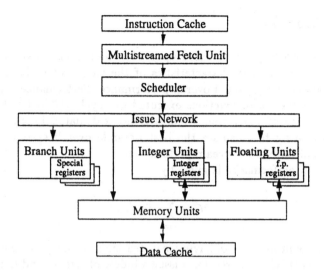

Fig. 1. Functional block diagram of a multistreamed, superscalar processor.

instruction takes 17 cycles to complete. Figure 2 shows the basic structure of a multistreamed, superscalar processor model. The processor has N instruction streams. The functional units are shared between the streams and multiple copies of certain functional units can improve overall performance as the number of streams is increased. The configuration has two integer units (IU), two floating point units (FPU), one memory unit (Mem), and one branch unit (BR).

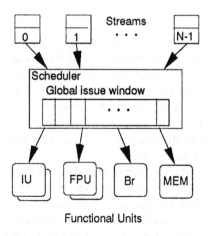

Functional Units

Fig. 2. Basic structure of a multistreamed, superscalar computer.

4 The Model

The model is stochastic and based upon Markov models. Given a description of the processor and the characteristics of the workload to be executed, the model calculates the expected overall performance. Performance is measured by the overall number of instructions executed per cycle (IPC) by the processor. The performance of individual streams is not calculated by the model since we are attempting to calculate the best possible overall system performance independent of individual stream performance.

The overall performance is given by

$$IPC = \sum_{w=1}^{G} P_w IPC_w \tag{1}$$

where P_w is the probability of having w instructions in the global window, IPC_w is the expected IPC measured for a issue window of size w, and G is the size of the global window. The model considers the effect of structural hazards and the effect of control and data hazards. IPC_w models the effect of structural hazards while P_w is a scaling factor that models the performance degradation of the data and control hazards. The rationale behind this division is that the modeling of structural hazards is primarily dependent on the architectural (hardware) configuration while the modeling of control and data hazards is primarily dependent on the workload. In addition, since we employ a Markov chain in our technique, combining the structural hazards as well as dependencies into a single chain results in an extremely large number of states. The division simplifies the chain and makes the problem more tractable.

4.1 Architecture Characterization

The architecture of the processor is specified by a number of streams N, a stream issue window size S, and a functional unit configuration vector **C**. The stream issue window is the buffer consisting of instructions that can be issued within a given cycle. Issued instructions are removed from the window and new instructions are brought in the buffer to maintain it full. We assume that the stream windows have the same size, i.e., $S_1 = S_2 = .. = S_N = S$. Thus, the total number of instructions in the machine considered for issue is equal to $G = N \times S$ (the sum of the stream windows).

An instruction is ready-to-issue if it passes the data and control dependencies constraints. A ready-to-issue instruction is issued if there is any functional unit that can accept it in a given cycle. We assume that there is a perfect *fair* scheduler that dispatches instructions from the streams, i.e., the scheduler gives a fair opportunity to each stream to send instructions to the functional units. The functional unit are assumed perfectly pipelined so each functional unit is capable of accepting an instruction every cycle. The parameter T describes the number of distinct functional unit types. The functional unit configuration vector **C** describes the number of units of each type. Each element of the configuration vector, c_i, contains the number of functional unit of type i. For example, the IBM RS/6000, which can issue a maximum of 4 instructions per cycle (integer, branch, floating point, and condition register), has the following architectural specification: $S = 4, T = 4, \mathbf{C} = (c_{integer}, c_{branch}, c_{floating-point}, c_{cr}) = (1, 1, 1, 1)$. Table 1 summarizes the architectural specification parameters.

Table 1. Architectural specification parameters.

N	Total number of streams.
S	Stream issue window size.
$\mathbf{C} = (c_1, c_2, .., c_T)$	Functional unit configuration vector.

4.2 Workload Characterization.

The workload is defined as the threads that run on the multistreamed processor. The workload is characterized by the type and number of threads loaded and running in the hardware streams of the machine. We assume that each stream is loaded with a thread, thus the number of threads is equal to the number of streams.

The workload is characterized by its runtime instruction mix vector $\mathbf{V}$, the number of active streams N, and a histogram vector $\mathbf{H}$. We view each thread as a stochastic stream of instructions of different types, and we use the instruction mix to characterize the thread. The instruction mix specifies the percentage of instructions of each type. For example, the instruction mix for the benchmark hanoi is: 56% integer, 12% branch, and 32% memory instructions, e.g., $\mathbf{V} = (v_{integer}, v_{branch}, v_{memory}) = (0.56, 0.12, 0.32)$. The vector $\mathbf{H}$ that models the data and control dependencies will be presented in Section 4.4. Table 2 summarizes the workload characterization parameters.

Table 2. Workload characterization parameters

$\mathbf{V} = (v_1, v_2, .., v_T)$	Runtime instruction mix vector.
$\mathbf{H}$	Histogram vector.

4.3 Model of Structural Hazards

The expected IPC of a workload for a given processor configuration is computed as a function of the global issue window size (IPC_w). The global issue window contains all the instructions that are ready to be dispatched to a functional unit, i.e., that have no data or control dependencies. Thus, only the effects of structural hazards are considered; data and control dependencies are not considered.

We assume that there are w ready-to-issue instructions in the global window, where $0 \leq w \leq NS$. The state of the instructions is represented by a vector $\mathbf{M}_w = (m_1, m_2, .., m_T)$, where each element m_i describes the number of instructions of type i. The sum of the elements of $\mathbf{M}_w$ is equal to w. Given a global issue window of size w and T functional types, the total number of states for the window is equal to the number of w-selections of a T-set, or

$$\binom{w + T - 1}{T - 1}$$

For a global window state $\mathbf{M}_w$, the number of instructions that can be dispatched in a given cycle is determined by the number and type of ready-to-issue instructions in the global window and the number and type of functional units. Figure 3 shows an example of a global window of size three containing zero floating-point instructions and three integer instructions ($\mathbf{M}_9 = [0, 3]$); there are one floating-point unit and two integer units ($\mathbf{C} = (1, 2)$)). For the example, only two integer instructions can be dispatched in that cycle. In general, for a given functional unit configuration and global window state, the number of instructions of type j dispatched, referred to as i_j, is the smaller of the number of functional units of type j and the number of instructions of type j within the window. We define I_{M_w}, the total number of instructions that can be dispatched given the global issue window is in state $\mathbf{M}_w$, as the sum of i_j over all functional unit types:

$$I_{M_w} = \sum_{j=1}^{T} i_j, \quad i_j = min(c_j, m_j) \tag{2}$$

The expected IPC given a window of w instructions is calculated as the sum of the instructions issued, times the probability of being in state $\mathbf{M}_w$, defined as Q_{M_w}, for all states of the window:

$$IPC_w = \sum_{\mathbf{M}_w} I_{M_w} Q_{M_w} \tag{3}$$

We obtain Q_{M_w} by calculating the steady state probabilities of a Markov chain involving the states of the global issue window. The states are not independent due to the relationship between the window size and configuration of functional units. For a given processor configuration and global window state, all possible next states are determined by the number of instructions that can execute. For the example in Figure 3, two integer instructions are dispatched and one remains. Therefore, the next state must contain at least one integer instruction ($[2, 1], [1, 2],$ or $[0, 3]$). Table 3 lists all states, the possible next states, and the number of instructions dispatched in each state (I_{M_3}) for the example

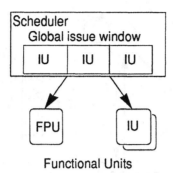

Fig. 3. Example with $w = 3$, $M_3 = [0, 3]$, and $C = (1, 2)$.

Table 3. State table for the example in Figure 3.

Current State $\mathbf{M}_3$	Possible Next States	I_{M_3}
$[3, 0]$	$[3, 0], [2, 1]$	1
$[2, 1]$	$[3, 0], [2, 1], [1, 2]$	2
$[1, 2]$	$[3, 0], [2, 1], [1, 2], [0, 3]$	3
$[0, 3]$	$[2, 1], [1, 2], [0, 3]$	2

in Figure 3. The next state depends on both the current state of the global issue window ($\mathbf{M}_w$) and the functional unit configuration ($\mathbf{C}$).

The state transition probabilities are based upon the runtime instruction mix for the workload ($\mathbf{V}$). Our next discussion restricts the workload to a single instruction mix even though more than one stream may be executing. Later, we discuss workloads of different instruction mixes.

Instructions of different types are assumed uniformly distributed throughout the execution of the workload. For the example in Figure 3, two instructions are executed in state $[0, 3]$. Thus, the probability of a transition to state $[2, 1]$ is equal to the probability of filling the global issue window with two floating point instructions. This probability, obtained from the runtime instruction mix ($\mathbf{V} = (v_0, v_1)$), is just the square of the floating point probability, or v_0^2. Figure 4 shows the Markov chain model and the state transition matrix STM for the example in Figure 3.

After solving for the steady state probabilities ($Q_{[x,y]}$), we calculate the expected IPC for a window of size w, IPC_w, using equation (3). The expected IPC in our example for a window size of 3 is given by

$$IPC_3 = I_{[3,0]}Q_{[3,0]} + I_{[2,1]}Q_{[2,1]} + I_{[1,2]}Q_{[1,2]} + I_{[0,3]}Q_{[0,3]}$$

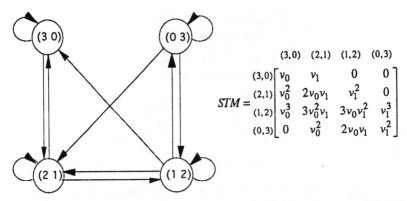

Fig. 4. Markov chain and state transition matrix (STM) for the example in Figure 3, assuming homogeneous workloads.

4.4 Model of Data and Control Hazards

The model of the data and control hazards uses an estimated degradation factor to scale the performance obtained by the structural hazards model. The instruction-level parallelism of a workload is degraded by data and control hazards. This section describes a measure of the instruction-level-parallelism of a workload, i.e., the capacity to issue a number of instructions in a cycle.

The histogram vector $H = (h_0, h_1, h_2, .., h_S)$ characterizes an active stream issuing part or all of its window of instructions in a cycle. Each element h_i describes the probability that i instructions are issued in a given cycle. The single stream measure H for the benchmark can be obtained using performance monitoring hardware, as in the new POWER2 RS/6000 architecture [16]. In our case, we use simulation to extract the histogram vector.

We define P_w as the probability of having w instructions ready-to-issue in a cycle. For the case of two homogeneous workloads (and therefore the same histogram vector), the probability P_0 of zero ready-to-issue instructions is equal to the probability of the two workloads having zero ready-to-issue instructions, or h_0^2. For our example, the following coefficients are obtained with a stream window size of two ($N = 2, S = 2$):

$$P_0 = h_0^2$$
$$P_1 = 2h_0h_1$$
$$P_2 = h_1^2 + 2h_0h_2$$
$$P_3 = 2h_1h_2$$
$$P_4 = h_2^2$$

For convenience of notation we group all the histogram vectors of the streams in a matrix $\mathbf{H}$ of dimensions $N \times S$. An element h_{i,x_i} in $\mathbf{H}$ contains the probability that x_i instructions from stream i are issued in a given cycle. For a given number of streams N and a matrix $\mathbf{H}$, the probability that w instructions are ready to be issued in a cycle is obtained from the probability that each stream i has x_i ready-to-issue instructions, such that $x_1 + x_2 + .. + x_N = w$, and $0 \leq x_i \leq S$:

$$P_w = \sum_{\substack{x_1 + x_2 + .. + x_N = w, \\ 0 \leq x_i \leq S}} h_{1,x_1} h_{2,x_2} .. h_{N,x_N} \tag{4}$$

Estimating the histogram is important. Often, the actual values cannot be obtained and must be predicted. For example, the executable code, actual machine configuration, or tools to extract data may not be available. However, since we had the tools to extract this information, we chose to execute the programs and measure the values. A machine configuration with an infinite number of functional unit of each type is used to collect the values. The rationale behind this method is that the infinite functional units will negate any effect of structural hazards and we can more accurately measure the data and control hazards.

4.5 Rationale

One of the main problems faced by the analyst is the large state space cardinality of the Markov chain associated with models, which precludes not only the model solution, but also the generation of the transition rate matrix [11]. The storage and time needed for a large non-sparse Markov chain of N states is $O(N^2)$ and $O(N^3)$ respectively[2]. Ways must be devised to cut down the number of states while keeping the accuracy of the solution.

In our case, a Markov chain could be built with all states for all ready-to-issue instructions. The total number of states would be equal to the sum of the number of states for all possible w ready-to- issue instructions, i.e.

$$\sum_{w=0}^{NS} \binom{w+T-1}{T-1} = \binom{NS+T}{T}$$

In contrast, the maximum number of states in our approach happens when the number of ready-to-issue instructions is equal to the window size, i.e.

$$\binom{NS+T-1}{T-1}$$

Thus, the ratio of the total number of states to the maximum number of states is $NS/T + 1$. For example, for four streams, four instruction types, and stream window sizes of four ($T = 4, N = 4, S = 4$), the total number of states is 4844 requiring 187.7 Mbytes of storage in double precision. In our approach the maximum number of states is 968 states, requiring 7.5 Mbytes of storage in double precision[3].

---

[2] Recent methods have reduced the time complexity to $O(N^{2.4})$ approximately
[3] Storage is proportional to the square of the number of states in a non-sparse matrix

In practice, the number of states can be further reduced by considering only the most significant instruction types. Thus, the number of instructions types (T) is reduced by selecting the most significant frequency/number (v_{unit}/c_{unit}) values since the less frequent instruction types do not often experience structural conflicts. This is specially true for the bigger window sizes where saturation effects are observed.

Our method applies a decomposition method of large Markov chains [3]. The Markov chain is decomposed into smaller chains that are solved independently. The probability of each subchain is computed from the characteristics of the workload. From the way we decompose the chain, we expect the method to be more accurate for streams that don't have enough instruction-level parallelism to saturate the configuration of functional units, or for configurations with many functional units.

5 Validation

We validated the analytic technique by comparing its estimations to the results of a multistreamed hardware simulator. We begin this section by describing our simulation environment and benchmark suite. Next, we present the results from both the analytical technique and the hardware simulator. Finally, we quantify the differences between the model and the simulator and discuss the discrepancies.

5.1 The Simulation Environment

Our hardware simulator is a program written in C that emulates a multistreamed version of the IBM RS/6000 instruction set and architecture [6]. We studied three different hardware configurations in terms of the number of functional units, from a restricted configuration with only one functional unit for each instruction type, to a expanded configuration with several functional units of each type. Table 4 shows the number of functional units of each type for each configuration ($C1$, $C2$, and $C3$). Table 4 also shows the latency in number of cycles of each instruction type. We assume that the functional units are independent and perfectly pipelined (i.e., capable of accepting an instruction on every cycle). The integer unit executes all integer arithmetic and logic instructions except multiply and divide which have a longer latency and are executed in the multiply and divide units, respectively. Memory instructions execute in the load/store unit in a single cycle. All branch outcomes are predicted correctly and no branch delays exist. The CR (condition register logic) unit executes the condition register instructions of the RS6000 instruction set.

The simulator marks instructions which have passed the data and control dependencies checks as ready-to-issue. In the simulations, the scheduler may dispatch instructions within the stream issue window out-of-sequence in the absence of data hazards. An instruction is removed from the stream issue window as soon as the scheduler dispatches it to a functional unit. The window is then compacted

Table 4. Configurations evaluated.

Functional Unit	Latency	Configuration		
		C1	C2	C3
integer	1	1	2	6
float	2	1	2	3
multiply	5	1	1	1
divide	17	1	1	1
cr logic	1	1	1	1
branch	1	1	1	2
load/store	1	1	2	3

and refilled with new instructions. The scheduler employs a fair scheduling algorithm. In this algorithm, priority is alternated among the streams on every cycle in a round-robin fashion. The scheduler dispatches as many instructions from the high priority stream's window as possible in the current cycle. If free slots in functional units still exist, the scheduler attempts to dispatch instructions from the other streams to fill the slots. For our experiments we used a stream issue window of four. This means that the maximum number of instructions that a stream can issue is four, resembling the original RS6000 architecture [5].

Each instruction stream is interpreted as a random stream of instructions of each type. The instruction stream is characterized by its instruction mix, i.e. each instruction type has a probability derived from the instruction mix. Instructions are generated from a binomial distribution using the instruction type probabilities. The accuracy of the model depends on the assumptions that 1) the instruction mix is a stationary process, and 2) instruction types appear in the instruction stream in a random manner. We used two estimators (σ, δ) to measure the degree of closeness to these assumptions:

- σ: The standard deviation of each instruction type computed from samples collected periodically from the simulations.
- δ: The skew measure of the degree of randomness of the workload:

$$\delta = \sum_{i=1}^{T} \sum_{j=1}^{T} v_i |v_j - t_{i,j}|$$

where v_i is the probability of the instruction type i, as derived from the instruction mix, and $t_{i,j}$ is the probability of transition from instruction type i to instruction type j, measured by the simulator. The skew measure is zero for a perfectly random workload because v_j must be equal to $t_{i,j}$. The skew measure approaches one for a deterministic workload. Thus, the skew is a good indicator of the fraction of randomness of a workload.

5.2 Benchmarks

A set of common benchmark is the workload. We selected the small benchmarks *hanoi, dhrystone, fibonacci, savage,* and *fft.* From the SPEC'89 benchmark suit we selected *doduc* and *li.* All the benchmarks were run to completion, with the exception of *doduc* and *li* which were run with traces of 150 million of instruction each [4]. Table 5 shows the characteristics of the workloads evaluated. The numbers indicate the mean and standard deviation of each instruction type (mean$\pm\sigma$) in percentage, and the skew (δ) of each benchmark. values were sampled at intervals of 1600 cycles. Integer multiply and divide instruction types are not shown since they constitute a very small percentage of all instructions.

Table 5. Workload characteristics.

	hanoi	dhrystone	fibonacci	savage	fft	doduc	li
skew	0.134	0.143	0.165	0.199	0.132	0.314	0.198
integer	56.1±0.1	46.5±0.0	48.4±0.0	15.1±0.2	23.2±0.9	17.3±4.0	36.4±6.5
float	0	0	0	38.1±0.2	16.8±3.5	27.9±4.8	0
cr logic	0	0.9±0.0	0	2.5±0.0	0	3.3±1.4	1.3±0.0
branch	12.2±0.0	20.0±0.0	22.6±0.0	9.2±0.1	18.9±1.6	10.2±3.1	21.2±1.1
memory	31.7±0.0	31.5±0.1	29.0±0.0	35.1±0.1	58.1±1.3	41.2±3.8	41.0±6.9

5.3 Comparison to the Simulation Results

The first experiment consisted in running homogeneous workloads, with multiple copies of the same thread loaded in all streams of the machine. For example, we ran *hanoi* with one, two, three, and four copies of the benchmark. Table 6 compares the predicted and simulation results in number of instructions-per-cycle executed (IPC) for one, two, three, and four streams. Table 6 also shows the percentage of error of the prediction. As seen from the results, the benchmarks with the smallest error are *hanoi* and *dhrystone.* They correspond to the benchmarks with the smallest skew and standard deviation values. In any case, the peak error does not exceed eight percent for all the configurations and benchmarks. The average error is smaller for configuration $C3$ where the number of resources is bigger than what an individual stream may use.

In general, combining heterogeneous workloads makes better use of the functional units. Higher saturation values could be achieved. For example, consider the extreme case of two heterogeneous thread types: one with only integer instructions, and the other with only floating point instructions. The saturation value of the first thread type is $c_{integer}$ (the number of integer units), while the saturation value of the second thread type is c_{float} (the number of float units).

Table 6. Results for Homogeneous workloads.

benchmark	streams	C1			C2			C3		
		predict	simul.	error%	predict	simul.	error%	predict	simul.	error%
hanoi	1	1.55	1.52	1.94	2.34	2.37	-0.93	2.56	2.56	-0.10
	2	1.76	1.76	-0.17	3.33	3.31	0.44	5.02	5.13	-2.05
	3	1.78	1.78	-0.04	3.53	3.53	-0.12	7.05	6.96	1.38
	4	1.78	1.78	0.00	3.56	3.56	-0.03	8.28	8.14	1.77
dhrystone	1	1.59	1.64	-3.29	2.15	2.21	-2.58	2.34	2.36	-0.89
	2	2.01	2.03	-0.98	3.35	3.41	-1.73	4.54	4.55	-0.40
	3	2.12	2.12	-0.05	3.89	3.90	-0.36	6.35	6.39	-0.63
	4	2.14	2.14	0.05	4.12	4.12	0.00	7.62	7.67	-0.59
fibonacci	1	1.86	2.02	-7.75	2.58	2.72	-5.18	2.96	2.99	-0.73
	2	2.05	2.07	-0.97	3.62	3.86	-6.26	5.61	5.70	-1.58
	3	2.07	2.07	-0.07	3.91	4.08	-4.16	7.42	7.61	-2.45
	4	2.07	2.07	0.00	4.02	4.12	-2.61	8.30	8.63	-3.79
savage	1	1.43	1.51	-5.53	1.78	1.81	-1.42	1.83	1.83	-0.15
	2	2.07	2.16	-4.43	3.18	3.30	-3.57	3.57	3.60	-0.66
	3	2.34	2.38	-1.69	4.07	4.21	-3.15	5.04	5.16	-2.41
	4	2.46	2.48	-0.74	4.59	4.65	-1.43	6.13	6.29	-2.61
fft	1	1.16	1.22	-5.16	1.48	1.44	2.70	1.52	1.51	0.26
	2	1.57	1.66	-5.23	2.55	2.61	-1.95	2.93	2.94	-0.26
	3	1.68	1.72	-2.07	3.11	3.23	-3.85	4.00	4.09	-2.26
	4	1.71	1.73	-0.82	3.33	3.42	-2.41	4.65	4.81	-3.48
doduc	1	1.56	1.57	-0.64	2.04	2.01	1.49	2.12	2.17	-2.30
	2	2.15	2.11	1.90	3.50	3.44	1.74	4.08	4.05	0.74
	3	2.34	2.30	1.74	4.29	4.16	3.13	5.59	5.53	1.08
	4	2.41	2.36	2.12	4.65	4.50	3.33	6.54	6.41	2.03
li	1	1.77	1.68	5.36	2.42	2.32	4.31	2.73	2.67	2.25
	2	2.22	2.06	7.77	3.65	3.44	6.10	5.03	4.88	3.07
	3	2.35	2.19	7.31	4.16	3.90	6.67	6.40	6.19	3.39
	4	2.40	2.26	6.19	4.41	4.26	3.52	7.01	6.79	3.24

Combining the two workloads give a saturation value of $c_{integer} + c_{float}$, equal to the sum of the individual saturation values. We expect that the model gives results closer to the simulations, since the behavior of a combined workload of different types is more random than a homogeneous workload.

Although the model was explained with homogeneous workloads, the model can be extended to model heterogeneous workloads. This is accomplished by using an averaging technique over the heterogeneous workload to obtain average homogenous workload characteristics ($V_{average}$). These characteristics are entered into the analytical model to obtain performance estimates of heterogeneous workloads. Table 7 shows the results for heterogeneous workloads. We picked pairs of benchmarks and run simulations for one, two, three, and four pairs, i.e. two, four, six, and eight streams. For example, for the experiment with four streams, two copies of *hanoi* and *dhrystone* were loaded in the four streams of the machine.

Table 7. Results for Heterogeneous workloads.

benchmarks	streams	C2			C3		
		predict	simul.	error %	predict	simul.	error %
hanoi -	2	3.41	3.40	0.17	4.79	4.81	-0.50
dhrystone	4	3.90	3.87	0.90	8.07	8.04	0.41
	6	3.93	3.91	0.69	9.19	9.09	1.12
	8	3.94	3.92	0.49	9.44	9.37	0.69
hanoi -	2	3.31	3.30	0.34	3.99	3.94	1.17
fft	4	4.24	4.25	-0.24	6.60	6.63	-0.46
	6	4.45	4.50	-1.11	7.24	7.29	-0.65
	8	4.54	4.59	-1.21	7.31	7.35	-0.63
dhrystone -	2	3.19	3.27	-2.25	3.77	3.78	-0.14
fft	4	4.37	4.40	-0.80	6.35	6.49	-2.17
	6	4.69	4.65	0.91	7.19	7.22	-0.39
	8	4.80	4.73	1.49	7.33	7.29	0.49
fibonacci -	2	3.90	3.99	-2.13	4.70	4.75	-1.09
savage	4	5.27	5.29	-0.39	7.99	8.18	-2.33
	6	5.68	5.76	-1.44	9.13	9.28	-1.59
	8	5.85	5.99	-2.39	9.38	9.44	-0.62
fibonacci -	2	3.55	3.77	-5.68	4.37	4.43	-1.28
fft	4	4.55	4.80	-5.15	7.19	7.40	-2.81
	6	4.80	4.95	-3.04	7.70	7.78	-1.11
	8	4.90	4.99	-1.81	7.74	7.76	-0.29

5.4 Discussion

Our results show that the analytical technique produces estimates very close to those of the simulation results. Over the 168 total estimates, the average deviation from the simulation results was 2.3%. On average, the analytic technique produced fairly even results across the various configurations and number of streams employed within a workload.

Table 8 shows the average percentage of error for different configurations and workloads for homogeneous workloads. While the model makes many generalizations about the architecture and workload that may not seem to be representative of most programs, our results show the contrary. A closer examination of the results for each benchmark shows the predicted performance to be very close to the simulation results for all configurations in six of the seven benchmarks (*hanoi, dhrystone, fibonacci, savage, fft, doduc*). Also, the error is lower for configuration $C3$ than for configuration $C1$, suggesting that the error diminishes as the configuration grows in number of functional units.

Table 8. Percentage of error for homogeneous workloads.

Benchmark	$C1$	$C2$	$C3$	Average
hanoi	0.54	0.38	1.33	0.75
dhrystone	1.09	1.17	0.63	0.96
fibonacci	2.20	4.55	2.14	2.96
savage	3.10	2.39	1.46	2.32
fft	3.32	2.73	1.57	2.54
doduc	1.60	2.42	1.54	1.85
li	6.66	5.15	2.99	4.93
Average	2.64	2.68	1.66	-

Table 9 summarizes the errors for heterogeneous workloads. When benchmarks are combined the instruction mix behaves more randomly, thus the errors we obtain are smaller. For example, the combination of *hanoi* and *fft* produces less error than either benchmark considered individually. We can also see that the performance of heterogeneous workloads in some cases is better than the individual benchmarks. For example, when we run four streams in configuration $C2$, the combination of *hanoi-fft* has an IPC of 4.25 compared to 3.56 and 3.42 for *hanoi* and *fft* respectively.

Table 9. Percentage of error for heterogeneous workloads.

Benchmarks	$C2$	$C3$	Average
hanoi - dhrystone	0.56	0.68	0.62
hanoi - fft	0.72	0.73	0.73
dhrystone - fft	1.36	0.80	1.08
fibonacci - savage	1.59	1.41	1.50
fibonacci - fft	3.92	1.37	2.65
Average	1.63	1.00	-

An examination of the IPC versus the number of streams (Tables 6 and 7) reveals the overall performance gain levels out as the number of streams increases. In our simulations, this is due to the contention for functional unit resources. The phenomenon is most obvious in configuration $C1$ where saturation occurs when executing as few as two streams. For the larger configurations ($C2$ and $C3$), saturation occurs when executing a larger number of streams. The results show that the gain from multistreaming is largest in configuration $C3$ where the largest number of functional units is present. In contrast, when the number of functional units is not large enough to support the execution of multiple streams, the gain from multistreaming is minimal. Figure 5 summarizes the performance improvements as the number of streams is increased.

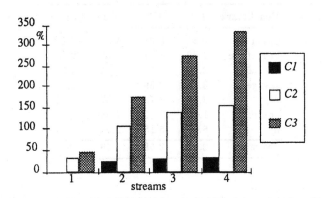

Fig. 5. Average percentage of improvement, compared to $C1$, single-stream.

6 Conclusions

We have presented an analytic technique for evaluating the performance of multistreamed, superscalar processors. Our results demonstrate that the technique produces accurate instructions-per-cycle (IPC) estimates of the overall performance through comparison with a multistreamed, RS/6000 simulator.

The analytical technique provides a quick way to examine the many variations of an architecture at a high level of abstraction. The technique only requires simple descriptions of the workload and architectural configuration as input parameters. These parameters are easily measured or estimated using tools that are commonly available. In addition, the simplicity of the model makes it easy to implement and much faster to execute than a hardware simulator.

As the trend moves towards integrating more functional units within the processor, designers face the challenge of utilizing these additional resources effectively. Superscalar processors cannot effectively exploit the additional functional units due to the inherent limit of instruction-level parallelism within a single stream. Multistreaming provides a technique to overcome the superscalar limitations. Recent operating systems and parallelizer compilers provide multiple threads of computation. Integrating multistreaming within superscalar architectures is an effective method for maximally exploiting the additional functional units needing for this new generation of sofware. Figure 5 shows that a multistreamed, superscalar processor executing four streams can improve the overall machine throughput by factor of three over an equivalent single stream machine.

References

1. A. Agarwal, B. Lim, D. Kranz, and Kubiatowicz, "APRIL: A Processor Architecture for Multiprocessing," Proc. of the 17th Symposium on Computer Architecture, May 1990, pp. 104- 114.
2. T. M. Conte, "Systematic Computer Architecture Prototyping," Ph.D Thesis, Electrical Engineering, University of Illinois at Urbana-Champaign, 1992.
3. P. J. Courtois, Decomposability. Queuing and Computer System Applications. Academic Press. 1977.
4. S. I. Feldman, D. M. Gay, M. W. Maimone, and N.L. Schryer, "A Fortran-to-C Converter," Computing Science Technical Report No. 149, AT&T Bell Laboratories, Murray Hill, NJ, 1991.
5. G. F. Grohoski, "Machine Organization of the IBM RISC System/6000 processor", IBM Journal of Research and Development, Vol. 34, No. 1, January 1990, pp. 37-58.
6. J. L. Hennessy, and D. A. Patterson, Computer Architecture, A Quantitative Approach, Morgan Kaufmann Publishers, 1990.
7. W. J. Kaminsky, and E. S. Davidson, "Developing a Multiple-Instruction-Stream Single-Chip Processor," IEEE Computer Magazine, Dec. 1979.
8. J. S. Kowalik, ed., Parallel MIMD Computation: HEP Supercomputer and its Applications, MIT Press, 1985.
9. M. D. Nemirovsky, F. Brewer, and R. C. Wood, "DISC: Dynamic Instruction Stream Computer," Proceedings of the 24th ACM/IEEE International Symposium and Workshop on Microarchitecture, Albuquerque, NM, Nov. 1991, pp. 163-171.
10. M. Serrano, M. D. Nemirovsky, and R. C. Wood, "A Study on Multistreamed Superscalar Processors," Technical Report #93-05, Department of Electrical and Computer Engineering, University of California, Santa Barbara, March 1993.
11. E. de Souza e Silva, P. M. Ochoa, "State Space Exploration in Markov Models", Performance Evaluation Review, Vol. 20, No. 1, June 1992.
12. C. A. Staley, "Design and Analysis of the CCMP: A Highly Expandable Shared Memory Parallel Computer," Ph.D Dissertation, University of California, Santa Barbara, August 1986.
13. B. Smith, R. Alverson, D. Callahan, D. Cummings, B. Koblenz, A. Porterfield. "The Tera Computer System". Proceedings of Supercomputing'90. pp. 1-6.
14. K. B. Theobald, G. R. Gao, and L. J. Hendren, "On the Limits of Program Parallelism and its Smoothability". IEEE Micro 25, Oct. 1992. pp. 10-19.
15. J. E. Thornton, "Parallel Operation in the Control Data 6600," Proceedings Spring Joint Computer Conference, 1964.
16. E. H. Welbon, C.C. Chan-Nui, D.J. Shippy, and D.A. Hicks, "The Power2 Performance Monitor", IBM RISC System/6000 Technology: Volume II. Sept. 23, 1993.

Accelerating the Evaluation of Parallel Program Performance Models Using Distributed Simulation

A. Ferscha[1] and G. Chiola[2]

[1] Institut für Angewandte Informatik und Informationssysteme, Universität Wien, Lenaugasse 2/8, A-1080 Vienna, Austria
[2] Dipartimento di Informatica, Università di Torino, Corso Svizzera 185, I-10149 Torino, Italy

Abstract. Petri nets with an explicit notion of time have proven to be a good modelling tool for the qualitative and quantitative study of asynchronous concurrent systems with parallelism, synchronization and resource sharing. The formalism could be successfully applied in performance modelling and evaluation of concurrent programs, parallel systems and mappings among the two. However, performance models of realistic systems are in general very costly to evaluate.

In this work we develop a Time Warp based distributed discrete event simulation scheme to accelerate the evaluation of timed Petri net performance models. The particular model structure that typically results from performance modelling of parallel systems explicitly provides partitioning possibilities which are very cumbersome to identify in general Petri net models. With our implementations of a distributed simulation engine running on an Intel iPSC/860 multiprocessor we show that real speedup over sequential simulation model evaluation can be achieved depending on the inherent model parallelism.

1 Introduction

Due to the technological progress of parallel processing systems and their increasing availability for scientific and commercial applications, the appropriate development of optimum parallel programs is a challenging task for software engineers of parallel systems. Besides other aspects, the objective to design and implement parallel programs showing optimum performance when being executed on a parallel processing system, is the most crucial one in the whole development cycle. This follows from the fact, that the performance of a parallel program depends on many more factors, compared to a program executed on a conventional computer system.

Petri nets (PNs) [17] with an explicit notion of time are accepted as a good modelling tool for the qualitative and quantitative study of asynchronous concurrent systems with parallelism, synchronization and resource sharing (conflicts). Previous work on modelling parallel programs and parallel systems include [10], [2], and [12]. A terse graphic representation and a solid mathematical background, supporting a high degree of automation and tool integration in the performance oriented development process of parallel programs [11], are some of their key characteristics. The use of contemporary Petri net modelling and analysis tools [9], however, shows that complex simulation models are in general very costly to evaluate. We have frequently stumbled against the limits of practical tractability of the net evaluation as soon we considered parallel program models of "realistic" size. Nevertheless, due to the intractable complexity of analytical model solutions and their applicability to restricted classes of nets (e.g. SPN, GSPN [4] or SRN [8], etc.,) simulation often remains the only practical technique to obtain performance indices. One hope to escape from this evaluation dilemma is to apply *parallel* and *distributed* simulation techniques, trying to alleviate these shortcomings by devoting multiple processors to a large simulation task in order to reduce elapsed simulation time.

Our motivation for studying the applicability of parallel and distributed simulation techniques to timed Petri net models of parallel programs can be summarized as follows:

- Practical experience with using Petri nets in their role as parallel program/system performance models has brought us to the limitations of analytical techniques: firstly, only net models of moderate size turn out to be practically tractable; secondly, only certain classes of Petri nets can be treated analytically. Hence the use of Petri nets for *simulation* modelling with *parallel* evaluation seems to be an approach to escape from the shortcomings of "traditional" Petri net performance modelling.
- The availability of multiprocessors with high degree of integration into user environments allows the scheduling of computation intensive tasks on parallel processors (after decomposition), and thus offers the possibility of (elapsed) execution time improvement without involving the user in the parallel computation process [7] [6].
- The Petri net formalism and the related analysis knowledge gives a strong basis for an *automated* problem decomposition process (parallelization), i.e. the verification of structural properties of Petri nets required for the decomposition of the simulation task can be invoked automatically without preservation of additional user knowledge upon the model.
- Finally the availability of a parallel/distributed simulation tool achieving actual speedup when applying an increasing number of parallel processors could have considerable impact on the use of the Petri net formalism in performance engineering of parallel application programs, since the parallel execution devices required are already in the environment.

The paper is organized as follows. In Section 2 we develop a framework for the distributed simulation of TTPNs combining the isomorphism among the execution behavior of TTPNs and discrete event systems with the Time Warp distributed simulation strategy. The problem of decomposing a TTPN into a set of simulation submodels is addressed and studied using empirical observations of simulation performance on the iPSC/860. Partitioning criteria for parallel program performance models are worked out in Section 3, meeting requirements for performance efficient distributed simulation. Section 4 uses a performance model of parallel N-body molecular dynamics simulations to verify the suitability of partitioning rules and describes the performance improvement gained by distributed simulation. Conclusions are drawn in Section 5.

2 Distributed Simulation of Timed Petri Nets

2.1 Discrete Event Simulation

Simulation analysis of Petri nets where time delays are associated with transition firings (timed transition Petri nets, TTPN) is based on the consideration of a discrete event system underlying the execution behavior of the net. A *simulation engine* is generally responsible for repeatedly processing the occurrence of events in simulated (virtual) time by maintaining a time ordered *event list* (EVL) holding time stamped events scheduled to occur in the future. A (global) *clock* indicates the current simulated time and *state variables* define the current state of the system. The occurrence of an event in the simulation system obviously corresponds to the firing of a transition in the TTPN model. The EVL hence carries transitions and the time instant at which they will fire, given that the firing process is not interrupted in the meantime. The state of the system is represented by the current marking M of the TTPN, which may change due to the processing of an event, i.e. the firing of a transition. A state change is simulated by withdrawing the transition with the smallest time stamp from the EVL, by removing tokens from its input places (modification of M) and by depositing an appropriate number of tokens in its output places (again a modification of M). Finally the virtual time (VT) is set to the timestamp of the transition whose occurrence was just simulated. The new marking, however, potentially enables new transitions and/or disables previously enabled transitions, so that the EVL has to be updated accordingly: new enabled transitions are scheduled while occurrences of disabled transitions are removed from the EVL.

2.2 Discrete Event Simulation by Logical Processes

To accelerate simulation time of discrete event time dynamic systems, asynchronous, distributed and parallel strategies have been developed along two different front-lines: the *conservative* (Chandy/Misra) [16] and the *optimistic* (Time Warp) [15] approach. A variety of amendments to these schemes is at hand today [13]; most of them have been developed to improve the performance of implementations by exploiting domain intrinsics of the simulated (*physical*) system. Common to both approaches is a partitioning of the simulation model into a set of submodels being simulated by communicating *logical processes* (LPs) (see Figure 1).

An LP comprises two kinds of functionality: first, a simulation engine SE which performs the repetitive execution of only those event occurrences (thus progressing a *local clock*), related to only the part of the simulation model assigned to that LP. Obviously, events having causal impact only on the local state

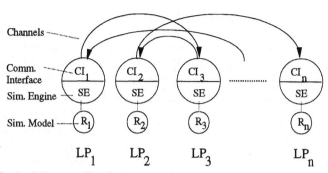

Fig. 1. Logical Process Simulation

of some LP (*internal events*) can be handled without interaction with other LPs. Should an LP produce an event affecting the context of another LP (*external event*), an appropriate message has to be sent, and the target LP has to incorporate it in its local simulation without violating global causality. Hence, the second functionality that an LP has to implement is a communication interface to other LPs which performs the distribution and collection of messages exchanged among the LPs in order to propagate the effect of external events. The exchange of messages is necessary to maintain causality relations among events across all participating LPs. Therefore the communication interface can be considered as a mechanism to synchronize the progress of work in the LPs with respect to those event causalities. The synchronization mechanism employed is what is referred to as the distributed discrete event simulation (DDES) *strategy* (or protocol).

The Chandy/Misra protocol does not allow an internal event with timestamp t to be processed if there is the chance of the notification of an external event with timestamp $s < t$. The Time Warp protocol, on the other hand, allows LPs to compute under the assumption of absence of external events; in the case that such an external event occurs actually, the LP invalidates the preliminary processing of internal events (*rollback*) and starts over again taking the external event into account. Since the propagation of external events in DDES is implemented in terms of time stamped messages exchanged among LPs, the Chandy/Misra protocol forces an LP to block the processing of internal events at time t, if it cannot be sure that there are no pending messages with timestamp $s < t$. Certainly this blocking yields a waste of CPU cycles and can cause considerable simulation performance degradation in cases where there are *actually no* pending messages of that kind (Chandy/Misra is overly pessimistic for simulation models in which the worst case causality situation seldom occurs actually). Time Warp would process internal events without respecting any bound on local time, and rolling back upon the receipt of a message timestamped in the local past. Although Time Warp could – in contrast to Chandy/Misra – gain from 'lucky' causality situations, there is the danger of progressing local event simulation too far ahead in virtual time, such that the probability of running into a causality violation reaches an unacceptably high level.

2.3 Logical Process Simulation of Petri Nets

The isomorphism among the execution behavior of TTPNs and discrete event systems can be exploited for the development of a logical process (distributed) simulation in the same way as for (sequential) discrete event simulation [3] [5]. The principal idea is to topologically decompose the TTPN graph into smaller subnets [7] which we call TTPN *regions* $R_1, R_2, \ldots R_n$. An LP_i is statically assigned the region R_i, whose events are simulated by a local simulation engine. SE is identical to the (sequential) discrete event simulation engine, and replicated in every LP. The communication interface CI_i realizes the DDES protocol applied (e.g. the Chandy/Misra or the Time Warp protocol), and the interconnecting channels follow the interconnection of the underlying TTPN regions. The set of LPs along with their linking channels defines the graph of logical processes, which is statically mapped to individual processing elements and communication links of the target parallel hardware platform. Each and every LP is assigned to a dedicated processor, and resides there for the whole simulation run. Channels are assigned to directed, reliable, FIFO hardware links.

In a more abstract way we can look at a logical process to as being a tuple $LP_k = (CI_k, SE_k, R_k)$ where R_k is a topological "region" of some TTPN (i.e. a subset of the places, transitions and arcs), CI_k is the LP's communication interface implementing the protocol handling message exchange among LPs, and SE_k is the simulation engine simulating events in R_k. The set of all LPs $LP = \bigcup_{k=1}^{n} LP_k$ together with communication channels $CH = \bigcup_{k,i\ (k \neq i)} ch_{k,i} = (LP_k, LP_i)$ constitute the *Graph of Logical Processes GLP* $= (LP, CH)$.

Partitioning into Topological Regions In order to decompose a TTPN simulation model topologically, consider the definition of a $TTPN = (P, T, F, W, \tau, M_0)$ where: $P = \{p_1, p_2, \ldots, p_{n_P}\}$ is a finite set of places, $T = \{t_1, t_2, \ldots, t_{n_T}\}$ is a finite set of transitions with $P \cap T = \emptyset$ and $P \cup T \neq \emptyset$. $F \subseteq (P \times T) \cup (T \times P)$ is a finite set of arcs among places and transitions denoting input flows $(P \times T)$ to and output flows $(T \times P)$ from transitions. $W : F \mapsto \mathbf{N}$ assigns weights $w(f)$ to elements of $f \in F$ denoting the multiplicity of unary arcs between the connected nodes. An explicit timing function $\tau : T \mapsto \mathbf{R}$ assigns firing delays τ_i to T-elements $t_i \in T$. $M_0 : P \mapsto \mathbf{N}$ is the marking $m_i = m(p_i)$ of P-elements $p_i \in P$ with tokens in the inital state of $TTPN$ (initial marking).

A trivial topological decomposition of TTPN is the one where every place $\{p_1, p_2, \ldots, p_{n_P}\}$ and every transition $\{t_1, t_2, \ldots, t_{n_T}\}$ establishes a region to be hosted exclusively by one LP, thus generally giving a very large number of LPs ($\mid LP \mid = \mid P \mid + \mid T \mid$) of smallest possible grain size (R is either a single place or a single transition). Although this partitioning fully exploits the 'potential' model parallelism, a series of practical arguments arise for avoiding it. First, different simulation engines would have to be created for place LPs and transition LPs, the former realizing a token deposit, the latter implementing

---

[3] The terms *logical process simulation* and *distributed (discrete event) simulation* are used interchangeably throughout this work.

enabling and firing semantics of transitions. This in turn would yield unbalanced simulation efforts in the different kinds of LPs. Second, transition LPs would require a double handshake message exchange protocol to verify local enabling based on distributed token locations, causing a (possibly tremendous) amount of messages [19] [1]. Thirdly, for distributed memory target architecture the ratio of computational requirements and communication load would practically nullify any chance of execution speedup over sequential simulation.

The minimum grain size of regions practically useful is the one where every transition within some region can determine enabling based on local information, i.e. avoiding any communication for enabling verification. We call the partitioning that prohibits the assignment of transitions and all of their input places to different LPs, avoiding the implementation of a costly distributed conflict resolution policy the *minimum region decomposition* [7]. An arbitrary TTPN can always and uniquely be partitioned into minimum regions defined upon the topological structure of the TTPN graph (R is the set of all transitions sharing input places, together with all their input places). A very useful implication of the minimum region partitioning is that $SE_k = SE$ for all k, i.e. the simulation engines in every LP are identical replicates of the same piece of program code.

Apparently the decomposition into regions has a strong impact on the DDES performance. Small sized regions naturally rise high communication demands, while large scale regions can potentially clump local TTPN behavior inside one LP. The minimum region partitioning will generally be too fine grained and induce substantial interprocessor communication overhead. A grain packing mechanism is employed to merge minimum regions into larger ones, therewith reducing communication overhead to attain better performance, but also to match the number of resulting LPs with the number of physically available processors. Rules for packing grains rely on information stemming from the TTPN topology, but also from behavioral properties of the net. Moreover, the DDES strategy coincidentially affects the partitioning with respect to optimizations that can be performed on the CI.

A Time Warp based Communication Interface The local simulation within LP_i can generate an external event if an internal transition is fired and the corresponding output place is located in a region corresponding to a remote LP_j. The instant of time the token was produced (the *local virtual time* LVT of LP_i at firing time) is an essential piece of information for LP_j to preserve causality; LP_j must "integrate" the token into the local event structure at exactly the (virtual) time it was created. To provide for this, the communication interface creates a message out of the token, containing an identifier of the destination place and a timestamp equal to the creation (virtual) time (*timestamped token messages*), which is then sent to the destination LP. CI_j (in the receiving LP) then has to take care for a proper consideration of the token in its own virtual time. Two situations can occur: LVT of LP_j is less than the timestamp of the received token, meaning that the external event is in the local future. In this case the token is stored to be considered when LVT has progressed up to the token timestamp. The other, most unfortunate, situation is when the token received is in the past of LP_j (*straggler message*). In this case all the simulation work between the timestamp of the token and LVT has to be invalidated (rollback), and SE has to restart simulation from that state on, where the 'old' token has changed the local event structure.

Since token messages cannot be guaranteed to arrive in (increasing) time order due to different LVT progression speeds in the different LPs, the Time Warp rollback mechanism takes care for proper synchronization respecting causalities. However, it requires a record of the LP's history with respect to the simulation of internal *and* external events. Hence every CI has to keep sufficient internal state information in a *state stack* SS (for internal events) which allows the restoring of a past state, and has to administer an *input queue* IQ for intermediately storing (future) tokens (incoming external events). Moreover an *output queue* OQ for token messages already sent (outgoing external messages) is necessary, since those might have to be invalidated due to local rollback. A direct consequence of this is that token messages have to be distinguished in: *positive* (token) messages $(m = \langle \#, P, TT, '+' \rangle)$ that are used to propagate external events, and *negative (anti-) messages* $(m = \langle \#, P, TT, '-' \rangle)$ that indicate a request to annihilate a previous positive message. In the original Time Warp an LP receiving a message with timestamp less than LVT initiates sending antimessages immediately (*aggressive cancellation*) when executing the rollback procedure. Since the impact of antimessages might be that also succeeding LPs are forced to roll back, a *chain of rollbacks* that eventually terminates is generated. Reducing the size of the rollback chain (that can cause overwhelming communication overhead in real implementations), can be attempted by postponing erroneous message annihilation until it turns out that they are not reproduced in the repeated simulation (*lazy cancellation*).

The CI_i drives the computations within an LP_i. For every 'input' channel $ch_{i,k} \in CH$ directed towards LP_i an *input buffer* IB[i] is used for collecting arriving messages. When a token from LP_i has to be sent to some LP_j via $ch_{k,j} \in CH$, it is 'packed' into a message and deposited into an *output buffer* OB[j] (one for every channel) before actual sending is invoked. The operation of CI_i is the following. After an initialization in which the local simulation state is started from virtual time zero and the initial state of R_i, CI_i repeats the following steps until the predefined end time is reached: take all messages from the IB and insert the arriving tokens in IQ (Step 1). If the smallest token time observed is less then LVT then invoke rollback (Step 2), else compare the timestamp of the next future token with the time of the next transition firing scheduled in the event list EVL. Related to that comparison, either process the first token (Step 3.1) (i.e. change the marking) or fire the first scheduled transition (Step 3.2) (i.e. remove the transition from EVL, schedule new enabled transitions (internal events), generate and send output messages (external events), deschedule obsolete transitions) and set LVT to the instant of the firing. Finally the outputmessages in OB are transmitted and the processed event is logged into SS (Step 4). The protocol and relevant data structures have been worked out in detail in [6].

Performance To give an idea of the Time Warp communication interface we first consider a trivial TTPN example in Figure 2 (right), describing two customers cycling among two service stations to obtain service in exponentially distributed time (the firing delay of T1 (T2) is $exp(\lambda_1)$ $(exp(\lambda_2))$). The maximum degree of model parallelism in this particular example is 2: while the firing of T1 is simulated by one LP on one physical processor, T2 could be simulated on a second one. The minimum region partitioning for the model along with the LP structure for its distributed simulation is also given in Figure 2 (left).

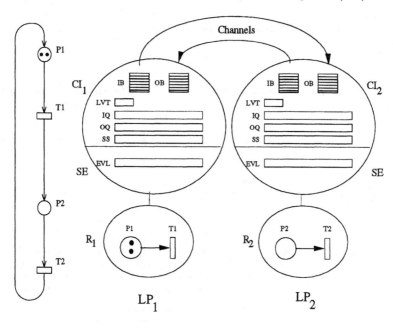

Fig. 2. Logical Process Architecture for a small TTPN

Figure 3 shows execution performance of the example in Figure 2, simulated by two logical processes with a *lazy cancellation* Time Warp communication interface mapped onto two nodes of an Intel iPSC/860, and is meant to explain two serious performance influence phenomena related to partitioning, namely *model parallelism* and *load balance*.

Model Parallelism With two customers (tokens) initially in P1 we have very little model parallelism and no hope to gain practically from a distributed simulation. The system with one customer does not have model parallelism at all. Figure 3 shows the impact of the model parallelism (for now only look at numbers where $\lambda = 1$). With one customer the management effort in the CI significantly overwhelms the real simulation work (the ratio of execution time used for processing events is less than 5%; the rest is wasted for communication, data structure manipulations and blocking due to the lack of events to process if the customer is currently in service at the other

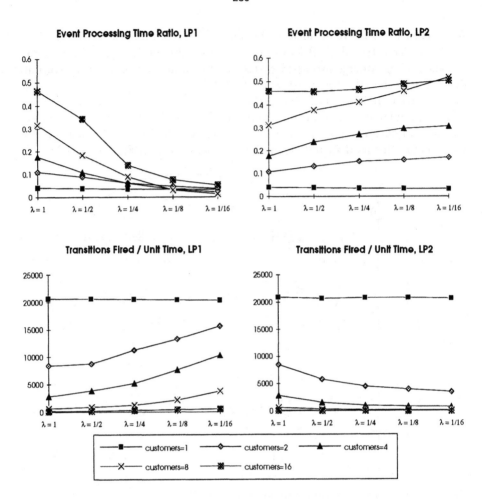

Fig. 3. Execution Profile of the Distributed Simulator on the iPSC/860

LP). The situation changes when more customers are in the system; with 16 customers the LPs can already use about 50% of the execution time for event processing. So we might conclude that for partitioning we have to take care for sufficient model parallelism, in this case increment the number of customers. However, looking at the figures showing the absolute number of transitions fired per elapsed time unit, we must recognize that the more customers there are in the system, the slower the simulation engines become due to augmented data structure manipulations (average EVL size increases, state saving becomes more complex).

Load Balance Causality violations in Time Warp are always due to inhomogeneous LVT increments in communicating LPs. An LP with small LVT increments followed by an LP with high LVT increment will frequently force its successor to rollback, given they work at the same event processing speed.

Rollback in turn can cause dramatic performance losses. In our example we can control the balance of LVT progress by the parameter λ_2; with $\lambda_2 = 1(= \lambda_1)$ we have the balanced situation, where service at T1 takes on average as long as at T2 . Setting $\lambda_2 = 1/2$ makes T2 twice as fast (with respect to LVT progression) than T1, i.e. the enabling time is twice as long; the higher the expected service time of T2 (which is $\frac{1}{\lambda_2}$), the more customers will reside in P2 (in steady state) enabling T2. We can see that an imbalance in the LVT progression can also cause a shift of load among LPs. The effect of this load shift causes the absolute number of transition firings in LP_1 to increase with an increase of parameter λ_2, since it shifts load to LP_2 (λ in Figure 3 stands for λ_2), and has the opposite effect on the 'simulation speed' of LP_2.

What we can conclude from this empirical observation is that performance influences in distributed simulation are highly interweaved. We must never follow a strategy that maximizes model parallelism while neglecting load balance effects caused by the timing of the model; on the other hand a perfect load balance is useless if there is no model parallelism. Thus, in general (practical) situations, partitioning must be considered a hard problem.

In the next section we shall work out partitioning criteria for Petri net performance models typically arising in the context of parallel systems (concurrent programs, parallel hardware). We shall see, that those models – due to their structural properties – extensively simplify the partitioning problem.

3   Partitioning Parallel Program Models

The PRM-net (Program Resource Mapping-net) modelling technique [10] for parallel systems has been developed to serve as an *integrated performance model*, combining performance characteristics of parallel programs expressed as a Petri net (P-net), of parallel hardware (R-net) and the assignment of programs to hardware (Mapping). The P-net of a concurrent program is developed by describing its execution skeleton in terms of a (timed) Petri net, using a modular, compositional approach. Parallel programs are modeled by hierarchical (de-)composition of *process building blocks* for concurrency, sequential execution, iteration, nondeterministic choice, communication and synchronization. The execution of atomic software processes is related to the firing of transitions, where input places and output places to that transition are used to model the current state of the process; an arriving token in the input place potentially starts the process (enables the transition), whereas depositing tokens in the output places after the process has terminated is interpreted as passing control to the next process. In a bottom-up modelling approach complex process structures are created by composition of other process structures using composition operators like parallel, sequential etc. Vice versa, top-down modelling is supported by process (transition) refinement in terms of substitution by structure templates. Every process, atomic or complex, is preceded by a single input place, the *entry place*

p_E, and followed by an *exit place* p_A. Some *composition/substitution* templates are recalled graphically in Figure 4. Note that any process transition represented as an empty box in Figure 4 can be substituted by any other of the templates as a matter of refinement (except synchronization).

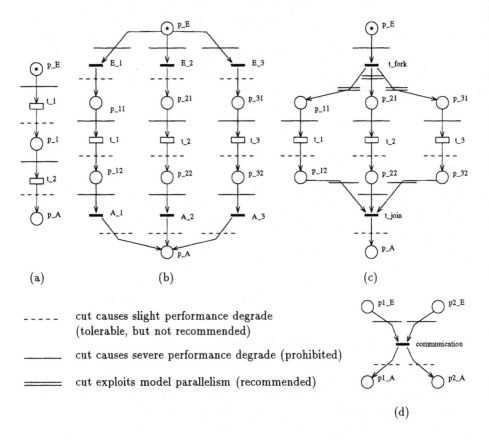

(a) (b) (c)

- - - - - cut causes slight performance degrade
(tolerable, but not recommended)

——— cut causes severe performance degrade (prohibited)

≡≡≡ cut exploits model parallelism (recommended)

(d)

Fig. 4. Process Templates: (a) sequential (b) alternative (c) parallel (d) synchronization

P-net models constructed by hierarchical composition or substitution using these templates are partitioned for logical process simulation as follows:

Sequential In a *sequential* process composition (Figure 4 (a)) two consecutive processes t_i, t_{i+1} have to be executed in the order of their appearance, i.e. the termination of t_i enables the process t_{i+1}. Generally, the component transitions in a sequential composition are *mutually exclusive* (ME). $t_i, t_j \in T$ are said to be ME, denoted by $(t_i\ ME\ t_j)$, if and only if they cannot be simultaneously enabled in any reachable marking, which is obviously the case.

We can state the partitioning rule, that ME transitions should go into a single region R, since they do not bear any model parallelism. Hence there is no speedup gain when assigning components of a sequential process composition to different LPs. (We have annotated in Figure 4 by dashed lines potential candidates for arc cuttings, i.e. separating into different regions, that would cause "tolerable" performance degradation.)

Alternative The *alternative* process template (Figure 4 (b)) models a nondeterministic choice among potential branches. By definition of the minimum region partitioning we must not separate the entry place from its output transitions in order to avoid distributed conflict resolution. Doing so could cause a severe performance loss. From the arguments for the sequential composition on the other hand we should also not partition the template in any other way, since all the transitions there are pairwise ME. (A sufficient condition for $(t_i\ ME\ t_j)$ is that the number of tokens in a P-invariant out of which t_i and t_j share places prohibits a simultaneous enabling. Making a strongly connected component out of the template and solving for the minimum support P-invariants verifies this property.)

Parallel In the *parallel* process composition the component processes $(t_1, t_2, t_3 \ldots)$ become active concurrently if the fork process t_{fork} has terminated. t_{join} (the join process) can start executing as soon as the last $t \in \{t_1, t_2, t_3 \ldots\}$ has terminated. The process composition terminates if t_{join} terminates. All the component processes between t_{fork} and t_{join} are *concurrent* transitions. ($t_i, t_j \in T$ are said to be *concurrent* denoted by $(t_i\ CN\ t_j)$, if they are causally independent (firing t_i might not cause an enabling of t_j), not conflicting (firing t_i might not cause a disabling of t_j), and not ME).

The inherent model parallelism expressed by a parallel process composition can easily be exploited. All the component processes should go into different regions, given they have sufficient grain size. Otherwise a packing of the concurrent grains is recommended. From the local load point of view, the partitioning of one parallel component into a single LP is ideal since the size of EVL cannot become higher than 1 (almost no data structure manipulation overhead). With every additional component that we pack into the LP, the average EVL size (and local load) increases.

Synchronization and Communication Processes having a fork-transition in common are allowed to communicate with each other. Interprocess communication is expressed by matching **send** and **receive** processes in concurrent programs, modelled by a *synchronization transition* as in Figure 4 (d). In this example synchronous message passing is modelled, i.e. the **send** process is blocking. Using an additional place to model a message buffer where **send** deposits and **receive** reads from would help to describe asynchronous message passing in the same simple way.

Since communication also causes synchronization among components of a parallel process composition, our partitioning rule has to be revised. If there is frequent synchronization, *arc-degree reduction* should be the driving force for partitioning. A static measure of the potential communication induced by LP_i is the *arc-degree (of connectivity)*, i.e. the number of arcs $f \in F$ originating in R_i and pointing to places in adjacent R's and vice-versa: $AD(R_i) =$

$| \bigcup_{k \in \text{adjacent\_regions}(R_i)} (P_i \times T_k) \cup (T_k \times P_i) |$. The partitioning rule can be formulated as follows: Among all the possible partitionings of a parallel process composition into a set of a fixed number of regions, choose the one with minimum average arc-degree $AD(R_i)$ to minimize potential message traffic intensity.

4   Example: Modelling Molecular Dynamics Simulations

The SPLASH [18] set of parallel applications for the evaluation of (shared-memory) multiprocessing systems provides a suite of *realistic* applications in the area of high performance computing. To demonstrate the potential of simulation evaluation acceleration via distributed simulation we have selected an example from the SPLASH suite, namely *Water*. *Water* is an N-body molecular dynamics application which evaluates forces and potentials in a system of water molecules in the liquid state. The principal computation is evaluating the intermolecular force-law performed on a vector of N water molecules over a user-defined number of time-steps in order to reach steady-state. Newtonian gravitational N-body simulations encounter a similar situation with respect to the infinite range force-law. Thus *Water* stands for a class of algorithms characterized by a running time proportional to the number of pairs of molecules (or more general: particles), posing a significant challenge for multiprocessors if the problem size exceeds a few thousands of particles.

The necessary data and parallel program structures in [18] are tuned for a shared memory target architecture. If one intends to implement *Water* in a distributed memory environment, different tuning choices for the data structures, data partitioning and program structure would be made. In a performance oriented development approach [11] one would not implement the various possibilities of program variations, but develop a performance model of the application in the early design phase of the application, and reason on variations of models rather than full implementations to find the optimum parallel application design choice. In what follows we demonstrate for such a performance model – which can become overwhelmingly complex – how our distributed model evaluation improves analysis.

For our purpose here it is sufficient to consider only the subproblem of evaluating the potential energy of the particle system in parallel, rather than the more complex determination of forces on all particles. Let P be the number of available processors, N the number of particles and $N_i = N/P$ the number of particles statically assigned to processor P_i. A particle is an aggregate data structure describing the state of the physical particle (position, mass, charge, etc.; 672 bytes of storage per water molecule). The total potential (energy) of the particle system is:

$$E_{\text{potential}} = \frac{1}{2} \sum_{i=1}^{P} \sum_{j=1}^{P} \phi(\mathbf{x_i}, \mathbf{x_j})$$

procedure N-body_Potential_Energy (Π)

   partition $\Pi = \cup_{(0 \leq i \leq P-1)} \pi_i$ into particle sets π_i $(0 \leq i \leq P - 1)$
   distribute particle sets π_i $\pi_{(i+1) \bmod P}$ to all processors P_i
   (in P_i set $\pi_{\text{local}} = \pi_i$, $\pi_{\text{inbuffer}} = \pi_{i+1}$, $E = 0$)
   for $k = 1$ **to** $P - 1$ **do**
      par for all P_i, $(0 \leq i \leq P - 1)$ **do**
         for all pairs_of_particles **do**
            $E_{i,j} = \phi(x_i \in \pi_{\text{local}}, x_j \in \pi_{\text{inbuffer}})$
            $E = E + E_{i,j}$
         od
         $\pi_{\text{outbuffer}} = \pi_{\text{inbuffer}}$
         par
            send $\pi_{\text{outbuffer}}$ to $P_{(i+1) \bmod P}$
            receive π_{inbuffer} from $P_{(i-1) \bmod P}$
         od
      od
   od
   collect E_i from all P_i, $(0 \leq i \leq P - 1)$
   $E_{\text{potential}} = \sum_{i=0}^{P-1} E_i / 2$

end

Fig. 5. Parallel Evaluation of N-body Potential Energy

where the pair potential function ϕ returns the potential energy contribution of the two argument particles $\mathbf{x_i}$ and $\mathbf{x_j}$ (For simplicity we here intentionally overlook the computationally crucial fact that $\phi(\mathbf{x_i}, \mathbf{x_j}) = \phi(\mathbf{x_j}, \mathbf{x_i})$).

Let the set of particles be Π. To compute $E_{\text{potential}}$ on e.g. a ring of P processors with a static assignment of N_i particles per processor (π_i), processor P_i must somehow obtain data attributes of $N - N_i$ particles and must evaluate ϕ for $N_i N$ times. In other words, every pair of particles must meet in some processor at some time, and once there, it must execute pairwise interaction with *all* other particles available in the host processor. Crucial to this approach is that communication can be done concurrently. With the ring topology we can think of P_1 steps – where each processor sends and receives (in a clockwise or counterclockwise direction) exactly one particle – to be required for one particle to meet all the P processors. After an appropriate setup of the ring, particle sets are circulating through the ring meeting each other for pair potential computation. Processors concurrently work on sets of ϕ computations between consecutive particle set exchanges. The algorithm is sketched in Figure 5.

A performance issue of interest would be the one clarifying whether it is more desirable to implement the algorithm in a way that only single particles circulate in one message exchange step (as one extreme), or that particle sets of size N/P are circulating (as the other extreme), or that any intermediate particle set between should circulate. A Petri net performance model devised for that issue is

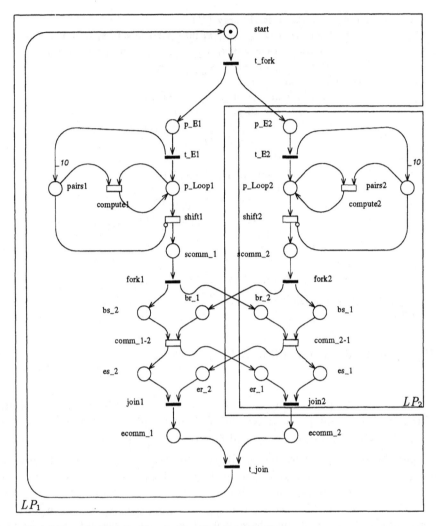

Fig. 6. Petri Net Model of a Simple Parallel N-body Program

constructed from the algorithm structure as follows (see Figure 6 for a two processor model): Independence of the particle set size is achieved by not explicitly modelling the **for** $k = 1$ **to** $P - 1$ **do**-loop, since the number of iterations here will vary from $P - 1$ to $(P - 1)N/P$. Neglecting the initial setup of the ring and the initial data distribution, the model is developed by applying a parallel process composition for the two concurrent processors, whose behavior is modelled between t_{fork} and t_{join}. Depending on the size of circulating particle sets, every branch establishes a loop on ϕ calculations, depositing an appropriate number tokens in the places *pairs1* and *pairs2* respectively. Pair potential computation and energy cumulation are embedded in the transitions *compute1* and *compute2*. Note that rather large subnets expand beyond those transitions (subnets not shown), causing a large number of events to be simulated during the evaluation, which fortunately do not affect events outside their domain. After all pairs have been computed, transition *shift* becomes enabled, which stands for shifting the contents of *inbuffer* to *outbuffer*. Finally particle sets are exchanged in parallel, which is modelled again by two parallel process compositions, containing a single communication/synchronization each. The extension of the model for a number of processors larger than two is straightforward.

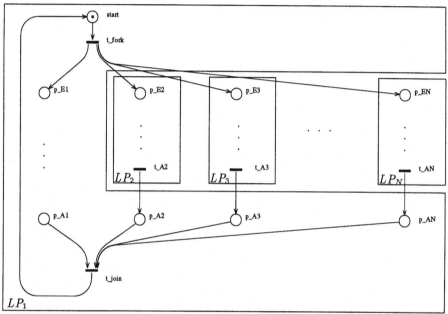

Fig. 7. Partitioning Parallel Process Compositions for Distributed Simulation

Distributed simulation now recommends a partitioning of the model in a way that subcomponents of the parallel compositions go into separate LP (see the recommended arc cutting in Figure 4 (c)). For the special case of two processors, the proposed partitioning is annotated in Figure 6. A general model partitioning would be as in Figure 7.

For the particular model a slight imbalance on behalf of LP_1 results from the proposed arc cutting, forcing LP_1 to manipulate three additional places and two transition enablings. An equally balanced situation would be the one where LP_2 takes the transition t_{join}, but this would mean cutting the arc ($eccom_1$, t_{join}), which would require a distributed enabling protocol. Compared to the performance loss due to such a protocol the imbalance is negligible. Figure 8 reports the execution traces of the distributed simulation of the model in Figure 6 using Time Warp on the iPSC/860. The three ParaGraph [14] gantt chart displays show the behavior of LP_1 mapped on Processor 0 and LP_2 mapped on Processor 1 when the number of particle pairs LP goes from 1 to 10 to 100 (for representational reasons we have intentionally chosen unrealistically small values) during 0.3755 sec of their execution. The different tasks are color-coded

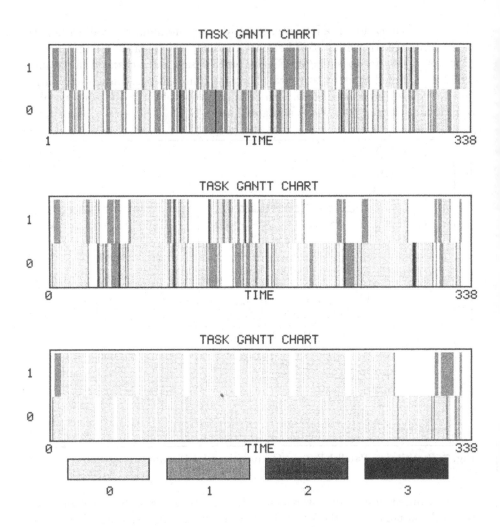

Fig. 8. Execution Behavior of Time Warp for the N-body Program Model

(top: 1 particle pair, middle: 10 pairs, bottom: 100 pairs per processor)

where 0 represents the event processing time (time that SE spent to fire transitions), 1 stands for communication time (time that CI spent for manipulating data structures and sending/receiving messages), 2 is the total execution time not displayed in the charts and 3 is the time used to invoke and execute rollback (state restoration). White bars indicate that an LP is blocked due to the lack of local events to be processed, i.e. waiting for a token message, or the intrusion introduced to take these traces (thin white bars). The charts show the typical phenomenon of Time Warp distributed simulation: LPs with insufficient local event processing suffer from overwhelming communication and rollback overheads. The frequent occurrence of external events in the first chart (1 particle pair) reveal that Time Warp simulation cannot gain speedup over sequential discrete event simulation. The situation improves slightly with 10 particle pairs per processor (second chart) where overlap of real simulation work (event processing) can be recognized, still not sufficient to dominate communication and rollback overhead. The 100 particle case (third chart) shows that Time Warp perfectly gains from our partitioning that clumps local behavior (i.e. event processing without external effects) as soon as the problem size increases. Note that a realistic parametrization of the model would be with 5000 to 10000 particle pairs, and that distributed simulation can in fact accelerate the evaluation of performance models using multiple processors instead of a single one.

4.1 Partitioning Parallel Compositions with Communication

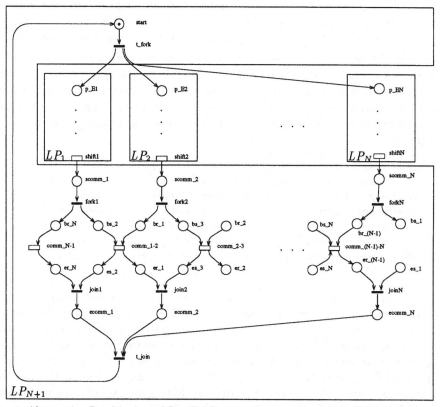

Fig. 9. Alternative Partitioning of Parallel Process Compositions with Communication

In the previous example we have seen the suitability of the proposed partitioning for Time Warp based communication interfaces of LPs, which is related to the structural property of the net model (Figure 6). In the loop of ϕ computations on particle pairs the events generated within one LP have no external effect. For firing the communication transitions $comm_{1-2}$ in LP_1 and $comm_{2-1}$ in LP_2, however, the two LPs have to synchronize. This synchronization is the reason why rollback will definitely occur when execution reaches that point, either in

LP$_1$ (if LP$_2$ has progressed LVT farther than LP$_1$) or in LP$_2$. Fortunately, this rollback cannot invoke the sending of antimessages, since no external events have been created in the loop simulation, and both LPs entered their loop simulations with identical LVTs (firing time of t_{fork}). A direct consequence of this is that no "rollback chain" can ever occur, and we can consider this kind of rollback as inexpensive (with respect to elapsed time). Looking now at the firing of $join1$ and $join2$ we recognize the same situation as with $comm_{1-2}$ and $comm_{2-1}$: for the case $\tau(comm_{1-2}) \neq \tau(comm_{2-1})$ again one rollback will occur, but now with the chance of having to send antimessages (potentially expensive rollback).

Generally, when parallel compositions comprise communicating components and partitioning is done the way as proposed in Figure 4, rollbacks will be the consequence of synchronization, the number of which can be determined from the net structure: Let I_P be the incidence matrix representation of the net model. Then every vector $\mathbf{i}$ that solves $I'_P \cdot \mathbf{i} = 0$ is called a place invariant, i.e. the number of tokens $\sum_{j=1}^{|P|} c_j$ (where c_j is the number of tokens in p_j as denoted by the j-th component of $\mathbf{i}$) remains constant in the set of places $p_j \in P$ with $c_j \neq 0$ for all possible markings reachable from the initial marking. $\mathbf{i}$ is a minimal place invariant if it cannot be constructed by linear combination of other place invariants. Looking now at the minimal place invariants for the model in Figure 6:

$$start + p_{E1} + pairs1 + p_{Loop1} + scomm_1 + bs_2 + es_2 + ecomm_1 = 1$$
$$start + p_{E1} + pairs1 + p_{Loop1} + scomm_1 + bs_2 + er_1 + ecomm_2 = 1$$
$$start + p_{E1} + pairs1 + p_{Loop1} + scomm_1 + br_2 + er_2 + ecomm_1 = 1$$
$$start + p_{E1} + pairs1 + p_{Loop1} + scomm_1 + br_2 + es_1 + ecomm_2 = 1$$
$$start + p_{E2} + pairs2 + p_{Loop2} + scomm_2 + bs_1 + es_1 + ecomm_2 = 1$$
$$start + p_{E2} + pairs2 + p_{Loop2} + scomm_2 + bs_1 + er_2 + ecomm_1 = 1$$
$$start + p_{E2} + pairs2 + p_{Loop2} + scomm_2 + br_1 + er_1 + ecomm_2 = 1$$
$$start + p_{E2} + pairs2 + p_{Loop2} + scomm_2 + br_1 + es_2 + ecomm_1 = 1$$

in conjuction with the (single) minimal transition invariant (t_{fork}, t_{E1}, t_{E2}, #_pairs(compute1), #_pairs(compute2), $shift1$, $shift2$, $fork1$, $fork2$, $comm_{1-2}$, $comm_{2-1}$, $join1$, $join2$, t_{join}), saying that firing each of these transitions with the corresponding multiplicity brings back the net to the initial marking, we recognize that in one iteration of the algorithm 4 messages are exchanged among LP$_1$ and LP$_2$ (generated by $fork1$, $fork2$, $comm_{1-2}$, and $comm_{2-1}$) for the communication part of the algorithm model. Precluding the case of evolving to identical LVTs in both LPs the resulting number of rollbacks is 2, one of which could have been avoided if $comm_{1-2}$ and $comm_{2-1}$ were in the same LP.

The consequence for partitioning of communicating parallel compositions therefore is that hardware performance has to be considered when constructing TTPN regions. With an alternative decomposition of our TTPN model into regions where the whole communication part of the algorithm goes into a single LP (see Figure 9), pairwise message exchanges among adjacent LPs can be avoided, reducing the absolute number of messages from $4(N-1)$ (where $N-1$ are created by t_{fork}, $2(N-1)$ by $fork(i)$ and $comm_{i-(i+_{modN}1)}$ respectively, and $N-1$ by $join(i)$) to $2N$ (N created by t_{fork} and N by $shift(i)$). Moreover the

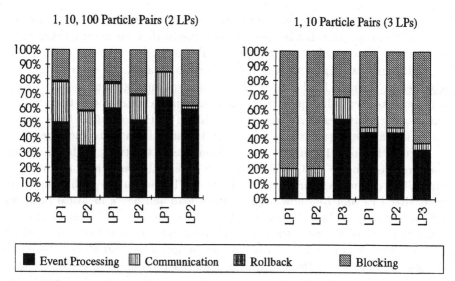

Fig. 10. Time Warp Performance on iPSC/860: Execution Profile

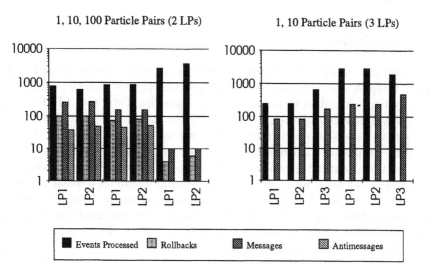

Fig. 11. Time Warp Performance on iPSC/860: Events and Messages

upper bound on the number of rollbacks is reduced from $2(N - 1)$ to $N - 1$, since all the "potentially expensive rollbacks" (invoked by $comm_{i-(i+_{modN}1)}$) are precluded. The remaining $N - 1$ (potential) rollbacks invoked in LP_{N+1} have neither local nor external effects, since every transition in LP_{N+1} is *persistent*, i.e. once enabled it cannot become disabled by the firing of another transition in LP_{N+1}. So whatever the timestamp of a token message entering LP_{N+1} via the firing of $shift(i)$ is, we can schedule and fire it without calling rollback at all (we have introduced this optimization to the original Time Warp protocol in [6]). Note also that t_{fork} can never produce causality violations in the respective

LVTs as was seen from the place invariants by the fact that place *start* is covered by all of them, i.e. *start* is a homeplace. This means that whenever a token meets *start*, a new time epoch starts for all the LPs involved.

With this packing of all the communication transitions $comm_{i-(i+_{modN}1)}$ together in a single LP however, the parallelism among them is lost; LP_{N+1} will simulate them all sequentially. An appropriate choice among the proposed partitioning strategies hence can only bee made on the execution time ratio of processing an event (transition firing) and transmitting a token message.

Figures 10 and 11 compare the performance of the two possible partitionings executed on the iPSC/860. The 3 LPs case is the one where the model in Figure 6 is decomposed according to Figure 9. The execution profile clearly reflects that LP_1 and LP_2 now have identical load and hence identical behavior. LP_3 is mutually exclusive to the other two LPs, causing significant blocking to them and vice versa. Increasing the load can make the event processing part of LP_1 and LP_2 dominate LP_3, such that the performance degradation caused by the sequentialization in this partitioning strategy is overwhelmed. (Note that no rollbacks, and as a consequence no antimessages, occur in the 3 LP simulation of the model.)

5 Conclusions

We have developed a distributed discrete event simulation protocol based on Time Warp to accelerate the evaluation execution of performance models of parallel systems described in terms of timed transition Petri nets (TTPNs). The necessary topological decomposition of TTPNs into a set of submodels, which appears to be a hard problem for general net models, was related to the model structures typically applied when modelling parallel systems. In this way heuristics for the partitioning process could successfully be developed, which were empirically validated using an N-body molecular dynamics parallel application example. The distributed simulation tool developed for the Intel iPSC/860 target architecture is integrated into the GreatSPN [3] graphical input interface, providing the user with a powerful environment as the frontend for the design and analysis of performance models.

References

1. H. H. Ammar and S. Deng. Time Warp Simulation of Stochastic Petri Nets. In *Proc. 4th Intern. Workshop on Petri Nets and Performance Models*, pages 186 – 195. IEEE-CS Press, 1991.
2. G. Balbo, G. Chiola, S.C. Bruell, and P. Chen. An Example of Modelling and Evaluation of a Concurrent Program using Coloured Stochastic Petri Nets: Lamport's Fast Mutual Exclusion Algorithm. *IEEE Transactions on Parallel and Distributed Systems*, 3(2):221–240, March 1992.
3. G. Chiola. GreatSPN1.5 Software Architecture. In *Proc. of the 5th Int. Conf. on Modelling Techniques and Tools for Computer Performance Evaluation. Torino, Italy, Feb 13-15, 1991*, pages 117 –132, 1991.

4. G. Chiola, M. Ajmone Marsan, G. Balbo, and G. Conte. Generalized Stochastic Petri Nets: A Definition at the Net Level and its Implications. *IEEE Transactions on Software Engineering*, 19(2), February 1993.

5. G. Chiola and A. Ferscha. Distributed Simulation of Petri Nets. *IEEE Parallel and Distributed Technology*, 1(3):33 – 50, August 1993.

6. G. Chiola and A. Ferscha. Distributed Simulation of Timed Petri Nets: Exploiting the Net Structure to Obtain Efficiency. In M. Ajmone Marsan, editor, *Proc. of the 14th Int. Conf. on Application and Theory of Petri Nets 1993, Chicago, June 1993*, Lecture Notes in Computer Science 691, pages 146 – 165, Berlin, 1993. Springer Verlag.

7. G. Chiola and A. Ferscha. Exploiting Timed Petri Net Properties for Distributed Simulation Partitioning. In *Proceedings of the 26th Hawaii Int. Conf. on Systems Sciences*, pages 194 – 203. IEEE Computer Society Press, 1993.

8. G. Ciardo, A. Blakemore, P. Chimento, J. K. Muppala, and K. S. Trivedi. Automated Generation and Analysis of Markov Reward Models using Stochastic Reward Nets. In C. Meyer and R. J. Plemmons, editors, *Linear Algebra, Markov Chains, and Queueing Models*. Springer-Verlag, 1992.

9. F. Feldbrugge. Special Volume: Petri Net Tools Overview 92. *Petri Net Newsletter*, (41):2 – 42, 1992.

10. A. Ferscha. Modelling Mappings of Parallel Computations onto Parallel Architectures with the PRM-Net Model. In C. Girault and M. Cosnard, editors, *Proc. of the IFIP WG 10.3 Working Conf. on Decentralized Systems*, pages 349 – 362. North Holland, 1990.

11. A. Ferscha. A Petri Net Approach for Performance Oriented Parallel Program Design. *Journal of Parallel and Distributed Computing*, 15(3):188 – 206, July 1992.

12. A. Ferscha and G. Kotsis. Optimum Interconnection Topologies for the Compute-Aggregate-Broadcast Operation on a Transputer Network. In *Proceedings of the TRANSPUTER '92 Conference*, pages 307 – 326, Amsterdam, 1992. IOS Press.

13. R. M. Fujimoto. Parallel Discrete Event Simulation. *Communications of the ACM*, 33(10):30–53, October 1990.

14. Michael T. Heath and Jennifer A. Etheridge. Visualizing Performance of Parallel Programs. Technical Report ORNL/TM-11813, Oak Ridge National Laboratory, May 1991.

15. D. A. Jefferson. Virtual Time. *ACM Transactions on Programming Languages and Systems*, 7(3):404–425, July 1985.

16. Jayadev Misra. Distributed Discrete-Event Simulation. *ACM Computing Surveys*, 18(1):39–65, March 1986.

17. T. Murata. Petri Nets: Properties, Analysis and Applications. *Proceedings of the IEEE*, 77(4):541–580, April 1989.

18. J. P. Singh, W.-D. Weber, and A. Gupta. SPLASH: Stanford Parallel Applications for Shared Memory. Technical report, Computer Systems Laboratory, Stanford University, CA 94305, 1993.

19. G. S. Thomas. Parallel Simulation of Petri Nets. Technical Report TR 91-05-05, Dep. of Computer Science, University of Washington, May 1991.

An Algorithm for Off–Line Detection of Phases in Execution Profiles [*]

Thomas D. Wagner[1] and Brian M. Carlson[2]

[1] Vanderbilt University, Nashville TN 37235, USA
[2] Dakota State University, Madison SD 57042, USA

Abstract. This paper describes a method for the detection of phases in time–series data. The specific application considered is time–series data which describes an execution profile of a parallel program. The algorithm is described and examples are presented. This paper presents the problem in the broader sense of general off–line time–series analysis. In analyzing time–series data, off–line algorithms enjoy the luxury of more computationally intensive approaches than possible with on–line algorithms. This paper describes an off–line method for the analysis of time–series data which detects periods of maximum homogeneity in time–series data, specifically phases in execution profiles. The technique is applied to execution profiles from program executions on hypercube computers. This is used to provide another means of workload characterization, and has been used to predict the speedup of parallel programs.

1 Introduction

The execution profile of a parallel program graphically describes the number of processors in a multiprocessor environment which are simultaneously utilized. Kumar [1] defines an execution profile as the number of processors kept busy as a function of time for a specific application on a specific architecture. Sevcik [2] defines a parallelism profile as the number of processors kept busy by a specific application as a function of time where the number of available processors exceeds the maximum parallelism of the application. For the purposes of this paper, the two terms are used synonymously. A graph of the execution profile looks much like a random walk. There are frequently, however, subsequences in the execution profile which are visually distinct and can be mapped back to the underlying parallel algorithm. These periods of roughly homogeneous processor activity are called *phases* in the execution profile. Identification of phases implies a partitioning of the time–series data into subsequences. In previous work it has been demonstrated that the identification of phases in execution profiles

[*] This research was supported by the Applied Mathematical Science Research Program of the Office of Energy Research, U. S. Department of Energy via subcontract 19X-SH008V from the Oak Ridge National Laboratory and by the Scientific Computing Program of the Office of Energy research, U. S. Department of Energy via subcontract 19X-SL131V from the Oak Ridge National Laboratory.

improves the prediction of the speedup of a parallel program if more processors are allocated [3]. In that work, the phases were identified manually. This paper presents a method of automatic detection of phases off–line. The algorithm is presented in terms of the C language structures that were used to implement it along with a narrative description of the steps involved. The analysis method is cast as a solution to a problem of off–line time–series data analysis. Two definitions motivate the remainder of the discussion.

Definition 1 (Stationary Phase). *A stationary phase is a contiguous subsequence of an execution profile which has roughly uniform processor activity.*

Definition 2 (Transitional Phase). *A transitional phase is a contiguous subsequence of an execution profile which constitutes an abrupt change in processor activity.*

It follows from the two definitions given, that a description of the stationary and transitional phases, is a complete description of the execution profile. Figure 1 illustrates these definitions. In that figure, subsequences A, C, and E represent stationary phases, and subsequences B and D represent transitional phases. (Figure 1 is reprinted from Carlson *et. al.* [3].) In this simplified example, Phase A can be viewed as the fraction sequential. Phases C and E represent periods of homogeneous processor utilization at different levels of utilization. The speedup characteristics of these two phases will likely be different. For example, if Phase E represents a period of full utilization, it will scale quite differently than Phase C which is a period of lesser utilization. Phases B and D are typical of periods of synchronization and will scale quite differently than Phases C and E.

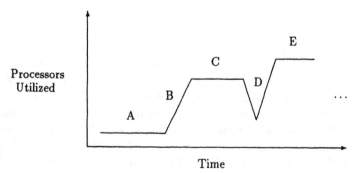

Fig. 1. Idealized execution profile

Sevcik [2], takes a different view of phases by disassociating processor activity with time. In his characterization, called *parallelism shape* the periods of identical behavior are all collected together. He shows that this can be used to compute average parallelism which may be a useful metric upon which to base heuristics for processor allocation for a multiprogrammed multiprocessor. In this paper we treat phases of activity as being associated with time as described in Figure 1.

Section 2 describes an implementation of a method of off–line phase identification based upon the metric of least squares error. Section 3 steps through an idealized example, illustrating the evolution of the data structures in the algorithm. Section 4 presents additional examples to show how the implementation supports the definitions of phases given above.

Variations of this approach are found in the literature. Fisher [4] introduces the basic approach employed here as a method of time–series analysis. He applies the algorithm to a series of monthly measurements of the depth of lake Michigan. The algorithm given here is essentially a greedy implementation of Fisher's approach. More recently, Berry and Hellerstein [5] employ a variation on Fisher's work which detects changes in customer arrival rates to an M/M/1 queueing system from the time–series data which is average customer response times. Their approach divides the time–series data into two partitions and then recursively subdivides the two partitions into subpartitions. The difficulty with this approach is knowing at what point to terminate the recursion. Berry and Hellerstein observe that by employing knowledge of the probability distribution which generates the response time data, a termination criterion can be constructed. Unfortunately, the underlying distribution which generates a parallelism profile is unknown and differs between parallel programs and system architectures. It is reasonable to assume that the distribution is not available to the analyst. The implementation given here, while using the binary splitting of the time–series data that Berry and Hellerstein employ, always creates a subpartition which maximally reduces the total least squares error. Consequently, we refer to the algorithm as taking a *greedy* approach.

2 The Algorithm

The basic idea underlying the algorithm presented here is that the data set can be iteratively decomposed into subsequences. Once a partition (or phase) boundary is added, it is never changed. This section presents a description of the data structures and the steps in the algorithm. (Notationally, physical arrays are designated using the C convention of $A[x \cdots y]$ whereas functions which manipulate data over the domain x to y are depicted as $A_{x,y}$.) The algorithm makes a number of passes over the time–series data. Therefore, the data is read into the array Series$[1 \cdots \texttt{NumPoints}]$. (Note: Series$[0]$ is never used.) Another array, Sum$[0 \cdots \texttt{NumPoints}]$ is generated from the input as well. In particular,

$$
\texttt{Sum}[i] = \begin{cases} \displaystyle\sum_{j=1}^{i} \texttt{Series}[j] = \texttt{Sum}[i-1] + \texttt{Series}[i] & \text{if } 0 < i \leq \texttt{NumPoints} \\ 0 & \text{if } i = 0 \end{cases}
$$

Sum$[0 \cdots \texttt{NumPoints}]$ provides a quick way to compute the mean of an arbitrary subrange of Series$[\,]$. Let Mean$_{a,b}$ denote the mean of the data values in Series$[a \cdots b]$.

$$\text{Mean}_{a,b} = \frac{\displaystyle\sum_{i=a}^{b} \text{Series}[i]}{b-a+1} = \frac{\text{Sum}[b] - \text{Sum}[a-1]}{b-a+1} \tag{1}$$

The result of Equation 1 is needed for computing the least squares error of a partition over $\text{Series}[a\cdots b]$. The algorithm proceeds by partitioning the time–series data into 2 sets. The point at which the time–series data is split is the point which minimizes the sum of the least squares error of the two sub-partitions. In other words, a point t is found such that $\text{LeastSquares}_{1,t-1} + \text{LeastSquares}_{t,\text{NumPoints}}$ is minimized. Let $\text{Error}_{1,\text{NumPoints}}$ be this minimum sum of least square error, then

$$\text{Error}_{1,\text{NumPoints}} = \min\left(\text{LeastSquares}_{1,t-1} + \text{LeastSquares}_{t,\text{NumPoints}}\right) \tag{2}$$

Equation 2 is compute intensive since all values $1 < t \le \text{NumPoints}$ must be examined. The algorithm accomplishes the splitting by repeatedly calculating the two values $\text{LeastSquares}_{1,t-1}$ and $\text{LeastSquares}_{t,\text{NumPoints}}$ and finding the point t which minimizes their sum. The result is stored in a structure of type PartitionPoint.

```
typedef struct {
        int Location;           .
        double LeftError, RightError;
        } PartitionPoint;
```

where the values stored in the structure are:

Location $\leftarrow t$,
LeftError $\leftarrow \text{LeastSquares}_{1,t-1}$, and
RightError $\leftarrow \text{LeastSquares}_{t,\text{NumPoints}}$.

As the algorithm proceeds, the partition which decreases the overall error the most by splitting it into two subpartitions is next divided. Thus, the partitioning problem is turned into a greedy best first search algorithm.

Accounting for where partition points have been found, and how much benefit there is in subdividing the partitions further is accomplished in a straightforward manner. This is the purpose of the array $\text{PointSet}[0\cdots\text{SetSize}-1]$, where SetSize is the current number of partitions. $\text{PointSet}[]$ should be regarded as a completely unordered set. For programming convenience, whenever the algorithm decides that $\text{PointSet}[i]$ should be split into two pieces, the left piece is put back into $\text{PointSet}[i]$ and the right piece is put at the end of $\text{PointSet}[]$ (*i.e.*, $\text{PointSet}[\text{SetSize}]$). Obviously, a linked list based implementation would employ a different ordering of $\text{PointSet}[]$.

```
struct {
    int Start, End, Location;
    double Error, LeftError, RightError;
    } PointSet[MaxNumberOfPartitions];
```

The fields PointSet[i].Start, and PointSet[i].End indicate the left and right sides of the ith partition. The least squares error of this partition is stored in PointSet[i].Error. The remaining three fields indicate the result of further subdividing this partition. By immediately subdividing a newly added partition, the algorithm avoids redundant computations.

3 Stepping Through an Idealized Example

Following is a trace of the algorithm given the idealized time–series data found in Table 1. This idealistic execution profile is used to illustrate each step in the algorithm and is taken from [6]. Examples using real execution profiles are presented in Section 4.

Table 1. Time series data

```
 1  2  1  2  1  2  1  2  1  2  1  2  1  2  1  2
11 12 11 12 11 12 11 12 11 12 11 12 11 12 11 12
21 22 21 22 21 22 21 22 21 22 21 22 21 22 21 22
31 32 31 32 31 32 31 32 31 32 31 32 31 32 31 32
```

The time–series data given in Table 1 is graphically illustrated in Figure 2. The mean of this data is 16.50 and the least squares error is 8016.00.

Visually it is clear that the data shown graphically in Figure 2 should be partitioned into 4 phases $(1 \cdots 16)$, $(17 \cdots 32)$, $(33 \cdots 48)$ and $(49 \cdots 64)$. As described above, a single partition point is found which splits the graph into two pieces. Figure 3 graphically illustrates the sum of the least squares errors of the two subpartitions as a function of t (i.e., the partition point). The sum of errors is minimized at $(1 \cdots 32)$, $(33 \cdots 64)$. (The graph illustrated in Figure 3 is smooth and symmetric due to the data given as input. In general, it is observed that this function $t \leftarrow \text{LeastSquare}_{1,64}$ can be quite noisy.)

Once $t = 33$ has been determined, both $(1 \cdots 32)$ and $(33 \cdots 64)$ are partitioned and the results are stored in PointSet[]. Thus, PointSet[] is initialized as the following.

$$
\text{PointSet[0]} = \begin{cases}
\text{Start} = 1 \\
\text{End} = 64 \\
\text{Error} = 8016.00 \\
\text{LocationPoint} = 33 \\
\text{LeftError} = 808.00 \\
\text{RightError} = 808.00
\end{cases}
$$

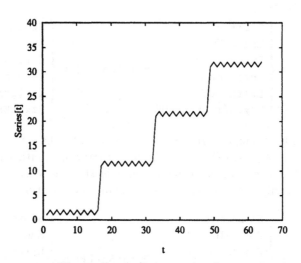

Fig. 2. Graph of time–series data

PointSet[0] denotes a single partition over Series[1 ⋯ 64] whose least squares error is 8016.00. It also indicates that if this partition is subdivided at (1 ⋯ 32), (33 ⋯ 64) the least squares error in (1 ⋯ 32) is 808.00 and the least squares error in (33 ⋯ 64) is 808.00. Partitioning at $t = 33$ leads to the following modifications in PointSet[].

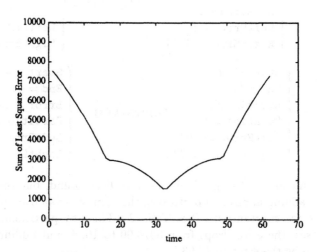

Fig. 3. Graph of least squares error

$$\text{PointSet}[0] = \begin{cases} \text{Start} = 1 \\ \text{End} = 32 \\ \text{Error} = 808.00 \\ \text{Location} = 17 \\ \text{LeftError} = 4.00 \\ \text{RightError} = 4.00 \end{cases} \quad \text{PointSet}[1] = \begin{cases} \text{Start} = 33 \\ \text{End} = 64 \\ \text{Error} = 808.00 \\ \text{Location} = 49 \\ \text{LeftError} = 4.00 \\ \text{RightError} = 4.00 \end{cases}$$

The sum of least squares error for the data set is $808.00 + 808.00 = 1616.00$. This is where best first search begins. Because partitioning either PointSet[0], or PointSet[1] leads to the same reduction in least squares error, PointSet[0] is chosen arbitrarily. This leads to the following modification of PointSet[].

$$\text{PointSet}[0] = \begin{cases} \text{Start} = 1 \\ \text{End} = 16 \\ \text{Error} = 4.00 \\ \text{Location} = 2 \\ \text{LeftError} = 0.00 \\ \text{RightError} = 3.73 \end{cases} \quad \text{PointSet}[1] = \begin{cases} \text{Start} = 33 \\ \text{End} = 64 \\ \text{Error} = 808.00 \\ \text{Location} = 49 \\ \text{LeftError} = 4.00 \\ \text{RightError} = 4.00 \end{cases}$$

$$\text{PointSet}[2] = \begin{cases} \text{Start} = 17 \\ \text{End} = 32 \\ \text{Error} = 4.00 \\ \text{Location} = 18 \\ \text{LeftError} = 0.00 \\ \text{RightError} = 3.73 \end{cases}$$

On the next iteration, PointSet[1] is partitioned at 48 resulting in the following.

$$\text{PointSet}[0] = \begin{cases} \text{Start} = 1 \\ \text{End} = 16 \\ \text{Error} = 4.00 \\ \text{Location} = 2 \\ \text{LeftError} = 0.00 \\ \text{RightError} = 3.73 \end{cases} \quad \text{PointSet}[1] = \begin{cases} \text{Start} = 33 \\ \text{End} = 48 \\ \text{Error} = 4.00 \\ \text{Location} = 34 \\ \text{LeftError} = 0.00 \\ \text{RightError} = 3.73 \end{cases}$$

$$\text{PointSet}[2] = \begin{cases} \text{Start} = 17 \\ \text{End} = 32 \\ \text{Error} = 4.00 \\ \text{Location} = 18 \\ \text{LeftError} = 0.00 \\ \text{RightError} = 3.73 \end{cases} \quad \text{PointSet}[3] = \begin{cases} \text{Start} = 49 \\ \text{End} = 64 \\ \text{Error} = 4.00 \\ \text{PartitionPoint} = 50 \\ \text{LeftError} = 0.00 \\ \text{RightError} = 3.73 \end{cases}$$

At this point all of the "real" phases have been found, but the algorithm continues. By adding additional partitions, the sum of errors over all partitions does not decrease substantially. (See Table 2.) Thus, by partitioning the data into four phases, the error drops from 8016.00 to 16.00, but adding four more phases only drops the error to 14.93.

Table 2. Least squares error as number of partitions increases

#	Partitions	$\sum$ L. S. Error
1	$(1\cdots 64)$	8016.00
2	$(1\cdots 32), (33\cdots 64)$	1616.00
3	$(1\cdots 16), (17\cdots 32), (33\cdots 64)$	816.00
4	$(1\cdots 16), (17\cdots 32), (33\cdots 48), (49\cdots 64)$	16.00
5	$(1\cdots 16), (17\cdots 17), (18\cdots 32), (33\cdots 48), (49\cdots 64)$	15.73
6	$(1\cdots 1), (2\cdots 16), (17\cdots 17), (18\cdots 32), (33\cdots 48), (49\cdots 64)$	15.47
7	$\vdots$	15.20
8	$\vdots$	14.93

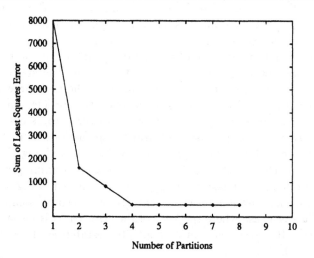

Fig. 4. Graph of least squares error versus number of partitions

Notice that iterations 5 and 6 create nonsensible partitions. Iteration 8 creates a single point partition $(49\cdots 49)$ and similarly, successive iterations create single point partitions. This suggests empirically that a criteria for stopping the search is when the knee of the sum of the least squares error curve is reached. Consider Figure 4 which is a graph of the data in Table 2. Again, it is clear visually that the knee of the least squares curve occurs when there are 4 partitions. There is simply no reduction in the least squares error to be achieved by further partitioning. In order to automate the determination of when to stop partitioning, consider how the human visual system makes the decision. What is considered to be the knee is the point where two lines could be drawn on the graph, one from the leftmost point of the curve to the knee and one from the knee to the rightmost point of the curve. These two added lines should be the best least squares fit of the entire curve of all possibilities of two lines. Therefore, as a heuristic to determine the knee, for each number of partitions, two lines are calculated that best fit the data from the leftmost curve point to the proposed knee and from the proposed knee to the rightmost curve point. The least squares error of the data from those two lines is then calculated. The number of partitions that minimizes the least squares error is taken as the knee and, therefore, the optimal number of partitions. In this case, the point that minimizes that error is at four partitions.

4 Further Examples

In this section the algorithmic approach of identifying phases discussed in the previous section is applied to the execution profile of three parallel applications. The execution profiles examined were extracted from PICL trace files [7]. Consider first the execution profile in Figure 5 which represents a parallel implementation of a Fast Fourier Transform on 16 nodes of an iPSC/2 hypercube.[3] This profile is free of noise and it is visually apparent that there are 17 stationary phases in the execution of this program. The transitional phases that separate the stationary phases are of very short duration and can be considered to be part of the stationary phases surrounding them. The stationary phases are of two types. One type represents the time that the multiprocessor is fully utilized and are depicted by the peaks in the execution profile. The other type is the periods that represent time that only one or zero processors are utilized. These periods of very low processor utilization are periods of communication between the processors when the processors are idle. Applying the algorithm described above, including the heuristic approach to determining when to stop partitioning, produces the phase boundary points identified in Figure 6 by vertical dashed lines. The least squares error generated by the partitioning is graphed in Figure 7. This is a case where the knee of the error curve is visually more difficult to identify. However, the heuristic adopted establishes 17 as the number of partitions at which to stop.

As another example of the application of this partitioning algorithm, consider Figure 8. This execution profile represents the execution of a parallel sort on 16 nodes of an iPSC/2 hypercube. Again, noise in this profile is minimal. In Figure 8 the dashed lines represent the first ten phase boundary points that the algorithm selects. The numbers above the dashed lines represent the order in which the boundaries were chosen. Figure 9 is a graph of the least squares error as more partitions are added. This execution profile is more difficult to visually partition than the previous FFT example. Application of the stopping heuristic to the error curve indicates that four phases are appropriate. The first four phases (separated by the first three boundary points) that are identified represent (from left to right):

1. a period of initialization while all processors are busy,
2. a period of ramping up in processor utilization,
3. a period of variable processor utilization as the sort algorithm executes, and
4. a period of post processing while, again, all processors are busy.

---

[3] Both this example and the next are taken from PICL trace files distributed with ParaGraph [8] through netlib.

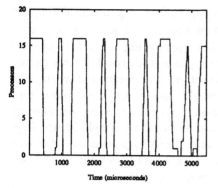

Fig. 5. Execution profile of Fast Fourier Transform

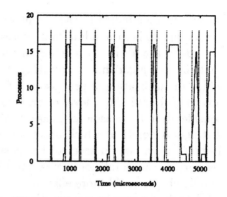

Fig. 6. Phases identified in execution profile of FFT

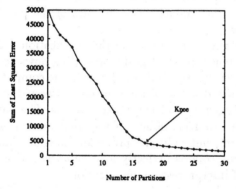

Fig. 7. Least squares error for FFT

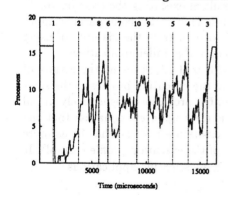

Fig. 8. Execution profile of parallel sort

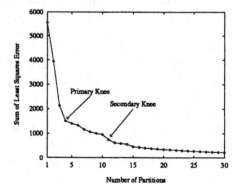

Fig. 9. Least-squares error of phases in parallel sort

Inspecting the error curve it is clear that the heuristic stopping criterion has chosen a knee in the curve. The choice of this stopping point does, in fact, make sense. If a finer granularity of phases is desired, the portion of the error curve to the right of the first stopping point can be further processed. Upon doing so, the secondary knee indicates that 11 phases are appropriate. Again considering the execution profile, the additional phases become apparent. The new phases that are added are all in the portion of the profile representing the period of time that the sort algorithm is executing. Each new phase is a period of time during which the utilization of processors is fairly constant. The decision of whether or not to go to this finer level of granularity would have to be made taking into consideration the reason for phase identification. If phases are being identified to distinguish points at which processor reallocation might be beneficial, the first stopping point would seem to be most suitable. The two phases at either end are periods that a full complement of 16 processors should be allocated. The two middle phases could more efficiently utilize fewer processors.

As a final example, consider the execution profile of a parallel implementation of a program that solves nonlinear shallow water equations on the sphere using the spectral transform method. This program constitutes an algorithmic kernel of a spectral global weather model.[4] The specifics of the parallel program are briefly described below.

The spectral transform method used in this application code approximates the solution on both a tensor product grid representing physical coordinates and a triangular grid representing spectral coordinates. The computational kernal of the method is the transformation of the solution between the two representations, requiring both Legendre and Fourier transforms. The parallel implementation decomposes the two grids in such a way that all processors participate in a Legendre transform, but only one processor calculates any given Fourier transform. The resulting code is perfectly load balanced, with all inefficiencies due to the overhead of interprocess communication.

Three execution profiles for this program are graphed in Figures 10, 11, and 12. These three execution profiles are the result of executing the code on 16, 32 and 64 processors of an iPSC/860 hypercube. This is the same parallel application that was used as an example in [3] to demonstrate the viability of using phases to improve the prediction of the speedup of parallel applications.

The dashed vertical lines in Figures 10, 11, and 12 represent the results of applying the phase identification algorithm to the execution profiles of the shallow water application running on 16, 32, and 64, processors, respectively. In all three cases 12 phases are identified and it is clear upon examination that the phases are scaling individually as more processors are added. That is, as more processors are allocated, the phase structure of the execution profile is stable. It has been our experience that this is true of many scientific applications.

5 Summary and future work

The motivation of the work described in this paper was a search for an algorithmic technique to identify phases in execution profiles of parallel programs.

---

[4] The serial version of the program was provided by J. J. Hack at the National Center for Atmospheric Research. The distributed version was implemented by Patrick Worley at Oak Ridge National Laboratory.

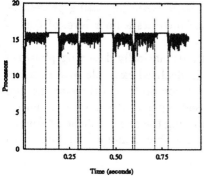

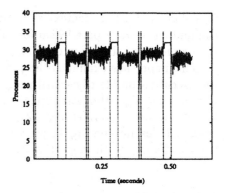

Fig. 10. Execution profile of shallow water code with 16 processors

Fig. 11. Execution profile of shallow water code with 32 processors

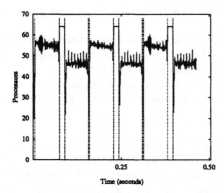

Fig. 12. Execution profile of shallow water code with 64 processors

Previous work demonstrated that identifying phases is worthwhile because it can improve the prediction of speedup of parallel applications when more processors are allocated. The algorithm presented here has realized the original objective. It is effective at automating the identification of phases in execution profiles.

There are other applications of this approach. One example is in the area of traditional workload characterization. Often, the load presented to a system varies dramatically over time. Both Haring [9] and Artis [10] have studied the time varying nature of workloads. It is often the case that in constructing workload models to be used for parameterizing system models, there are clear phases in the workload in the time dimension. Identifying those phases can provide a more accurate model of the workload. Automating that phase identification is desirable. The algorithm presented here can be applied to that task.

If a multiprocessor is to be multiprogrammed (*i.e.*, more than one application executed simultaneously), identifying phases could be useful for the scheduler. In particular, in a batch environment with real–time constraints, where applications are run repeatedly, having detailed knowledge regarding the size of phases would provide a basis for a schedule of effectively allocating and deallocating processors to applications. Sevcik [2], for example, discusses the use of average parallelism in a slightly different context as a useful heuristic when allocating processors. Computing average parallelism for each phase provides a finer level of control than a one time allocation decision.

This paper has examined the problem of automatically detecting phases in the execution profile of parallel programs. An idealized example was described and then the algorithm was applied to the execution profiles of three application traces obtained from Intel iPSC/2 and iPSC/860 hypercubes using PICL. Future work includes automating the process of identifying the knee of the least squares curve and applying the algorithm described here to shared memory architectures.

References

1. M. Kumar, "Measuring parallelism in computation intensive scientific engineering applications," *IEEE Transactions on Computers*, vol. 37, no. 9, pp. 1088–1098, 1988.
2. K. Sevcik, "Characterization of parallelism in applications and their use in scheduling," in *Proceedings of the ACM SIGMETRICS Conference on Measurement and Modeling of Computer Systems*, pp. 171–180, 1989.
3. B. Carlson, T. Wagner, L. Dowdy, and P. Worley, "Speedup properties of phases in the execution profile of distributed memory parallel program," in *Computer Performance Evaluation '92* (R. Pooley and J. Hillston, eds.), pp. 83–95, Antony Rowe Ltd., 1992.
4. W. Fisher, "On grouping for maximum homogeneity," *Journal of the American Statistical Association*, vol. 48, pp. 789–798, 1953.
5. R. Berry and J. Hellerstein, "An approach to detecting changes in the factors affecting the performance of computer systems," in *Proceedings of the ACM SIGMETRICS Conference on Measurement and Modeling of Computer Systems*, pp. 39–49, 1991.
6. B. Carlson, *Markovian analysis of deadline oriented queueing systems*. PhD thesis, Vanderbilt University, 1991.
7. A. Geist, M. Heath, B. Peyton, and P. Worley, "PICL: a portable instrumented communication library, C reference manual," Tech. Rep. ORNL/TM–11130, Oak Ridge National Laboratory, Oak Ridge, Tennessee, 1990.
8. M. Heath and J. Etheridge, "Visualizing the performance of parallel programs," *IEEE Software*, vol. 8, no. 5, pp. 29–39, 1991.
9. G. Haring, "On stochastic models of interactive workloads," in *Performance '83* (A. Agrawala and S. Tripathi, eds.), pp. 133–152, North–Holland, 1983.
10. H. Artis, "Characterizing the time varying nature of TSO workloads," in *Workload Characterization of Computer Systems and Computer Networks* (G. Serrazi, ed.), pp. 11–19, North–Holland, 1986.

Experiences on SIMD Massively Parallel GSPN Analysis *

S. Caselli, G. Conte, F. Bonardi, and M. Fontanesi

Dipartimento di Ingegneria dell'Informazione
Università di Parma
Viale delle Scienze - 43100 Parma - Italy
e-mail: {caselli,conte,bonardi,fontanem}@CE.Eng.UniPR.it

Abstract. Generalized Stochastic Petri Nets (GSPN) are a prominent tool for the performance analysis of many real systems, but are prone to a state space explosion phenomenon that limits the applicability of the methodology. In this paper we describe CM-2 based algorithms dealing with the most computing-intensive part of the solution of GSPN models. Their implementation is fully integrated with a well-known tool for GSPN analysis (GreatSPN) and allows the solution of nets with up to several 10^6 states. The efficiency of the approach varies with the structure of the model under analysis.

1 Introduction

Petri Nets (PN) [16] are a consolidated formal tool for the modeling and analysis of concurrent, asynchronous, non-deterministic systems. For performance analysis applications, Generalized Stochastic PN (GSPN) [2] represent a dialect in widespread use. In GSPN transitions are divided into timed transitions (to describe the execution of time-consuming activities) and immediate transitions (to describe some logical behavior of the model). Delays of timed transitions are stochastic variables with exponential distribution, and, if the net is bounded, the resulting models yield a state space isomorphic to a finite Markov chain.

While GSPN permit very detailed and semantically correct modeling of systems, they tend to generate large state spaces even from apparently innocuous models. In fact, the quantitative analysis of GSPN models of significant systems always results in a complex problem only solvable with automatic tools. Moreover, the state space explosion phenomenon is a serious drawback in several potential applications (e.g., design of concurrent architectures [10]).

There are three main approaches to the solution of large GSPN models:

- exact numerical solution [1];
- simulation with Monte Carlo methods [12];
- approximate techniques [15].

* This work was partially supported by MURST 40% funds (special program on Distributed Architectures), by MURST 60% funds and by CNR contracts n.91.02313.CT12 and n.92.00065.CT12.

The approximate solution techniques are based on reduction and decomposition of the net into simpler nets, in order to allow a simplified analysis. Unfortunately, this approach is only feasible in some particular cases and cannot be applied in the general ones.

The simulation techniques are cumbersome, from a computational point of view, if results with reasonable confidence levels are to be reached. Distributed simulation is being investigated to overcome the computational complexity. Promising results have been attained with massively parallel simulation, but only with reference to specific subclasses of nets (Marked Graphs) [4].

The exact numerical solution requires the construction of the space state representation, called Reachability Set (RS), followed by the construction and the solution of the infinitesimal generator of the associated Markov chain. Of course, the exact numerical solution must directly face the state space explosion. The sequential implementation (e.g., in the GreatSPN package [11]) is severely constrained in its applicability by the exponential increase in the time and the computational resources needed, especially of memory.

In this paper we illustrate our ongoing efforts toward a massively parallel, exact numerical solution of *large* GSPN models. In particular, our investigation has dealt with what we have perceived as the main bottleneck in the solution of large GSPNs, the construction of the state space. Algorithms for RS construction have been implemented and tested for a Connection Machine CM-2 massively parallel computer [17, 7].

The effort we have undertaken may be regarded as a brute force approach to the problem. However, the use of a massively parallel computer rather than a "conventional" supercomputer has led to new algorithms and implementations emphasizing the potential sources of parallelism. Sources for massive parallelism have actually been found in the problem, but they do not match well with the strictly SIMD type model of computation supported by the CM-2. Hence, execution times (so far) do not compare as well as hoped with those of the conventional workstation implementation.

In spite of its unsatisfactory performance, the CM-2 implementation significantly extends the range of solvable nets. To the best of our knowledge, we have achieved the largest GSPN RS cardinalities reported in the literature, in the order of 2 million tangible states. About 10 million states are actually achievable, but we have refrained from solving such "extra large" size models solely for testing purposes because of a fair resource usage policy. One should understand the system implications of solving large models, with over 40 MBytes of intermediate results generated per million states. In [8] a 1.6 million state space cardinality has been obtained by resorting to a hierarchical, colored approach (Hierarchical GCSPN) which is beneficial both from a structured modeling perspective and from a computational viewpoint. The drawback of such an approach is that hierarchical modeling is not always applicable or apparent to the designer. Moreover, plain GSPN represent a much more established and widespread modeling tool.

The parallel RS construction algorithm described in this paper is fully integrated with the well-known GreatSPN analysis tool [11], in that consistent internal representations are exploited. In fact, we have successfully completed the GSPN solution of our models by resuming the GreatSPN sequential algorithms for the construction and solution of the infinitesimal generator of the associated Markov chain. This compatibility requirement, which we have set as a mandatory feature, has further contributed to the difficulties met in parallelizing.

The paper is organized as follows. In section 2 the characteristics of the CM-2 system are summarized. In section 3 the RS construction problem is described and two potential sources for parallelism highlighted. In sections 4 and 5 CM-2–based implementations exploiting this parallelism are illustrated. Section 6 discusses some of the results achieved so far, while Section 7 concludes the paper.

2 The CM-2

The Connection Machine CM-2 is a fine grain massively parallel architecture consisting of up to 64K single bit processing elements (PE). A group of 16 PE is implemented on a single VLSI chip and forms a nucleus connected with other identical ones using a 12-dimensional hypercube interconnection structure. An optional floating point accelerator is associated with every 32 PEs. Each processing unit can directly access its associated memory (up to 4 Mbytes). The PEs operate in a strictly SIMD manner and the programming style follows the data parallel paradigm.

Data is much more parallelizable than instructions in most practical computing intensive tasks. However, there are relevant problems that have been reported to only partially benefit from the data parallel approach. Even in these cases it is not clear whether this behavior is really associated with the problem *per se* or merely with a solution technique which is inadequate but amenable to potential improvements.

Figure 1 shows the internal architecture of the CM-2, which can operate as a coprocessor associated with a sequential front-end. The flow of control is handled by the front-end, including storage and execution of the program and all interaction with the user.

The Connection Machine system was specifically designed to handle the largest computational problems, and can be efficiently used whenever the same operation can be performed simultaneously on many data objects. Though the CM-2 basically operates in SIMD mode, features at the hardware level allow different processors to behave very differently. For instance, conditional instruction execution is provided as follows: each processor has a state bit, called a context flag, which determines whether or not the processor will execute the commands issued by the sequencer. Data remain in the CM-2 memory during the execution of the program and are operated in parallel. Whenever this is not true the performance of applications running on the CM-2 is seriously affected.

The main problem in developing an application on the CM-2 is thus to carefully reconsider the algorithm that must be executed on the machine in order to define a procedure for the solution efficiently exploiting the computational

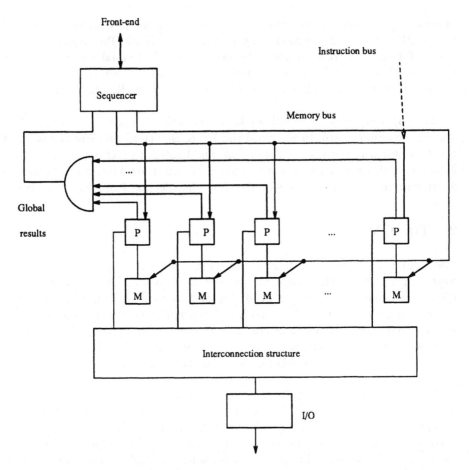

Fig. 1. The CM-2 parallel processing unit.

paradigm. The distribution of the data set into the memory of the machine must also be carefully considered, since the performance behavior of the interconnection network is largely influenced by the data flow pattern. In localized, regular data structures, the data access pattern is the same everywhere and the paths between the accessed data items are very short. In irregular data structures the access pattern varies from one part to another, and often determines very long data access paths.

The parallelism of the machine can be virtually increased by associating each physical processor with a part of its local memory. Thus the physical processors operate in sequence on all the subsets of data on which the global data structure have been partitioned. This behavior is transparent to the user, who can thus operate on a number of virtual processors (VPs) largely exceeding the physical parallelism of the machine. The number of VPs emulated by each PE is called the *VP ratio*. A *shape* is a logical *n*-dimensional organization given to parallel data structures.

The configuration of the CM-2 used in this research is the following:

- 8K processors with floating point units.
- 32K bytes of local memory per processor (256 Mbytes).
- SPARCstation2 front end with a 64 Mbyte main memory and a 2 Gbyte disk.

It should be noted that the PEs and the other components of the CM-2 are (deliberately) designed in a very conservative and somewhat obsolete 7MHz clock technology, since a major tenet of the architecture has been to promote the data parallel and massively parallel paradigms as alternative approaches to costly high-speed technology.

3 Problem analysis

We begin this section with some relevant GSPN terminology definitions. In GSPN the state notion is represented by the *marking* of the net. A marking is *reachable* if it can be obtained from an initial or current marking by firing a suitable sequence of transitions. If only timed transitions are enabled in a marking, the marking is said to be *tangible*. If at least one immediate transition is enabled, the marking is said to be *vanishing*. Each marking, vanishing or tangible, is characterized by a list of enabled transitions: timed transitions have zero priority, while immediate transitions have non-zero priority. The transitions to be fired are the highest priority ones; a tangible marking is recognized by an all-zero priority enabled transitions list. Markov chains isomorphic to stochastic Petri nets can be built by considering only the set of reachable tangible markings (Tangible Reachability Set – TRS). The graph whose nodes are the reachable tangible markings and whose arcs are labeled with the firing sequences is termed Tangible Reachability Graph (TRG).

Given the high cost of solving large nets, the quantitative analysis should always follow an extensive qualitative analysis, to ensure that all the basic correctness and logical properties are satisfied [16]. This is true for both the sequential and parallel implementations.

Analysis of large GSPN models on a workstation using GreatSPN demonstrates that the crucial computation is the construction of the state space, the TRG. This computation may require in the order of 1 hour on a Sun SPARCstation1 for a 300,000 state space, although there is considerable variability depending on the structural complexity of the net. Once the state space has been computed, the successive steps in a GSPN solution may still require a significant amount of time, but always substantially less than for the TRG. Moreover, whatever the time required, only TRG computation may prevent a successful solution, since the other steps are always feasible once the TRG is available. Thus, TRG computation is currently the true bottleneck in large GSPN model analysis.

The time required by TRG computation is mainly due to the lack of locality in the data access patterns. It is well known that the performance of today's computing systems is largely dependent upon the actual validity of the locality principle on the memory access patterns.

In the sequential implementation, a large TRG construction may fail according to one of two distinct patterns, *sudden death* or *thrashing*. In the former case, the computation of the state space proceeds in a regular way until the memory limit is reached, and then aborts. In the latter case a typical memory thrashing phenomenon occurs, and the TRG computation proceeds at an increasingly slower rate owing to continuous swapping. The thrashing type failure is particularily pernicious because of its global effects. What kind of failure will arise for a given GSPN model depends on the net structural complexity, vanishing states, and invariants, and also on workstation configuration at the system level. Examples of failure belonging to both categories will be provided later on in the paper.

The TRG construction algorithm consists of an iterative procedure starting from the initial marking M_0, and comprising the following steps: generate next state, verify if next state is a new state, insert new state in the state graph. In particular, the core loop of the algorithm is a breadth-first search and can be sketched as follows:

```
<let NEWMARKS be an ordered list of markings>;
<let RS be the reachability set of the net>;
NEWMARKS:= [MO] {*the initial marking*};
RS := [MO];
while <NEWMARKS is not empty> do
begin
     <let MARK be the first element of NEWMARKS>;
     <remove MARK from NEWMARKS>;
     for each <transition t enabled in MARK> do
     begin
         <let REACHEDMARK be the marking
              obtained by firing t in MARK>;
         if <REACHEDMARK is not in RS> then
         begin
              <insert REACHEDMARK at the end of NEWMARKS>;
              <add REACHEDMARK to RS>
         end
     end
end;
```

As potential sources of parallelism, attention has been devoted to the search and the insertion of a new marking into the state space and to the marking evolution from one state to the next (transition firing). The two problems embody the manipulation of an irregular, complex state space, whose growth is typically unpredictable and not uniform. Hence, neither of the problems exhibits the regularity in data structure required for easy SIMD implementation.

Since transition firing is substantially independent of search and insertion of new markings, descriptions of SIMD algorithms for the two problems are given next in separate sections.

4 Parallel transition firing

Firing enabled transitions represents a systematic state space exploration whereby new markings are computed and transition firing sequences leading from a tangible marking to the next are determined. Transition firing is similar to the state space analysis performed in strategy games (e.g. chess) [3], where the state space tree (sequence of possible game configurations) is generated in order to choose the best game strategy.

Efficiency in transition firing is achieved, in both sequential [5, 11] and parallel [6] implementations, by means of techniques such as:

- the ability to directly manipulate the markings in their particular encoded representation, without resorting to any code conversion;
- the test of only a reduced number of transitions in the "find enabled transitions in new marking" step, based on a preventive structural analysis of the net.

Let M_i be a new marking to be examined obtained by firing transition t_k in marking M_j. Given the preventive structural analysis of the net, the enabled transitions list associated with M_i is derived from the list associated with M_j (containing only a few of the whole set of transitions of the net) as follows: transitions disabled by the firing of t_k are removed from the list (a sufficient condition for a transition to be disabled is that it is either mutually exclusive or in structural conflict with the firing transition t_k), while transitions that have been enabled are added (a necessary condition is that these transitions are in causal connection with t_k).

As an additional feature, the sequential implementation of transition firing takes advantage of subnets analysis to manipulate only subnet-specific fields of vanishing markings [5]. However, this optimization in the parallel implementation would significantly increase the memory required at each PE. Moreover, in the parallel implementation we have elected not to maintain any representation for vanishing markings (not needed for TRG construction), and thus subnets analysis is not exploited.

The basic idea of the parallel implementation is to assign a marking with n firing transitions to n distinct processors: every processor containing a copy of the marking fires a different transition. While at the beginning of TRG construction only the initial marking M_0 must be examined, in the successive firing steps several markings are typically available in the NEWMARKS list (refer to the algorithm in Section 3). Hence a bi-dimensional shape is used on the CM-2 for parallel firing from as many markings as is feasible. Given the $8K$ processor configuration of the CM-2 at the University of Parma, a 256 row, 32 column grid was chosen. The 32 column dimension is a critical parameter, in that it must accommodate a full segment of firing history from a tangible marking until tangible markings along all possible next state evolutions are reached and until no immediate transition remains to be fired. The default value for this parameter was set to 32 following our empirical evidence. Furthermore, no VP emulation occurs, both to avoid incurring in the associated overhead and to maximize the memory available at each PE.

The major phases of a parallel firing run are outlined below.

1. Initial
   At each firing cycle, up to 256 tangible markings are taken from the **NEWMARKS** bag and inserted in the first column of the grid, one per row, together with all the information concerning the enabled transitions.

2. Spread
   Next, each marking is spread along the row to obtain as many replicas of the marking as the number of transitions to be fired. If the 32 column constraint is exceeded by spreading, an exception occurs and an appropriate error message is returned.

3. Fire
   All processors containing a marking fire the assigned transition, computing the new marking and its enabled transitions list. In the following we denote as *active* all processors in which a firing occurs. New markings replace the old ones (no longer useful) in the active processors, so that the active processors in a row now contain all the states (tangible or vanishing) reachable from the initial tangible marking of the row by firing a single transition. Meanwhile, the indices of fired transitions are stored in a parallel array for reconstruction of the firing sequence.

4. Test
   New markings are again identified as tangible or vanishing. Only processors containing vanishing markings are maintained active, because they have not yet terminated their tangible-to-tangible firing sequence.

5. Iterate and Terminate
   Processors containing a vanishing marking repeat the process of marking spreading and firing until all processors are found to contain only tangible markings and all possible firing sequences from the starting markings have been inspected.

A typical course in marking evolution is described in Figure 2. A single row of the grid is drawn, but the same kind of behavior is present in all rows for every fired marking.

The parallel implementation of transition firing has achieved somewhat disappointing performance, being about 1.5-2 times slower than the sequential one (depending on the specific net). Detailed trace analysis has shown that the main limitation of the parallel implementation is the time needed to compute the enabled transitions lists in the new markings, an operation requiring heavy access to global net transition information. These structures are not immediately available inside every processor, and their access involves several time-consuming

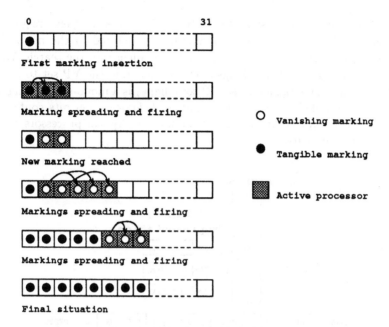

Fig. 2. Steps in parallel transition firing. A single row of the grid is drawn.

general communications inside the CM-2. This choice was initially made in order to minimize the local memory requirements, which would increase by several KBytes if this information were stored in every processor. Alternative possibilities exploiting the local insertion of this information are now being investigated in order to speed up the algorithm [6].

5 Parallel construction of the RG

RG construction is essentially based on a test of existence of a marking in the reachability set (search) followed by its possible insertion in the set. We have pursued two alternative RG construction algorithms. In the first one searching is carried out in parallel but only one marking at a time is considered. The second algorithm attempts to exploit the simultaneous availability of several markings from the firing algorithm (section 4) in order to concurrently insert them in the TRG. The former algorithm has been fully implemented and tested, while implementation of the latter is still in progress.

5.1 Parallel marking search

From the point of view of search and insertion, the state space is mapped onto the CM-2. Hence, the data structure organization is the key to an efficient algorithm implementation, in terms of both execution time and memory utilization.

New states are evenly distributed to the processors while they are generated. Thus, the resulting RS is split into subsets, each stored in the memory of a different physical processor to maximize the parallelism in the search. Working on independent subsets of markings does not constrain the VP set to assume a particular shape; thus the assumed VP set (shape) is a mono-dimensional array. Such a shape, predefined by the C* compiler (*physical*), permits configuration of the data structure according to the number of physical processors available during program execution.

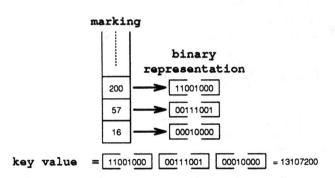

Fig. 3. Example of key generation.

Each subset is organized as a sorted list using a numerical value directly resulting from marking values. The markings are coded in arrays of bytes of fixed length, and a search key is created joining the last three bytes of the marking as shown in Figure 3. Of course, with this kind of key the markings are not univocally characterized. Only in the case of a "hit" in the hashing procedure are the marking values unused in the key checked. At the end of the search, two cases are possible:

- no marking coincides with the input one, which is added to the RS;
- the input marking is found to already belong to the RS.

The marking search is dichotomic, exhaustive and simultaneously executed by all physical processors in the locally ordered lists. Identical keys in different processors do not increase the complexity of the search algorithm, owing to independence of the lists. The worst case occurs when the same key is repeated in at least one list (processor): these processors must sequentially scan the markings with the same key as the input marking. Choosing a longer key reduces this effect but slows down key comparison. Moreover, the keys must be maintained twice: ranked first according to their value, and second according to the order of generation. Thus longer keys require more memory. Information about new reached markings is also stored into temporary structures for more effective searching and sorting mechanisms.

Figure 4 shows how new markings are stored into temporary structures. Such structures are sequentially filled by new markings; when the structures

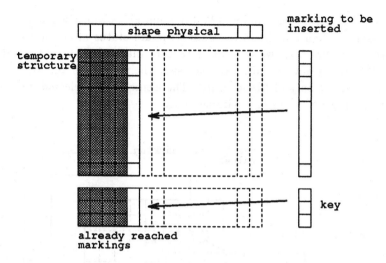

Fig. 4. Temporary structure for storage of markings.

are full, markings are simultaneously inserted into ordered lists. This approach also exploits a sort of locality property of the marking sequence. In fact, in a given step the probability of finding a marking already reached is higher around the last marking generated. Therefore a scan of the temporary structure often obviates the need for the scan of the complete RS.

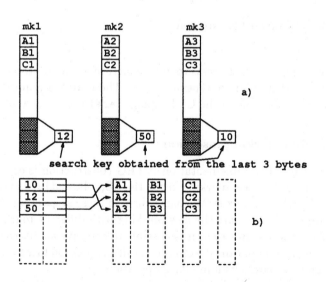

Fig. 5. A set of markings sequentially assigned to a processor (a) and their storage into the processor's memory (b).

Figure 5 shows how data are stored for a dichotomic search into each processor P_j. Let mk1, mk2 and mk3 be three markings which sequentially come to processor P_j: figure 5 shows the data structure status after the third marking insertion. Figure 6 shows the data structure mapped on the set of all processors. Processors have simultaneous access to their elements and every one works on one element only of the local list. The elements are selected addressing the parallel array by means of a parallel index.

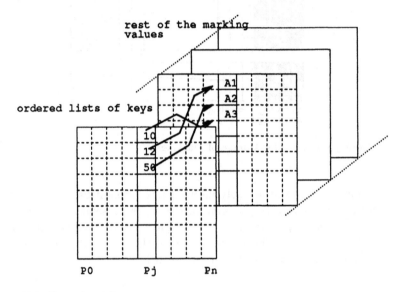

Fig. 6. Globally ordered list structure.

In order to obtain fast access to data structure, data are stored in special slice-wise arrays rather than conventional ones. These arrays require special PARIS functions, not available in the C* language (aref32(...), aset32(...)) [9].

5.2 Parallel marking insertion

The current parallel implementation of TRG construction is essentially a test of existence of a single new marking inside the RS, followed by its possible insertion in the set if it does not exist: the parallelism in these steps stems from the ability to simultaneously compare the new marking with different marking instances in all CM-2 processors. However, it is possible to conceive an alternative algorithm, based on the results obtained from the parallel transition firing (section 4), which yields a greater parallelism in CM-2 operations.

At the end of a parallel firing cycle, each processor that was active in the firing process holds a tangible marking obtained from an initial marking by following one of the possible firing sequences. Of course, not all available CM-2 processors have been involved in the firing sequence; however, a large number of processors (in the order of hundreds or thousands for large nets) contain tangible markings. The current search and insertion step of the algorithm only manages the comparison of one of these markings at a time and, even though efficient, it does not exploit the simultaneous availability of several markings for parallel comparisons. The basic problem is that it is not possible to forecast how the reached markings will be distributed among the processors in the course of the RS construction: this means that new markings must be compared with all those already in the RS.

In GreatSPN a marking is represented by a particular coding, which allows computation of the number of tokens contained in all the net places from a certain number of coding bytes. The number of coding bytes needed depends upon net complexity. By grouping the first two coding bytes (or any two coding bytes), a number in the range 0 to 65535 is obtained. Since a CM-2 may include up to 64k processors, this value can be used to univocally determine a processor into which the marking must be inserted before the existence test is made. In practice, the number obtained from these two bytes is the CM-2 processor index of the marking. For the 8K processor CM-2 at the University of Parma, for example, the processor index can be obtained by a *modulo* 8192 operation on the two coding bytes.

This hashing technique efficiently computes the index of the processor to which a certain marking is associated: consequently, only a local search must be made to determine whether this particular marking is already in the RS.

The large number of markings obtained with the parallel transition firing could now be managed simultaneously, inserting them in their associated processors before executing the comparisons with RS markings. Moreover, these comparisons can take place simultaneously and independently in every processor containing at least one new marking, thus the algorithm appears to more effectively exploit the SIMD capability of the CM-2.

A difficulty to be overcome for implementing this algorithm is the parallel management of a different number of comparisons inside the processors: the number of markings assigned to each processor is now unpredictable, because the RS would not be uniformly distributed to the processors, as happens in the current implementation.

6 Experimental results

A number of experiments have been carried out investigating the effectiveness and performance of parallel TRG generation [9, 13]. In this paper we only report selected results from two representative GSPN models: *benchnet* (figure 7) and *mesh2x2np* (figure 8). These results are typical of what has been obtained in general for TRG generation of large models, and are based on a sequential transition firing and the parallel search described in section 5.1.

The *benchnet* is part of the GSPN model of a complex manufacturing system [14], while the *mesh2x2np* is the GSPN model of a mesh of processors under multitasking and non-preemptive interaction policy [10]. Both models allow large

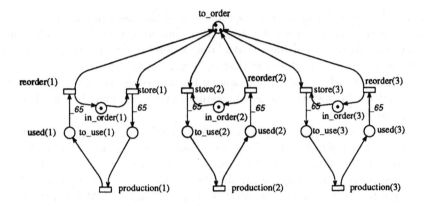

Fig. 7. The *benchnet* model.

state space generation by means of simple parametric changes, but they have structural features leading to different behaviors in state space generation.

- The *benchnet* model has a simple net structure and no immediate transitions. a large state space cardinality is obtained by modifying the arc weights, which in turn lead to high marking values in places. Hence, most of the memory is used to store the marking codes rather than the net structure.

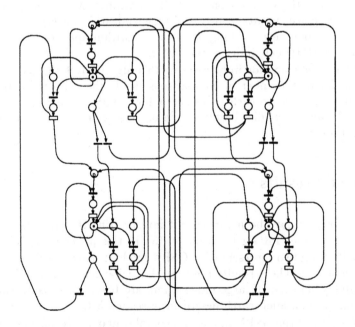

Fig. 8. The *mesh2x2np* model.

– In the *mesh2x2np* model the net structure is quite complex, and immediate and timed transitions coexist. In this case it is not necessary to inject many tokens to generate a large TRG. Memory is used more to store the net structure than the marking codes.

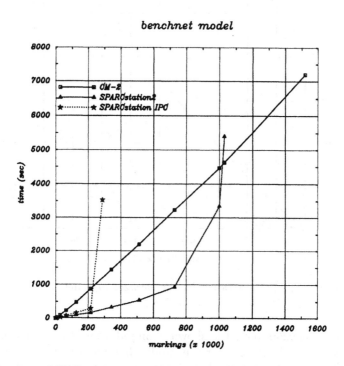

Fig. 9. Run time of TRG generation for the *benchnet* model.

Figure 9 shows the results obtained from the execution of the parallel and the sequential algorithms for TRG construction of the *benchnet* model. The best sequential run refers to a 64 MByte main memory, 40 MHz SPARCstation2 workstation. As shown, the parallel solution can guarantee a computation time almost linear in the number of states and carries out TRG computation of about $1,5 \cdot 10^6$ markings in 2 hours. Sequential execution time follows a linear law until the memory request exceeds memory capacity (this happens near 10^6 markings) and then grows exponentially owing to thrashing.

To validate that memory is the principal bound to the TRG construction problem, we have also included results obtained on the same GSPN model using a 32 MByte main memory, 25 MHz IPC SPARCstation workstation. In this case the cardinality of the TRG that can be computed is reduced to $2.5 \cdot 10^5$ markings and about one hour is required.

In figure 10 the comparison between parallel and sequential TRG construc-

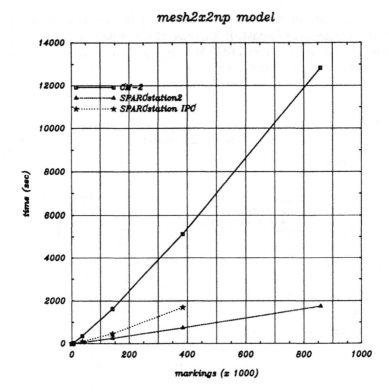

Fig. 10. Run time of TRG generation for the *mesh2x2np* model.

tion for the *mesh2x2np* model is shown. Computation of the same number of tangible markings takes longer for this model than for the *benchnet* model: this is due to the presence of immediate transitions that cause the algorithms to generate and store vanishing markings before computing tangible ones.

For values larger than the ones shown, the sequential implementations fail owing to sudden death, while the parallel one might be carried out up to about 10 million states. Up to the markings that can be generated, the sequential solution is faster than the parallel one. This behavior depends on the fact that memory requirements for storage of marking codes are less demanding, hence the memory bound has less influence on the sequential implementation. In the *benchnet* model the large number of tokens in the net requires a large amount of memory for storage of markings codes and the parallel machine can take better advantage of the larger available memory.

In summary,

- memory is the main bound to the sequential algorithm;
- both algorithms exhibit a computation time almost linear within their range of effectiveness, which is shorter for the sequential one;

– the CM-2 parallel implementation can manage larger state space cardinalities than the sequential workstation one, but unfortunately it is slower in the range where both are effective.

7 Conclusion

This paper reports an ongoing investigation of GSPN solution on a massively parallel SIMD computer. The specific problem addressed so far is TRG generation, but we also plan to investigate the parallel implementation of the steady-state numerical solution of the Markov chain generated from the reachability graph. In this case the problem reduces to the manipulation of a very large, almost triangular and very sparse matrix, and appears to lend itself more easily to parallelization.

The current CM-2 implementation of TRG generation has achieved somewhat disappointing results, in that larger state spaces are now feasible but at slower speed than on powerful workstations. Thus, a powerful sequential workstation equipped with a comparable primary memory would be in most cases a better choice than a CM-2 running the parallel algorithms. This is due both to technological aspects of the CM-2 and to the fact that the problem cannot be easily framed into a SIMD formulation. An alternative TRG construction algorithm, exploiting in a more integrated, synergistic fashion parallel transition firing and RS update, is now being investigated.

The implementation of parallel TRG construction is fully integrated into the GreatSPN package: the user can select between the sequential or the parallel solution according to the complexity of the problem.

References

1. M. Ajmone Marsan, G. Balbo, G. Chiola, and G. Conte. Applicability of stochastic Petri nets to performance modeling. In G. Iazeolla, P.J. Courtois, and O.J. Boxma, editors, *Proc. 2nd Int. Workshop Applied Mathematics & Performance/Reliability Models of Computer/Communication Systems*, Roma, Italy, May 1987. North-Holland.

2. M. Ajmone Marsan, G. Balbo, and G. Conte. A class of generalized stochastic Petri nets for the performance analysis of multiprocessor systems. *ACM Transactions on Computer Systems*, 2(1), May 1984.

3. S.G. Akl. *The Design and Analysis of Parallel Algorithms*. Prentice-Hall, Englewood Cliffs, NJ, 1989.

4. F. Baccelli and M. Canales. Parallel simulation of stochastic Petri nets using recursive equations. Technical Report 1520, INRIA, Sophia-Antipolis, France, September 1991.

5. G. Balbo, G. Chiola, G. Franceschinis, and G. Molinar Roet. On the efficient construction of the tangible reachability graph of generalized stochastic Petri nets. In *Proc. Int. Workshop on Petri Nets and Performance Models*, Madison, WI, USA, August 1987. IEEE-CS Press.

6. F. Bonardi. Tecniche parallele di analisi di reti di Petri stocastiche: sparo delle transizioni abilitate ed evoluzione delle marcature. Master's thesis, Univ. di Parma, July 1993.

7. F. Bonardi, M. Caselli, S. Caselli, and G. Conte. A parallel approach to stochastic Petri net analysis. In *Proc. Parma-CM Users Meeting*, Dip. di Ingegneria dell'Informazione, University of Parma, Italy, July 1993.

8. P. Buchholz. A hierarchical view of GCSPNs and its impact on qualitative and quantitative analysis. *Journal of Parallel and Distributed Computing*, 15(3):207–224, 1992.

9. M. Caselli. Tecniche parallele di analisi di reti di Petri stocastiche: costruzione del grafo di raggiungibilita'. Master's thesis, Univ. di Parma, March 1993.

10. S. Caselli and G. Conte. GSPN models of concurrent architectures with mesh topology. In *Proc. Int. Workshop on Petri Nets and Performance Models*, Melbourne, Australia, December 1991. IEEE-CS Press.

11. G. Chiola. *GreatSPN* 1.5 software architecture. In *Proc. 5th Int. Conf. Modeling Techniques and Tools for Computer Performance Evaluation*, Torino, Italy, February 1991.

12. G. Chiola. A simulation framework for timed and stochastic Petri nets. *International Journal of Computer Simulation*, 1991. Special Issue on Simulation of Multiple Processor Networks.

13. M. Fontanesi. Generazione del grafo di raggiungibilita' di reti di Petri stocastiche generalizzate: Confronto fra algoritmi sequenziale e parallelo. Working Paper DII-CE-WP003-93, Dip. di Ingegneria dell'Informazione, Univ. di Parma, Parma, Italy, September 1993.

14. M. Fontanesi. Modelli a reti di Petri di sistemi di produzione complessi. Working Paper DII-CE-WP002-93, Dip. di Ingegneria dell'Informazione, Univ. di Parma, Parma, Italy, July 1993.

15. M. K. Molloy. Structurally bounded stochastic Petri nets. In *Proc. Int. Workshop on Petri Nets and Performance Models*, Madison, Wisconsin, August 1997. IEEE-CS Press.

16. T. Murata. Petri nets: properties, analysis, and applications. *Proceedings of the IEEE*, 77(4):541–580, April 1989.

17. TMC. The CM-2 technical summary. Technical report, Thinking Machine Corporation, Cambridge, MA, 1991.

PAPS - The Parallel Program Performance Prediction Toolset *

H. Wabnig, G. Haring

University of Vienna
Institute of Applied Computer Science and Information Systems
Lenaugasse 2/8, A-1080 Vienna, Austria
Tel: +43 1 408 63 66 10, Fax: +43 1 408 04 50
e-mail: wabnig@ani.univie.ac.at, haring@ani.univie.ac.at

Abstract. This paper describes the PAPS parallel program performance prediction toolset which is currently based on an initial set of Petri net driven performance prediction tools. Flexibility, extendability and automation were the general goals in the design of the toolset. Parallel systems are described by acyclic task graphs representing typical workloads of parallel programs, by processor graphs describing MIMD distributed memory multiprocessor hardware, and a mapping function. The three specifications, program, resource and mapping, are separated to be able to change them independently from each other. Methods to make this separation feasible are described. The PAPS toolset is logically divided into four layers: The specification layer, containing tools for writing scalable, mapping independent specifications, the transformation layer, consisting of tools which generate performance models, the evaluation layer, that contains performance analysis tools, and the presentation layer, which consists of tools for the visualization and depiction of performance evaluation results.

1 Introduction

Multiprocessor computer systems have become a working alternative to conventional sequential computer hardware. The main advantage of the new computer architecture is the high, hopefully scalable system performance which can be provided at relatively low costs. Unfortunately the totally new hardware architectures demand new program development methodologies and software engineering considerations. In contrast to the sequential software development area where performance considerations are very often ignored, there exists a great demand for performance engineering of parallel computer programs. In almost every case parallel program development without performance engineering will lead to inefficient solutions. It must be kept in mind that a parallel program has to be suited to the system architecture. The parallel system characteristics have to be obeyed right from the beginning to write parallel programs that

* This work was funded by the Austrian Science Foundation (research grant S5303-PHY).

offer good performance on a particular architecture. But also load balancing, data-partitioning etc. are influencing the performance of the parallel system and hence should be taken into account during program development.

In addition to performance issues, the software engineering methods known from CASE tools for sequential software engineering should also be applied in the parallel software development area. Research in this direction has been extended over the last years. One approach is the CAPSE environment (Computer Aided Parallel Software Engineering) [5] which is under development at the University of Vienna. It defines which performance relevant activities have to be included into the traditional software lifecycle and how a complete set of tools supporting the whole software development process should look like. Other integrated development environments for parallel architectures are PIE from Carnegie Mellon University [24], Poker from University of Washington [26], PPSE from the Oregon Advanced Computing Institute [18], FAUST from the University of Illinois [11], IMSE [23], Tangram from UCLA Computer Science Department [10] and Paragon from Cornell University [2]. A good overview on performance tools rather than complete environments has been worked out by Nichols [17], a summary of existing design tools by Oman [19] and a summary of parallel processing tools by Harrison [20].

Our toolset is dedicated to MIMD computers with distributed memory. That means that accesses to data not stored in local memory have to be implemented by passing messages between processors. Because communication costs in current multiprocessor systems are high compared to the computational power they offer, the communication structure of the parallel program is an important performance factor. Another performance criteria are precedence relations between blocks of computation. Hence these aspects have to be described in the program specification. Directed acyclic task graphs [9] are the basis for the program specification. Task graph nodes are used to represent computational demands and arcs represent logical data flow between nodes. The hardware is described by a directed processor graph. Nodes in the processor graph represent processing elements and directed arcs communication facilities that allow to transfer messages from one processor to another processor. A mapping of task graph nodes onto processor nodes is required. The performance model generator reads the program, resource, and mapping information and automatically constructs a Petri net performance model out of it. Different Petri net performance models can be generated for various communication and computation models. Petri net simulation is used to evaluate the performance models. PICL (Portable Instrumented Communication Library [7] [8]) tracefiles can be generated during the simulation. They can be analyzed by the visualization tool ParaGraph [12] [13]. ParaGraph has a multitude of graphical displays giving insight into the execution of parallel programs.

Various different performance investigations, both relative and absolute ones, can be performed with the toolset. On the one hand side, it is possible to compare different parallel machines, different mappings and different parallel workloads without great efforts. Such investigations do not rely on accurate absolute execution times. In early program development phases where the execution time of program parts cannot be accurately determined, rough estimates have to be specified. Of course in that case the absolute execution time will not be meaningful, but, for example, a comparision of two different mappings of the hypothetical workload can be of interest. On the other hand, if the execution times of sequential program parts can be measured by benchmark programs, it is possible to determine meaningful absolute execution time predictions of a given parallel program that is mapped onto a specified parallel hardware. The optimum number of processors to be used when executing the parallel program depending on the problem size can be evaluated by generating execution time diagrams for different problem sizes. This information can then be used for scheduling purposes.

The paper is organized as follows. Section 2 describes the general guidelines, characteristics and goals of the toolset. Section 3 describes how parallel systems are specified and discusses some problems arising from the separation of the program, resource and mapping specifications. Section 4 describes the current structure of the PAPS toolset more detailed.

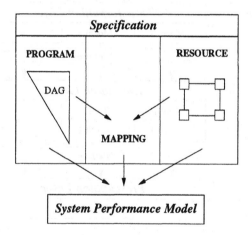

Fig. 1. The PRM methodology.

2 Characteristics and Goals of the Toolset

There are three major guidelines influencing the design of the toolset: Flexibility, extendability and automation.

Flexibility is mainly reflected by the PRM (*P*rogram, *R*esource, *M*apping) methodology [4] (see figure 1). The main idea of the PRM methodology is the separation of the three input specifications: program, resources, mapping. Due to the separation it should be possible to vary one of them without having to adapt the other specifications. That makes it possible to compare the performance of programs on different hardware architectures or with different mappings very quickly and efficiently. Flexibility is also reflected by the layer structure of the toolset (see figure 2). The transformation layer allows to generate different performance models according to the choosen methodology based on the same parallel system specification. The user is relieved from that task and can always work with the same specification environment.

Another characteristic is *extendability*. The toolset is open for including new methods and tools within the transformation, the evaluation, and the presentation layer. This is possible due to the modular structure of the toolset.

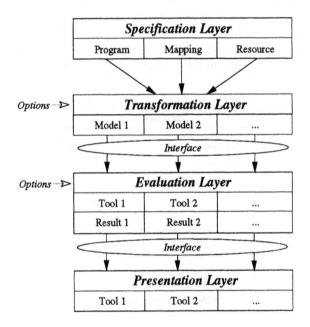

Fig. 2. The layer methodology.

There are four different layers:

- *The Specification Layer.* The current version of the toolset is based on textual input but a graphical scalable specification interface is under development.
- *The Transformation Layer.* The information derived from the specification layer is used to generate performance models. Various types of performance models can be generated (e.g. queueing networks, Petri net, deterministic models). In the current status of the toolset, Petri net performance models are generated.

- *The Evaluation Layer.* Performance analysis and prediction tools are contained in the evaluation layer. They can support the analysis of different kinds of performance models or realize different analysis methods. Appropriate performance models that can be evaluated by the particular tools have to be generated by the transformation layer. Interfaces might be necessary to convert data formats. Each tool can produce its own types of results. Some may be applicable for visualization and others for statistical purposes.
- *The Presentation Layer.* Sometimes it is necessary to preprocess the results before they can be accessed by a presentation tool. This adjustment is performed by the interface to the evaluation layer. The tools in the presentation layer have to provide the user with performance information about the system under consideration.

The third characteristic of the performance prediction toolset is *automation*. It is possible to automatically generate performance models out of the parallel system specifications. If the generation of the performance models would have to be done by hand, the reponsible person would have to have a deep knowledge of both, the performance evaluation methods themselves and the process transforming the specification into a performance model. Furthermore, Petri net performance models can become very huge. For example, Petri net performance models for bigger system specifications can consist of thousands of transitions and places. The generation of such huge performance models is very time consuming and error prone or even impossible if it is done by hand.

Different types of communication characteristics of the system, e.g. wormhole routing or store & forward communication networks, result in performance models of different complexity. The same is true when computation characteristics are considered, e.g. round robin or FIFO scheduling on the processor node. Performance investigations often require performance models for different problem sizes. All these arguments make it clear that the transformation of the system specification to an actual performance model has to be done automatically.

The goal of the toolset is to efficiently support the making of performance evaluation studies that experiment with the following specification elements:

- *Mapping.* Due to the PRM methodology the mapping information can be changed independently from the other input specifications. Different mappings can be compared to find the one that is most appropriate for the particular application.
- *Hardware Topology and Resource Characteristics.* It is possible to compare different systems by changing the following resource characteristics: The interconnection topology of the processing elements, different performance parameters for processors and communication hardware elements, the type of the communication network, or the scheduling strategy on a single computation node.
- *Implementation of Algorithms.* The system allows to compare different implementations of a parallel algorithm. For example the usage of synchronous and asynchronous communication primitives can be compared by adding ac-

knowledgement messages to the communication elements modelled in the program specification.

To demonstrate the applicability of the toolset a detailed performance study has been done, which is reported in [29]. In this study, for a parallel version of the Gaussian elimination algorithm, among other topics, the optimum number of processing elements depending on the problem size (i.e. the matrix dimension to compute) has been determined for various mapping strategies and various hypothetical hardware speed characteristics. See also [27] and [28] for performance studies based on an existing parallel system.

3   The Specification Layer

The specification layer consists of the program specification, the resource specification, and the mapping specification. At the moment all of them have to be specified textually. A description language has been developed to simplify the specification task. The language mainly consists of syntax constructs to define *model elements* and *connections* between them. An element is a task graph node, a processor node, or an unidirectional communication facility. A connection is an arc in a task graph, an arc in a processor graph, or an arc binding a node in the task graph to a processing node (i.e. mapping). The definition of a model element can either be done by specifying the name of the new element together with an element-dependent set of parameters or by assigning its characteristics to those of an already existing element. In the second case it is possible to specify multiple new element names and all of them are created with parameter values equal to the specified sample element. This is useful, for example, in resource specifications of homogeneous networks consisting of processors and communication facilities with equal parameters. The definition of a connection between model elements demands that the corresponding model elements are already defined. In the case that the connection is a task graph arc, also a parameter that specifies the size of the data that will be transferred across the arc has to be given. Additional language constructs are used to specify that a task graph arc is part of a global communication or that identical data is also sent across other arcs (see section 3.4). In the case of a mapping connection one or more task graph nodes have to be specified beside the processor node on which the task graph nodes are to be executed. Special language constructs are also available to describe routing paths. To define a routing path between two processor nodes a sequence of communication facilities has to be specified.

Although the description language simplifies the specification process, the specification by means of a textual description language is cumbersome and error prone if the models get bigger and more complex. For descriptions with thousands of task graph nodes and complex topologies with dozens of processors (and hundreds of routing paths between processor pairs), the typing of textual descriptions is also an enormous time consuming task. Usually, such big models do not consist of unstructured collections of model elements and connections.

In contrary, they are almost always very regularly structured. In that case it is possible to describe the elements and connections of a model not only on the basis of the simple description language. Increasing the expressiveness of the language by adding control-flow statements makes it possible to describe specifications by their regular structure rather than by single model elements or connections. This leads to the concept of *scalable specifications*, as it is realized in our tool. A scalable specification is a compact system description, which is parametrized by constant values, that specify, for example the problem size for the program specification generation or the number of processing nodes for the architecture specification. Scalable specifications can be used for all types of specifications. Using scalable specifications to describe a certain parallel system enables the quick generation of sets of specifications for different constant values. For example, it is very simple to make performance investigations with changing problem sizes or changing numbers of processors if scalable specifications are available.

3.1 Program Specification

The program description file contains the textual information about the task graph describing an actual workload of the program under investigation. Acyclic task graphs are an appropriate deadlock-free description for program workloads because task graphs explicitly model the communication and precedence structure of a workload but hide all the information about sequential program code within task graph nodes.

Task graphs consist of nodes and arcs. Task graph arcs represent precedence information between nodes. A node represents a block of sequential computation which is not interrupted by communication to other nodes. Hence communication is only possible at the end of a node. Nodes are activated if messages have been received from all input nodes. If a node has been executed, messages are sent to all output nodes. Logical data flows between nodes can be represented by arcs. It is important to recognise that arcs in the program description models program interdependencies in a general way. The decision whether a logical data flow is a real data flow is delayed until the mapping of task graph nodes onto processor nodes has been performed. Task graph arcs model asynchronous communication. If synchronous communication is required then acknowledgement messages have to be explicitly modelled within the program specification.

Besides the topological structure of the workload a set of parameters has to be specified. For each node the computation demand must be given. This can be done by specifying the absolute time to execute the corresponding sequential program or by specifying the number of scalar instructions and floating point instructions that are required to perform the computation. Dummy nodes are task graph nodes that have a computational demand of zero. They are used to represent model properties, for example the modelling of high level communications (see section 3.4). Because they do not require any computational power they cannot be delayed due to resource conflicts.

For an arc the communication demand in bytes and special properties of the arc have to be specified. Arc properties can be: The arc is a member of a bundle of arcs that represent a global communication or the arc is a member of a bundle of arcs that represents flows of identical data (see section 3.4).

3.2 Resource Specification

The resource specification is separated into two parts. One part describes the hardware topology which consists of the processing elements and the communication facilities. The other part defines system software properties. In the current version of the toolset the software is only described by the way routing paths are choosen to communicate between distant processors (i.e. processors that are not directly connected by a communication facility).

Hardware Description. The hardware description defines the processor topology and the speed of the processor nodes which are parametrized by instruction rates. A distinction between scalar instructions and floating point instructions can be made.

Communication facilities are characterized by the following parameters:

- *Startup time: T_s.*
- *Physical transfer rate: T_r.*
- *Constant overhead per communication hop: O_s.*
- *Message size dependent overhead per communication hop: O_l.*

The communication facility parameters are interpreted according to the specified communication network type (see section 4.1).

Software Description. In our tool, routing paths can be automatically generated by the Petri net performance model generator applying a shortest path algorithm. If there exists a sequence of communication facilities that connect two processor nodes then the shortest path is determined and stored according to the language syntax in the routing description file. If there already exists a routing description file for the topology then it is used in all subsequent model generations. This is important for topologies that consist of dozens or more processors because the computational effort to calculate shortest paths for all pairs of defined processors often can not be ignored. Furthermore in special topologies it is often predefined how messages are routed. Even if they are routed by a shortest path algorithm it can not be guaranteed that the automatic routing path generator will generate the right one because in the case that there is more than one shortest routing path, one is randomly selected. Such routing schemes are almost always regularly structured and hence the routing description file can be generated by a scalable specification.

3.3 Mapping Specification

In the mapping specification task graph nodes are assigned to processor nodes. This assignment either transforms logical data flow (see section 3.1) into physical communication or resolves a logical data flow in a pure precedence relation. Hence different mappings generate different communication patterns without requiring changes in the program specification.

3.4 Problems Arising from the PRM Methodology

The PRM methodology recommends a complete separation of the program, the resource and the mapping description. That should enable to change one of them without the necessity to adjust the others. But the separation can not be achieved completely. For example, each change of the number of processors can demand adaptions in the mapping description because processors may not be defined any more. By specifying general scalable mapping strategies, especially in the context of predefined topologies, this problem gets less serious, because the mapping description is automatically adjusted according to the number of specified processing nodes.

Another problem during model generation arises if multiple task graph nodes receive the same data from a single sender. This would be represented in the program description by connecting a single sender node with multiple receiver nodes. If each of the arcs are simply parametrized with the size of the data, problems can occur when the nodes are mapped onto processors. In general, a logical data flow between two task graph nodes mapped onto different processors is translated into a physical communication. If the program specification does not reflect, that on each of the corresponding arcs identical data is transferred, it would be possible that physical communications of the same data to the same processor are introduced into the performance model when mapping the receiver nodes onto the same processor. But physical communication of identical data to the same processor is redundant because sending the data once and duplicating it locally on the processor is much more efficient. To be able to automatically prevent these redundant physical communications it is necessary to explicitly specify that identical data is transferred across multiple arcs in the task graph program specification.

Furtheron, several programming languages and operating systems provide high level communication commands which are optimized for the particular interconnection topology of the parallel hardware. For example a tree communication structure can be used to perform a global broadcast [16]. The tree broadcast communication can be performed in less communication steps than simple sending messages to all processors in the topology. To model the high level communications in the right way it is necessary to translate them into usual low level communications that can be represented by ordinary task graph arcs.

In both cases additional (dummy) nodes are necessary to represent the adapted communication structure in the task graph model appropriately. If the topology of the system is changed then the mapping of the dummy nodes representing, for example a tree communication structure has to be adapted accordingly.

It is very cumbersome to explicitly describe each high level communication in terms of task graph arcs. Furthermore, if the communication pattern for high level communications is changed, the program specification has to be adapted. Hence a change in the resource characteristics directly affects the program and the mapping description.

The specification dependencies caused by identical data and high level communications can be automatically resolved (see section 4, mapping preprocessor). Hence different mappings and high level communication models do not affect the other descriptions and the PRM concept is fulfilled.

3.5 An Example for the Specification of a Simple Parallel System

To summarize the specification layer, a simple example is elaborated to demonstrate the specification process using the simple description language described at the beginning of this section 3. Figure 3 (a) shows a task graph representing a program that first has to execute a computation block (called P0), and then sends identical data (labelled x) to blocks P1 to P3 that can be executed independently from each other. Figure 3 (c) shows a processor graph for a line topology that consists of four processors. Dotted arcs describe the mapping function of task graph nodes to processor nodes.

Assume that the data size of x is 10 bytes and the computational demand of all processing nodes is 300 scalar instructions and zero floating point instructions.

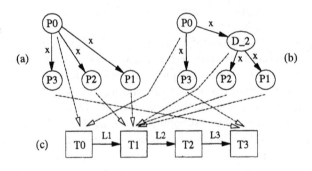

Fig. 3. A simple example for a parallel system specification.

The program description is as follows:

```
P0[inst(300)flop(0)]; P1 P2 P3 = P0;
x = 10;
P0 -> [x] P1; P0 -> [x] P2; P0 -> [x] P3;
```

Assume that the computational power of the processors in the topology is ten MIPS and two Megaflops and the parameters for the communication facilities are

25 microseconds startup time, about 12 million bits per second physical transfer rate, 23 microseconds constant overhead per communication hop and about 0.018 microseconds per bit message size dependent overhead per communication hop.

The resource description is as follows:

```
T0 [mips(10) mflops(2)];
T1 T2 T3 = T1;

L1 [startup(25 usec) rate(12.6229 mbps) overhead(23 usec 0.01815 uspb)];
L2 L3 = L1;

T0 -> L1 -> T1; T1 -> L2 -> T2; T2 -> L3 -> T3;
```

The automatically generated routing description is:

```
T0->T1 = L1;
T0->T2 = L1 L2;
T0->T3 = L1 L2 L3;
T1->T2 = L2;
T1->T3 = L2 L3;
T2->T3 = L3;
```

Assume the mapping function described in figure 3. The mapping description for such an assignment of task graph nodes to processor nodes is:

```
P0 -> T0; P1 P2 -> T1; P3 -> T3;
```

If the task graph description is taken directly as it is defined in figure 3 (a) then redundant communication is modelled because the same data (x) is transferred to the nodes P1 and P2 and both are mapped onto the same processor. In the mapping preprocessing step the redundant communication is removed by changing the program description according to the mapping information. Figure 3 (b) shows the resulting task graph. An additional task graph node, D_2, is required. It is mapped to processor T1 and locally "distributes" data item x.

4 The Current Structure of the Performance Prediction Toolset

The current structure of the toolset is described in figure 4. Ellipses represent integrated tools and squares describe results from and inputs to the tools. The user has to define a parallel system by providing mapping independent, scalable specifications for program, resource and mapping. The task graph viewer tool can be used to visualize the textual program description. The TxT-tool generates full size representations of scalable specifications and the mapping preprocessor resolves mapping dependencies. It is possible to skip the TxT- and/or the mapping preprocessor if the specification given by the user does not include scalable or mapping independent constructs (reflected by dotted arrows with empty arrow heads in figure 4). The performance model generator constructs Petri net performance models out of the specification either in a syntax that is suitable for the GreatSPN [3] tool or a syntax appropriate for the special Petri net simulator tool described in section 4.2. The Petri net performance model can be evaluated by the special Petri net simulator tool. Either statistical performance data for the whole simulation run or a tracefile is generated that contains detailed information about the hypothetical execution of the parallel system. The tracefile can be visualized by the ParaGraph visualization tool.

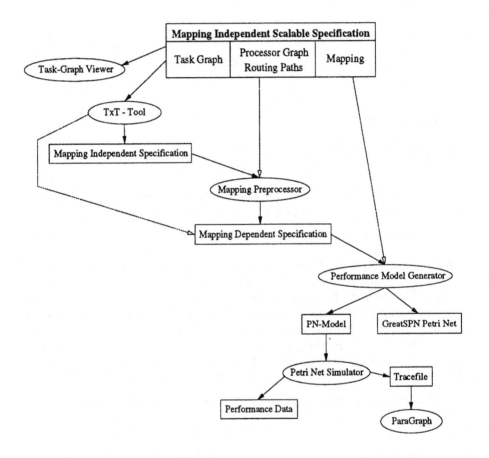

Fig. 4. The toolset structure.

The toolset consists of the following tools:

- *TxT-Tool:* The TxT-tool is used to convert scalable specifications (see section 3) into full text descriptions. The TxT-language is a procedural programming language that contains loops and conditionals. It is dedicated to the generation of textual output.

– *Task Graph Viewer:* The task graph viewer is a Xlib based tool for the visualization of the task graph structure of the program specification. At the moment the program specification has to be done textually. This is a time consuming and error prone task. The task graph viewer helps to debug program specifications. The viewer automatically generates a "good" layout, which can be adjusted by command line options. The layout can also be edited manually by using the mouse to move nodes to new positions. The arcs are adjusted automatically. Zooming, moving around, switching on/off labelling and other features are included. If a mapping is specified, the shape of the arc depicts whether the arc represents physical communication between processors (the nodes are mapped to different processors; the arc is drawn solid) or local communication within a processor (the nodes are mapped onto the same processor; a dotted arc is drawn).

The (edited) workload model can then be stored in PostScript format (encapsulated PostScript, including bounding boxes; this eases the task to import the picture into other commercial program packages). The picture is scaled to fit into A4 format.

Figure 5 shows an example for a task graph picture that has been generated by the task graph viewer. It is a Gaussian elimination task graph for matrix dimension 8 mapped onto a topology consisting of 8 processors.

– *Mapping Preprocessor for Resolving Mapping Dependencies:* It is not always possible to describe the program, resource, and mapping specifications completely independent from each other. Such cases have been explained in section 3.4. The mapping preprocessor automatically adaptes the task graph model to appropriately represent the desired model behaviour.

– *Performance Model Generator:* The performance model generator reads the input specifications and generates a Petri net performance model according to the specified system options (e.g. communication network type). Information about performance relevant elements in the Petri net are gathered during the construction process of the performance model. This information is required by the special Petri net simulator tool to produce correct and significant simulation results that can be interpreted on the level of the specification elements.

– *Special Petri Net Simulator:* The special Petri net simulator takes the information about the shape of the Petri net performance model and generates a special internal representation of the Petri net for simulation. During the simulation of the Petri net, information about performance relevant events are collected in a tracefile. The simulator also generates a report about the simulation run which includes total execution time, utilizations, link conflict delays etc.

– *Visualization Tool ParaGraph + Extensions:* The tracefile generated by the simulator contains information that is required to visualize the hypothetical execution of the program on the specified hardware under the given mapping. Several new displays have been added to ParaGraph to overcome the problems arising from the fact that ParaGraph assumes a system model

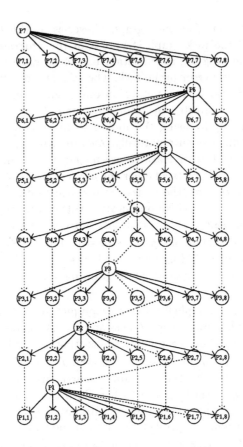

Fig. 5. An example of a task graph picture generated by the task graph viewer.

where there are no parallel processes executed in a time sliced fashion on a single node:

- The *utilization count display* visualizes the total number of processors that are busy at any instance of time.
- The *link delay display* shows the communication delay times caused by link conflicts, i.e. messages are waiting for communication links already in use by another message. The display can help to identify critical communication phases.
- The *link usage display* (see figure 6 (a)) shows the number of messages that are waiting to be transferred over a certain link or the number of waiting messages for the whole communication network. The link usage and the link delay display can be used to identify bottlenecks in the communication scheme of the application.
- The *active processes display* draws the number of active task nodes in the

task graph either for a certain processor or summarized for all processors. If task graph nodes represent real processes the number of parallel processes on a single processor indicates how big the overhead for the time slicing mechanism will be.

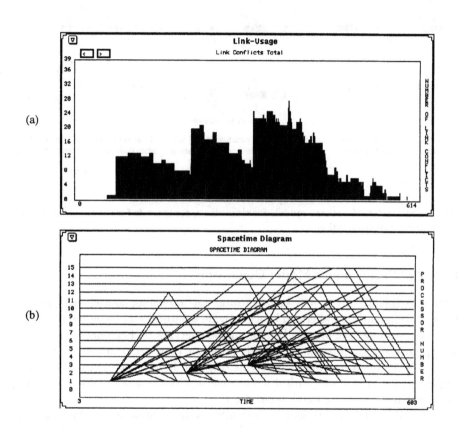

Fig. 6. Two sample ParaGraph displays.

A predefined display within ParaGraph is, for example, the *Spacetime Diagram* (see figure 6 (b)). The origin of a non horizontal line in the spacetime diagram marks a sending processor and the corresponding message sending time, and the destination of the line the receiving processor and the corresponding message arrival time.

Detailed information about all ParaGraph displays and the general handling of the ParaGraph tool is reported in [13].

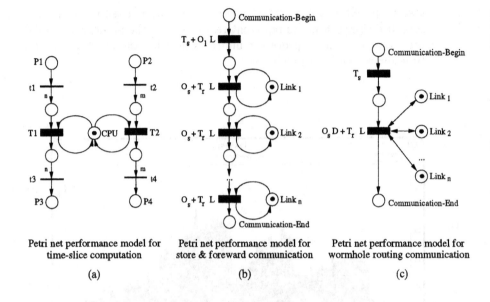

Petri net performance model for time-slice computation	Petri net performance model for store & foreward communication	Petri net performance model for wormhole routing communication
(a)	(b)	(c)

Fig. 7. Petri net performance models for time-slices, for store & forward and for wormhole routing communication.

4.1 The Performance Model Generator

Petri nets [21, 22] are a well known means for modelling concurrent systems. They are inherently parallel and hence provide a natural way to represent multicomputer systems. See Murata [15] for a good overview on Petri net terminology and application. By introducing a timing concept Petri nets can be used to evaluate the performance of parallel systems [25] [14]. In our Petri net performance models we use transitions with discrete deterministic time extensions to represent computation or communication efforts.

The performance model generator combines the information in the program, resource, and mapping description and builds a timed Petri net based performance model out of it. The Petri net performance model is constructed by plugging together small basic Petri net blocks. A basic Petri net block mainly consists of a deterministic timed transition that can require a resource token for execution. The transition represents a communication or computation demand, or a communication startup time. For example, a store & forward communication requires multiple communication substeps. Each substep is represented by a basic Petri net block that requires one link. The whole communication is then represented by a composite Petri net block that consists of all of them, plus a basic Petri net block representing the startup time (see figure 7 (b)). All task graph nodes and arcs specified in the program description are translated into basic or composite Petri net blocks and are connected according to the task graph structure.

An example of a basic Petri net block representing a task graph node applying time slice behaviour is shown in the left half of figure 7 (a). If a token arrives at place P1 the immediate transition t1 can be fired which generates n tokens in the output place. Hence the deterministic transtion T1 is then fired n times. Afterwards the immediate transition t3 reduces all n tokens from its input place to a single token in place P3. For each resource defined in the resource specification a resource place containing a resource token that represents the availability of the resource is defined in the performance model. Each firing of the timed transition T1 requires the resource token in the place, labelled CPU [6]. In general, if an element in the program specification needs a resource token to perform a task then the corresponding resource token has to be available. The right hand side of figure 7 (a) represents a second instance of a Petri net for a task graph node. The timed transition T2 represents the execution of the computation contained in the corresponding task graph node. The firing delay of the transitions T1 and T2 have to be equal to simulate a balanced sharing of the CPU resource between both transitions. The firing delay time is then called the common time basis of the task graph nodes. Because the selection of transitions that are enabled at the same time is done randomly, in the long term time slice behaviour is appropriately reflected. Transition T2 is fired m times. The actual values for n and m are determined by the absolute execution time of the corresponding task graph nodes divided by the common time basis. The accuracy of the Petri net representation is determined by the firing delay of transition T1 and T2. The smaller the common firing delay is, the greater gets the accuracy of the model, because the values of n and m grow. That improves on the one hand side the time slice effect and on the other hand side it reduces errors in the modelling of the absolute execution time of the task graph nodes, because only execution times that are multiples of the common firing delay time can be generated.[2] But the simulation effort increases because more transition firings are required to execute a task graph node. Hence a tradeoff between accuracy and simulation effort has to be found.

All task graph nodes that can potentially be executed in parallel define sets of potentially parallel tasks. For all nodes in a set of potentially parallel tasks a common time basis has to be defined if time slice execution of computation is specified. The common time basis is used as the basic firing delay of the timed transition that represents execution of node computations, e.g. transition T1 in figure 7 (a). Because the smallest execution time of a node in a set of potentially parallel nodes determines the value of the time basis, it is possible that a very huge number of transition firings is necessary to simulate the execution of a single task graph node. But the structure of the program task graph often allows optimizations of the performance model. The performance model generator has integrated routines to detect minimum sets of potentially parallel task graph nodes. The knowledge of such minimum sets can reduce the number of needed

---

[2] Of course, if the Petri net for task graph nodes is extended by an additional timed transition, which represents the lacking node execution time, then no such error can occur - but, the Petri net gets more complex.

transition firings substantially. Unfortunately the detection of such sets gets very time and memory consuming for task graphs consisting of more than about fifty nodes. But simple optimations of the Petri net, e.g. removing Petri net elements that do not affect performance results, can be applied without restrictions.

Communication can be handled by the tool in several different ways: The first possibility is to model the communication by a *store & forward* scheme. Whole messages are moved from one processor to the next on the predefined route until the destination processor is reached. Figure 7 (b) shows a Petri net performance model for store & forward communication. It is also described how the firing delays are calculated for all timed transitions based on the parameters for communication facilities explained in section 3.2. The second possibility is called *packet switching*. Messages are cut into small packets of fixed size and sent along the routing path as distinct messages that are reassembled at the destination node to the original message. In this kind of modelling, if there is no load in the network, after a filling phase of the communication path, each time one packet is sent, one packet reaches the destination node independently of the number of communication substeps necessary (similar to pipelining effect in computer architectures). The third possiblity to model communications is known as *wormhole routing*. In wormhole routing a path between the source and the destination processor is exclusively assigned to a message transfer. Hence there are no resource conflicts after the header of the message has reached the destination processor. Messages are split in small packets of fixed size. Similar to the packet switching networks this makes communication delays almost independent of the number of communication hops needed to perform the data transfer. The Petri net model for wormhole routing and the way the communication facility parameters are used in this kind of model, are shown in figure 7 (c).

For detailed information on Petri net models that are generated for different system characteristics see [27].

4.2 The Special Petri Net Simulator

The special Petri net simulator is dedicated to the simulation of timed Petri nets of the special shape generated by the performance model generator. The simulator is optimized for the constructs in this special kind of Petri nets. It is not possible to simulate general Petri nets with this simulator tool.

Simulation speed depends on the complexity of the Petri net. Transition firing rates up to 8000 transitions per second can be achieved on a Sparc ELC workstation. But the rate decreases for big and complex Petri nets down to about ten transition firings per second.

The special simulator generates two different kinds of results: The first kind is a tracefile in PICL syntax. Command line options determine which kind of events are recorded in the tracefile. Some ParaGraph displays rely on certain event information. Hence, if the information is not generated during the simulation it is not possible to use the displays in ParaGraph. On the other side, for big models a tracefile can become very huge if the accuracy of the gathered information is too big. The concept of options allows to analyze different performance aspects in different simulation runs. This reduces the size of the generated tracefiles.

Some examples for events that can be gathered in the tracefile are: Events which represent the begin and end of busy times of processors, events which mark begin and end of a communication, events which record the number of messages waiting for a certain link, and events which record the delay times due to such link conflicts.

The results generated by the Petri nets simulation are not deterministic if multiple enabled timed transitions occur during the simulation because the firing rules for Petri nets do not define which transition has to be taken first in such a case. But the choice influences the actual enabling time of transitions later in the simulation. Therefore, multiple simulation runs can be necessary to determine an average execution behaviour.

The second performance result is a text file, generated by the simulator. It contains information about each processor in the topology, e.g. utilization or the number of messages and bytes sent and received, and the total execution time for all communications, for all computations and for the parallel program as a whole. The text files containing the performance results can be used for statistical evaluation. Spreadsheet programs support these evaluations. For example, by using the macro definitions in the spreadsheet program Microsoft Excel [1], it is easy to generate graphical diagrams to visualize utilization depending on the problem size.

5   Conclusions and Future Work

This paper has presented the architecture of a performance prediction toolset based on Petri net performance models. The way parallel systems are specified has been described. The structure of the tools, their functionality and usage has been discussed.

The current toolset consists of several tools which can be assigned to different layers. The specification layer contains the TxT-tool, the mapping preprocessor and the task graph viewer. At the moment all system specifications have to be done by a textual language but the design and implementation of a graphical user interface supporting scalable hierarchical specifications is under development. The transformation layer contains the performance model generator tool. The evaluation layer consists of the special simulator tool. The presentation layer is built on top of the ParaGraph tool.

There are problems with the Petri net approach. For huge task graph models (more than 10000 nodes) the generation of the performance model fails due to memory restrictions. Even if the performance model can be generated, the time needed to simulate the performance model can be too long and hence evaluation is not possible. Faster tools based on other evaluation methods and performance models have to be integrated into the toolset to evaluate big parallel system specifications. A discrete event simulator is currently under development.

Another direction for future work is the extension of the expressiveness of the task graph model with conditional nodes. Conditional nodes can define boolean expressions among their input and output arcs, for example, allowing the node to be executed if it has only received messages from a subset of input nodes or generate only messages for a subset of output nodes.

References

1. Martin Althaus and Michael Ortlepp. *Das Excel 4 Buch*. SYBEX-Verlag GmbH, Düsseldorf, 1992.
2. Craig M. Chase, Alex L. Cheung, Anthony P. Reeves, and Mark R. Smith. Paragon: A parallel programming environment for scientific applications using communication structures. *Journal of Parallel and Distributed Computing*, 16:79–91, 1992.
3. G. Chiola. Greatspn1.5 software architecture. In *Proc. of the 5th Int. Conf. on Modelling Techniques and Tools for Computer Performance Evaluation. Torino, Italy, Feb. 13-15, 1991*, pages 117–132, 1991.
4. A. Ferscha. A petri net approach for performance oriented parallel program design. *Journal of Parallel and Distributed Computing*, (15):188–206, 1992.
5. A. Ferscha and G. Haring. On performance oriented environments for the development of parallel programs. *Kybernetika a Informatika, Proceedings of the 15th Symposium on Cybernetics and Informatics '91, April 3-5 1991, Smolenice Castle, ČSFR*, 4(1/2), 1991.
6. Alois Ferscha. Modellierung und leistungsanalyse paralleler systeme mit dem prm-netz modell. *Ph.D. Thesis, University of Vienna. To be published in OCG-Schriftenreihe, Oldenbourg Verlag*, 65, 1990.
7. G. A. Geist, M. T. Heath, B. W. Peyton, and P. H. Worley. Picl: A portable instrumented communication library. Technical Report ORNL/TM-11130, Oak Ridge National Laboratory, July 1990.
8. G. A. Geist, M. T. Heath, B. W. Peyton, and P. H. Worley. A users' guide to picl: a portable instrumented communication library. Technical Report ORNL/TM-11616, Oak Ridge National Laboratory, August 1990.
9. Erol Gelenbe. *Multiprocessor Performance, Series in Parallel Computing*. John Wiley & Sons Ltd., 1989.
10. Leana Golubchik, Gary D. Rozenblat, William C. Cheng, and Richard R. Muntz. The tangram modeling environment. In *Proc. of the 5th Int. Conf. on Modelling Techniques and Tools for Computer Performance Evaluation. Torino, Italy, Feb. 13-15, 1991*, pages 421–435, 1991.
11. V.A. Guarna Jr., D. Gannon, D. Jablonowski, A.D. Mallony, and Y. Gaur. Faust: An integrated environment for parallel programming. *IEEE Software*, 6(4), 1989.
12. Michael T. Heath and Jennifer A. Etheridge. Visualizing performance of parallel programs. Technical Report ORNL/TM-11813, Oak Ridge National Laboratory, May 1991.
13. Michael T. Heath and Jennifer A. Etheridge. Visualizing the performance of parallel programs. *IEEE Software*, 8(5):29–39, September 1991.
14. M. Molloy. Performance modeling using stochastic petri nets. *IEEE Trans. Comput.*, C-31:913–917, September 1982.
15. T. Murata. Petri nets: Properties, analysis and applications. *Proceedings of the IEEE*, 77(4):541–580, April 1989.

16. nCUBE. *nCUBE 2 Programmer's Guide*. nCUBE Corporation, 1990.

17. K. Nichols. Performance tools. *IEEE Software*, 7(3):21–30, May 1990.

18. OACIS. Parallel programming support environment research. Technical Report TR-PPSE-89-1, Oregon Advanced Computing Institute, 1989.

19. P. Oman. Case analysis and design tools. *IEEE Software*, 7(3):37–43, May 1990.

20. P. Oman. Tools for multiple cpu environments. *IEEE Software*, 7(3):45–51, May 1990.

21. C. A. Petri. Kommunikation mit automaten. *Bonn: Institut für Instrumentelle Mathematik, Schriften des IIM*, (3), 1962.

22. C. A. Petri. Communication with automata. Technical Report RADC-TR-65-377, vol.1, Suppl. 1, New York: Griffiss Air Force Base, 1966.

23. R. J. Pooley. The integrated modelling support environment, a new generation of performance modelling tools. In *Proc. of the 5th Int. Conf. on Modelling Techniques and Tools for Computer Performance Evaluation. Torino, Italy, Feb. 13-15, 1991*, pages 1–15, 1991.

24. Z. Segall and L. Rudolph. Pie: A programming and instrumentation environment for parallel programming. *IEEE Software*, 2:22–37, November 1985.

25. J. Sifakis. Petri nets for performance evaluation. In *Measuring, Modeling, and Evaluating Computer Systems*, pages 75–93. H. Beilner and E. Gelenbe, Eds. North-Holland, 1977.

26. L. Snyder and D. Socha. Poker on the cosmic cube: The first retargetable parallel programming language and environment. In *K. Hwang, S.M. Jacobs, E.E. Swartzlander (Editor): Proceedings of Int'l Conf. on Parallel Processing, IEEE Computer Society Press, Washington D.C.*, pages 628–635, August 1986.

27. H. Wabnig and G. Haring. Performance prediction of parallel systems with scalable specifications - methodology and case study (extended abstract). Accepted as Poster at the 1994 ACM SIGMETRICS Conference on Measurement and Modeling of Computer Systems, 16-20 May 1994, Nashville, USA, 1994.

28. H. Wabnig and G. Haring. Petri net performance models of parallel systems - methodology and case study. Submitted to PARLE'94 - Parallel Architectures and Languages Europe, 13-17 June 1994, Athens, Greece, 1994.

29. H. Wabnig, G. Kotsis, and G. Haring. Performance prediction of parallel programs. In *Proc. of the 7th GI/ITG Conference on Mearsurement, Modelling and Performance Evaluation of Computer Systems, 21-23 September 1993, Aachen, Germany*, pages 64–76. Springer Verlag, New York, 1993.

qcomp: A Tool for Assessing Online Transaction Processing Scalability

Neil J. Gunther

Pyramid Technology Corporation, San Jose, CA 95134
ngunther@pyramid.com

Abstract. We present a simple, semi-empirical, tool for gauging the possible throughput scaleup on shared-memory multiprocessors executing online transaction processing (OLTP) workloads. Unlike most scientific workloads, concurrent, multi-user, OLTP workloads exhibit a high degree of data sharing. The attendant sublinear scaling is found not to follow an Amdahl-like law due to simple seriality or single point of serialization. The concept of *super-seriality* is introduced to account for performance penalties due to compound serialization in OLTP workloads. Super-seriality has the effect of inducing a premature *maximum* in the scaling curve beyond which it is counter-productive to add further processors. This concept is incorporated into a performance tool called *qcomp* which has been used for both competitive benchmarking and system sizing.

1. Introduction

Ideally, a "scalable", multiprocessor architecture is one in which the cumulative efficiency [1] increases linearly with the number of processors (N). In reality, this ideal can only be approximated by computer systems running meaningful workloads. A more realistic bound on scalability is given by Amdahl's law [1] which states that any serial component in the workload, which could otherwise be executed in parallel on N processors, will cause the cumulative efficiency to grow only logarithmically with N.

Our interest, in this paper, is in modelling the throughput scalability of multiprocessor on-line transaction processing (OLTP) computer systems running commercially available relational database management system (RDBMS). We present a tool for providing realistic scalability bounds which are generally more pessimistic due to considerable resource sharing. As far as we are aware this is the first predictive tool of its type for RDBMS scalability.

Shared data and code abound in the currently available commercial RDBMS products. This use of sharing is driven by the desire for:

---

1. More formally, if S is the speedup at N processors and E = S/N the measure of efficiency, then the *cumulative efficiency*: $\sum_{k=1}^{N} E_k$, corresponds to the area under the efficiency curve. Throughout this paper we assume fixed processing power for each additional processor.

- A general purpose programming model in high-level database applications.

- Ease of production release and maintenance through the use of shared data structures.

Since the degree of sharing in RDBMS architectures is currently non-zero and is unlikely to improve in the near future, we must live with the constraint that linear scaling for transaction processing systems is as practically unattainable as it has been for massively parallel scientific workloads. Broadly speaking, database transaction workloads exhibit a higher degree of serialization than is the case for fine-grain parallelism. As we have discussed previously [2], Symmetric Multi-Processor (SMP) and Massively Parallel Processing (MPP) systems running workloads with a high degree of shared data can exhibit measureable retrograde performance [3], rather than just throughput saturation. The question then becomes: "What degree of sublinear scalability is acceptable for OLTP systems?"

From another vantage-point, sharing resources implies queueing behaviour. Typical queueing models of single bus MP systems comprise N delay centers representing the processors and a central server with fixed service rate representing the memory bus [4]. However, if we examine these queueing network models more closely we are faced with a discomforting paradox. Queueing models exhibit a plateauing in throughput scalability as the utilization of the bus service center reaches its saturation value [5]. Similarly, Amdahl scaling reaches a different saturation asymptote determined by the seriality parameter [6]. Multiprocessor systems running workloads with shared data, however, exhibit throughput degradation as a measureable effect. So, on the one hand we have the simplicity of an analytic model like Amdahl scaling which is overly optimistic while on the other hand, we have more sophisticated queueing models which are not consistent with typical scaling. We would like to maintain simple analyticity whilst including more realistic behaviour. How can we accomplish this goal?

We present a tool based on a model [2] that accounts for the increased amount of serialization due to the existence of shared data. As we shall see, the proposed model exhibits a in the scaling curve; the onset of which can occur long before the expected system saturation level! Moreover, since the models we present here are nowhere dependent on any assumptions about bus or interconnect technologies, they already draw attention to software inefficiencies.

The organization of the paper is as follows. In section 2 we develop the analytic scaling models. Section 3 describes a UNIX implementation of the scaling tool which incorporates the analytic models of section 2. In Section 4 we discuss the application of the tool which also provides validation of the models.

2. Fundamental Scaling Models

We study processor scalability from the point of view of the execution of a workload that is scaled to the number of available processors. DeWitt [7] refers to this as "scaleup" rather than "speedup".

2.1 Performance Metrics

Transaction scaleup is ideal for OLTP systems since each transaction is a relatively small, independent, task that can run on separate processors. It is for this reason that the TPC council has chosen a special set of scaling rules for database sizing as more processing power is applied to the workload. The capacity metric used in this paper is the throughput rate (X) measured in Transactions Per Second (TPS) and the workload is equivalent to that found in the TPC Benchmark B™ [7].

We compare the TPS rate achieved by N processors (X_N) relative to that achieved by a single processor (X_1). The transaction throughput scaleup is then given by the ratio: $\dfrac{X_N}{X_1}$. We assume that each processor is optimized for process concurrency i.e., enough transaction generation processes are present to utilize the available cycles on a processor and yet not cause a higher than necessary context-switch rate. Only one transaction generation process runs at a time on a single CPU. In general, the processor will not be doing any useful work if it must:

1. wait for a bus transfer from another processor or memory.

2. wait for I/O completion

3. serialize on an RDBMS lock or latch.

2.2 Dual-processor Scaleup

Suppose a uniprocessor completes t_1 transactions in an elapse time, τ. Its throughput is then given by:

$$X_1 = \frac{t_1}{\tau}.$$

Doubling the workload, at the same throughput, we would expect the uniprocessor to complete $2t_1$ transactions in twice the time, i.e., 2τ. Similarly, we would expect a dual processor (N = 2) system to complete $t_2 = 2t_1$ transactions in the same time it takes the uniprocessor to complete t_1 transactions.

Measurements on shared-memory multiprocessor systems show, however, that t_2 is generally less than $2t_1$ in time τ. In other words, the dual processor takes *longer* to complete $2t_1$ transactions! The two processors are said to "interfere" with each other's ability to execute independently.

Let us denote the additional time as a fraction σ of the uniprocessor time, i.e., $\sigma \tau$, where $0 \le \sigma \le 1$. Then, the dual processor throughput is given by:

$$X_2 = \frac{2t_1}{\tau + \sigma \tau}.$$

Dual-processor scaleup can then be expressed as:

$$X_2 = \left[\frac{2}{1+\sigma} \right] X_1,$$

where we have cancelled the common factors of τ in the quotient. The significance of this equation can understood in the following way.

Clearly, for any non-zero value of σ, the throughput capacity of the dual-processor system will be *less* than twice that of uniprocessor system. Suppose, for example, that $\sigma = 0.03$ then the scalability formula tells us that the effective throughput of the dual-processor will only be 1.94 times that of the uniprocessor. Therefore, if the uniprocessor is capable of 100 TPS, the dual-processor will achieve only 194 TPS - not 200 TPS! This will be the case if the two processors spend 3% of the elapsed time interfering with each other's ability to generate transactions. The parameter, σ, is a measure of this interference. But what is the nature of this interference?

In SMP transaction processing systems there are many points of serial contention in both the hardware (e.g., the memory bus), and the DBMS (e.g., critical code sections) where one processor must wait for the other to complete before it can continue to execute transactions. This time spent contending for shared resources is the dominant reason for dilating the dual-processor execution time. The impact of this effect may not be cause for concern in a dual-processor system but if the trend holds as more processors are added to the system then cumulative effects at the high-end (large-N) may have significant consequences. Recall that MPP hardware is often technically capable of supporting many hundreds and even thousands of processors. To determine the potential impact, for large-N systems, we need to generalize the dual-processor capacity equation.

2.3 Generalized Scaleup

Another way of looking at dual-processor scaleup is that it tells us about the effective number of processors doing real work i.e., actually executing transactions. Let us therefore define a generalized capacity function, $C(N)$ by analogy with our discussion in Section 2.2:

$$X_N = C(N) X_1.$$

At this point in our discussion, we do not know the form of $C(N)$ for an arbitrary number of processors. What we can expect, however, is that the following assumptions hold:

1. $C(1) = 1$, trivially.

2. The fraction of time, σ, will scale with the number of processors according to $\sigma(N) = \sigma f(N)$ where $f(N)$ is to be determined.

3. Any general capacity function should contain dual-processor scaling, $C(2) = \dfrac{2}{1 + \sigma}$, as a special case.

With these constraints in mind, we write a general capacity function for an arbitrary number of physical processors (N) as:

$$C(N) = \frac{N}{1 + \sigma f(N)}$$

where $f(N)$ is an undetermined function of the number of physical processors in the system. It should be clear by now that no matter what choice is made for $f(N)$, the scaleup will be nonlinear and in general it will be sublinear for non-zero values of σ. As in conventional queueing theory [5], it is reasonable to expect that the contributions to the time dilation are additive. We therefore assume, for simplicity, that $f(N)$ is polynomial in N. Elsewhere [2], we have considered two important forms for $f(N)$ in an attempt to model SMP transaction processing performance. We summarize those results here.

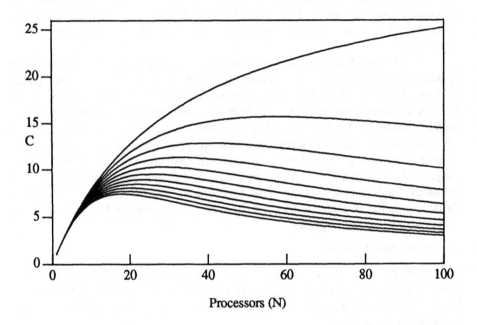

Figure 1. Characteristic curves for the super-serial model with σ held constant and λ ranging from 0.00 (top curve - Optimistic model), to 0.10 (bottom curve).

2.4 Optimistic Model

In this model[2], $f(N) = (N - 1)$, is linear in N. The explicit form of $C(N)$ for this model is:

$$C(N) = \frac{N}{1 + \sigma(N - 1)}$$

This Optimistic model produces scaleup in which $C(N)$ is monotonically increasing for all values of N and asymptotically approaches saturation at σ^{-1}. As we will demonstrate in the next section, this model is too optimistic in its prediction of multiprocessor throughput capacity. There are other effects contributing to the

2. The functional form of $C(N)$ is identical to that found in Amdahl's law [1] for parallel *speedup* of a fixed size, single task. The reader is cautioned against taking the analogy too far since our models are for multi-user scaleup.

execution time-dilation which are not accounted for in the Optimistic model.

2.5 Super-serial Model

In this model, $f(N) = (N-1) + \lambda N(N-1)$, is quadratic in N and $0 \leq \lambda \leq 1$ is another independent parameter. The explicit form of $C(N)$ is:

$$C(N) = \frac{N}{1 + \sigma(N-1) + \sigma\lambda N(N-1)}$$

This model has been found to provide a broader match with measured data, in part, because it approaches zero asymptotically as N^{-1} (see Fig. 1). Moreover, the Super-Serial model contains the Optimistic model as a special case when $\lambda = 0$. Both of these polynomial forms are expressed in terms of $(N-1)$ so as to satisfy the first of the assumptions in Section 2.3 viz, $C(1) = 1$ when $f(1) = 1$.

We can think of each term in the denominator as representing respectively one of the *Three C's*: Concurrency, Contention and Coherency. The first term represents the degree of concurrency available in the code. The second term represents contention for shared resources. The third term is more obscure but we have argued in [2] that it can be identified with additional latency due to some fraction, λ, of processor activity invalidating shared, writeable, data. We call this latter effect *super seriality* to emphasize its connection with the existence of serial contention (seriality) and the extension of its effects. This notion is captured by virtue of the coherency term vanishing when $\sigma = 0$.

Figure 1 depicts the general characteristic of $C(N)$ for N ranging from 1 to 100. σ is held constant while λ ranges from zero (Optimistic Model) to a nominal 0.1. Note the appearance of a broad maximum in the curve for $\lambda > 0$ and also how the peak narrows and moves toward the origin as λ increases. The most significant difference between the two models is the presence of a maximum at:

$$N_c = \left[\sqrt{(1-\lambda)/\lambda\,\sigma} \right]$$

processors. We have adopted the notation [.] to signify the *nearest integer* value. For a fixed value of σ, the position of the maximum is determined by the value of λ according to:

1. $N_c \to \infty$ as $\lambda \to 0$; (Optimistic model limit)

2. $N_c \to 0$ as $\lambda \to 1$

This, more complex, behaviour should be compared with the simpler asymptotic behavior of the Optimistic model depicted in Figure 1. In the remainder of the paper, we discuss a tool based on these models and apply it to measured SMP and MPP performance data.

3. Unix Implementation

The underlying models described in the previous section have been cast into a useable analytic tool in a Unix environment. The synopsis is:

```
qcomp [-g] [-m] [-s value] [-l value] [-d] data
```

where data is the file containing the measured TPS values.

1. The -g switch generates the estimated scaling data for each of the models in a format suitable for plotting.

2. The -m switch sets the maximum range of CPU values to be computed. This is useful for early projections beyond the range of measured data. The default setting is 12.

3. The -s switch allows the calculated seriality value to be overridden.

4. The -l switch allows the calculated super-serial value to be overridden.

5. The -d switch provides a debugging option.

The value of this approach is that the output of qcomp can be (Unix) piped to other utility programs such as statistical analysis programs like |STAT [8].

3.1 Polynomial Approximation

The qcomp program uses a standard linear least-squares fit [9] to the power series expansion:

$$C(N) = N - \sigma N(N-1) + \sigma^2 N(N-1)^2$$
$$- \lambda\sigma(N-1)N^2 + O(N^4)$$

to first calculate σ and then λ. It turns out that near N_c, the maximum in $C(N)$ can be predicted quite accurately from just the first two terms. This is a reasonable approximation because for $N \leq N_c$ the scaling curve is well-approximated by an inverse parabolic function, while for $N > N_c$ the curve is dominated by the coherency term carried with the factor λ.

The polynomial approximation also has a more intuitive interpretation. Considering just the first two terms in the above equation,

$$C(N) = N (1 - \sigma(N - 1))$$

we see that each additional CPU incurs a fixed contention penalty of σ % of *all* the existing CPUs in the system. Furthermore, since C(N) is a symmetric function with roots at 0 and N_{max} when $\sigma(N - 1) = 1$, N_c is now predicted to occur at the half-way point, namely:

$$N_c = \left[\frac{1+\sigma}{2\sigma}\right]$$

We refer to this second-degree polynomial approximation as the *Quad1* estimator. The numerical output for qcomp has the terminal screen format shown in Table 1. The corresponding graphical representation is depicted in Fig. 2.

Table 1. Terminal Output of qcomp

```
Estimated TPS Scaling for: "es.ora.7.0.tpcb.ipi.qdat"
===========================
```

CPU	Sample Data	Percent Sublin	Quad1 Estmt	Quad2 Estmt	Super- Serial	Optimistic Seriality
1	0	0.0	77.8	77.3	77.8	77.8
2	150	3.7	151.2	150.2	150.9	151.3
3	--	--	220.0	218.5	218.9	220.7
4	282	9.4	284.2	282.4	281.9	286.4
5	--	--	344.0	341.7	339.8	348.7
6	395	15.4	399.2	396.6	392.8	407.8
7	--	--	449.8	446.9	441.0	464.0
8	496	20.4	496.0	492.8	484.7	517.4
9	--	--	537.6	534.1	523.9	568.4
10	--	--	574.7	571.0	559.0	617.0
11	--	--	607.3	603.4	590.3	663.4
12	627	32.9	635.4	631.2	617.9	707.8
13	--	--	658.9	654.6	642.2	750.2
14	671	38.4	677.9	673.5	663.5	790.9
15	--	--	692.3	687.9	681.9	829.9
16	--	--	702.3	697.7	697.7	867.3
17	--	--	707.7	703.1	711.2	903.2
18	704	49.8	708.6	704.0	722.6	937.7
19	--	--	705.0	700.4	732.0	970.9
20	--	--	696.8	692.3	739.6	1002.9
21	--	--	684.1	679.7	745.7	1033.6
22	--	--	666.9	662.6	750.4	1063.3
23	--	--	645.1	641.0	753.8	1091.9
24	--	--	618.9	614.9	756.0	1119.6

```
Model parameters:
        alpha:      79.43609
        beta:        2.24441
        TPS[1]:     77.84352
        Sigma:       0.02908
        Lambda:      0.05000
        Nc (qd):    18
        Nc (ss):    26
```

The column labelled *Sample Data*, in Table 1, reports the measured TPS data provided as input to qcomp. The next column reports the calculated degree of sublinearity in the sample data. The next column reports the quadratic estimator based on the input data. The next column reports a 2-parameter quadratic estimator. The next column reports the super-serial estimator. The next column reports the optimistic estimator. The block of numbers at the end indicate some of the internal fitting values used in calculating the σ and λ parameters. The value of λ is calculated iteratively based on the computed σ value. Finally, the critical CPU values are calculated for both the quadratic and super-serial estimators.

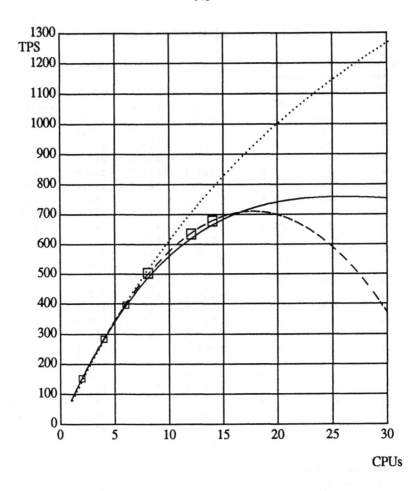

Figure 2. Graphical output of qcomp -g corresponding to the data in Table 1. The sample data is shown as squares, the dotted line is the projection for the optimistic model, the solid curve is the super-serial model and the dashed curve is the quadratic estimator.

4. Case Studies

We now turn to the validation and application of our scaling tool in the context of OLTP workloads. The only comprehensive data, available to us (both internally and externally), is for the TPC Benchmark B™ [7] and similar workloads. Although TPC-B is not typical of all transaction processing workloads, it does provide a common baseline by which to make certain scaling comparisons. It is sufficient for our discussion of latencies due to sharing because there are common functional parts of the RDBMS that must be exercised by all the benchmark transactions.

4.1 Multiprocessor Performance Tuning

Table 2 compares measured and predicted transaction throughput, R_N, for each of the models discussed. The column labelled *Sample TPS* shows measured performance data taken from a Pyramid MIServer™ running ORACLE 7™ with a workload similar to that found in the TPC-B™ benchmark. These data are accurate to about ± 3%. The next column is the degree of sublinearity (relative to a single CPU) expressed as a percentage.

Table 2. ORACLE7 OLTP Scalability						
Calculated parameters: $\sigma = 0.029$, $\lambda = 0.05$						
Measured TPS			Predicted TPS			
CPU	Sample TPS	% Sublin	Super Serial	% Error	Opti-mistic	% Error
1	–		77.8		77.8	
2	150.0	3.7	150.9	0.6	151.3	0.9
3	–		218.9		220.7	
4	282.0	9.4	281.9	0	286.4	1.6
5	–		339.8		348.7	
6	395.0	15.4	392.8	-0.6	407.8	3.2
7	–		441.0		464.0	
8	496.0	20.4	484.7	-2.3	517.4	4.3
9	–		523.9		568.4	
10	–		559.0		617.0	
11	–		590.3		663.4	
12	627.0	32.9	617.9	-1.5	707.8	12.9
13	–		642.2		750.2	
14	671.0	38.4	663.5	-1.1	790.9	17.9
15	–		681.9		829.9	
16	–		697.7		867.3	
17	–		711.2		903.2	
18	704.0	49.8	722.6	2.6	937.7	33.2
19	–		732.0		970.9	
20	–		739.6		1002.9	
21	–		745.7		1033.6	
22	–		750.4		1063.3	
23	–		753.8		1091.9	
24	–		756.0		1119.6	
25	N/A		757.2		1146.2	
26	"		757.5		1172.0	
27	"		757.0		1196.9	
28	"		755.7		1221.0	
29	"		753.8		1244.4	
30	"		751.4		1267.0	

To analyze the data[3], we used qcomp to assist in determining appropriate values of σ

---

3. The measured data samples in Tables 2 and 3 come from a variety of sources and therefore do not provide the complete set that otherwise would be desirable for our modelling discussion. Negative error values indicate an under estimation by the respective model.

and λ from sampled, low-end throughput data. The solid curves in Figures 2 and 3 were generated in this way.

Our approach assumes that single or dual processor measurements are generally more accurate than those taken on systems configured with a large number of CPUs. In the current MIServer™ ES-Series product line, a maximum of 24 R3000A processors are available. Up to 6 of these processors are usually dedicated to handling OLTP network traffic and RDBMS logging processes. Therefore, only 18 processors were available for executing database transactions. The official (client-server) TPC-A result was 645 tpsA [10]. The Super-Serial model predicts a capacity roll-off around $N_c = 26$; just beyond the number of CPU's that can be physically slotted into the ES backplane.

4.2 UNIFY 2000 Benchmarking

In this section, we briefly outline how qcomp was used to aid in the tuning effort of the Pyramid-UNIFY TPC-B benchmark [11]. Figure 3 depicts the *measure* → *estimate* → *tune* process of iterative performance improvement. The circles represent an early initial set of TPS measurements and the squares some early high-end TPS data.

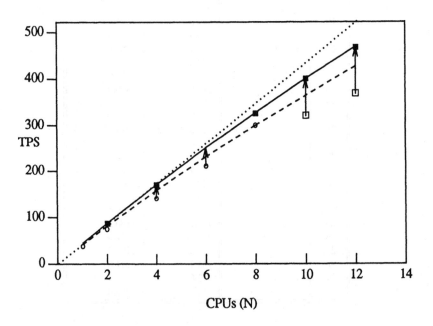

Figure 3. The *measure* → *estimate* → *tune* process. Circles represent an initial low-end data set. The dashed line is the qcomp projected scaling curve. Additional tuning problems in the the high-end data are shown as squares. The final data points are shown as boxes together with the corresponding qcomp estimated scaling (solid) curve. The dotted line represents linear scaling.

The dashed line shows the projected scaling curve based on the initial low-end data. Since the 8 CPU value is estimated to be the best value, it is clear that an attempt should be made to tune all the other data points, starting with the low-end measurements. By repeatedly applying this measurement-estimator-tune process, it was possible to tune the system to produce a competitive TPC-B benchmark result [11] at 12 CPUs. The final qcomp scaling curve is shown as the solid black line.

4.3 Massively Parallel Scalability

In this section we demonstrate how qcomp can be used to asses potential scalability in MPP systems. A more detailed discussion of MPP scalability can be found in [12-15].

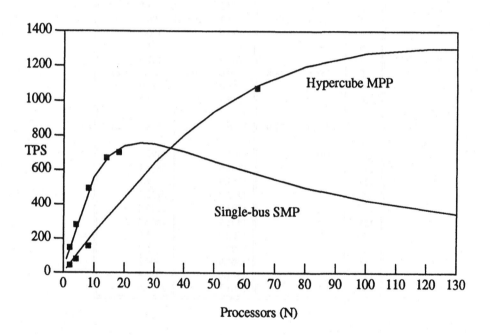

Figure 4. Comparison of predicted throughput capacity based on C(N) for the super-serial model (solid curves) with measured data (squares) for both SMP (from Table 1) and MPP (from Table 2) architectures.

Table 2 presents data for the nCUBE2™ running ORACLE 6.2 with the same workload as described in Section 4.1. The low-end data are informal and supplied by Oracle Corporation and probably responsible for error magnitudes that are higher than for the SMP case. The implication is that these were inferior TPS values that could have been improved on with additional tuning effort. The 64-node measurement is a bona fide TPC-B result [16]. The projections beyond 64 nodes were obtained using qcomp to calculate $C(N)$.

Figure 4 shows the SMP and MPP data (from Tables 2 and 3) together with capacity projections given by $C(N)$. The reader is encouraged to cover up the right-hand side of the figure and note how the initial perception of "linearity" must be mentally readjusted as the cover is removed and the roll-off appears around $N_c = 125$ processors for the nCUBE2. It should be noted in Figure 4 that σ is approximately an order of magnitude smaller than for SMP systems while λ about half the SMP value.

	Table 3. nCUBE2 OLTP Scalability					
	Calculated parameters: $\sigma = 0.0028$, $\lambda = 0.023$					
	Measured TPS		Predicted TPS			
CPU	Sample TPS	% Sublin	Super Serial	% Error	Opti- mistic	% Error
1	24.3	0.0	24.3	0	24.3	0
2	44.3	8.8	48.5	9.5	48.5	9.5
3	–		72.5		72.5	
4	80.5	17.2	96.3	19.6	96.4	19.8
5	–		120.0		120.2	
6	–		143.5		143.8	
7	–		166.9		167.3	
8	157.8	18.8	190.0	20.4	190.7	20.8
9	–		213.0		214.0	
10	–		235.8		237.1	
–	...					
64	1073.0	31.0	1086.1	1.2	1323.9	23.4
...	...					
120	N/A		1301.3		2192.4	
121	"		1301.6		2206.1	
122	"		1301.9		2219.7	
123	"		1302.0		2233.3	
124	"		1302.2		2246.8	
125	"		1302.2		2260.2	
126	"		1302.2		2273.6	
127	"		1302.1		2286.9	
128	"		1301.9		2300.2	
129	"		1301.6		2313.4	
130	"		1301.3		2326.6	

4.4 Non-Database Workloads

We have also been able to apply qcomp to the case of assessing different hardware cache polices. To demonstrate this use we draw on data available for the Sequent Symmetry™ which has been measured for different cache protocols namely, "write-through" and "copy-back" [3]. Both protocols refer to main memory transfers across the shared memory bus. The former protocol updates main memory (across the memory bus) on every update while the latter only does so on a read. In particular, our capacity function $C(N)$ accounts accurately for the maximum in the write-through data with $\sigma = 0.03$ and $\lambda = 0.15$. The copy-back data is fitted to $\sigma = 0.01$ and $\lambda = 0.02$. The data in Fig. 5 show an excellent fit to the Super-Serial model. Furthermore, since these data do *not* belong to a database workload, they lend support to the idea that super-seriality occurs in the presence of any shared, writeable data and therefore exists as a quite general phenomenon.

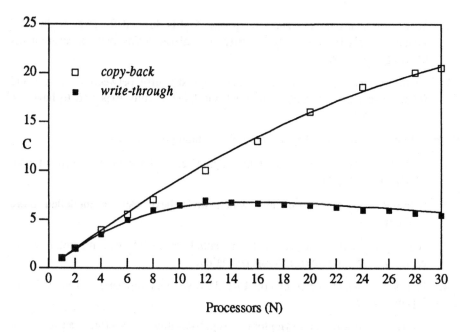

Figure 5. Architectural impacts on scaleup. Measured scaleup (squares) for different cache policies for a non-OLTP application with a significant degree of shared data. Solid curves correspond to Super-Serial C(N) for these data.

We have just indicated that hardware attributes such as cache protocols [17] and bus bandwidth can play a significant role on determining system scalability. In general, however, for a hardware platform with fixed protocols, and bus characteristics the values of σ and λ will thereafter be determined by software attributes [18] as reflected in: resident set size, cache footprint, lock management policies, contention for internal buffers and length of critical sections.

5. Conclusions

We have presented a simple analytic tool for assessing OLTP scalability of SMP and MPP systems in the presence of throughput degradation. The notion of super-seriality provides a simple, intuitively appealing model of the measured throughput degradation and the qcomp too, which incorporates this model, has demonstrated reliable predictive power. qcomp relies on small configuration measurements to iteratively determine the two model parameters, σ and λ. We reiterate that:

1. These modeling parameters are fitted in qcomp rather than calculated from some more fundamental model.

2. Validation of the models and the qcomp tool is difficult for large scale MPP systems simply because of the pausity of database performance measurements on these platforms.

3. At this stage, the qcomp tool does not distinguish between hardware and software resource contention although the latter is fully expected to have the greater impact in general.

Even with these limitations, the current implementation of qcomp:

1. Directs attention to the quantitative significance of resource sharing in both SMP and MPP architectures.

2. Requires only two parameters which we believe can be modelled more explicitly [12].

3. Alerts performance analysts to the potential for an intrinsic maximum in the capacity of the system due to super-seriality.

4. Places the onus on the commercial RDBMS vendors to ensure that the value of λ is minimized.

Performance roll-off is a symptom of super-seriality. This effect can arise, for example, when a processor is primarily serialized spinning on a lock to access shared data and is further serialized after obtaining the lock by the need to fetch cache lines (across the memory bus) that have been invalidated by the preceding action of other processors on that data. In SMP's, super-seriality is a latency effect due to the presence of a memory hierarchy e.g., local caches and global shared memory. It is similar in character, but different in time-scale, to the performance roll-off caused by virtual-memory paging ("thrashing") when the number of user-processes competing for finite main memory increases beyond the multi-programming level. At that point, additional latency is incurred because the current pages in memory must be replaced by pages that are in auxiliary memory - typically a slower disk. This effect has been modelled as a load-dependent central-server queueing network which has computable transient behaviour [18]. In fact, due to the presence of a performance optimum in both cases, the throughput curves for the virtual memory model show a remarkable, qualitative, similarity to the super-serial scaling curves presented here.

6. References

1. G. Amdahl, "Validity of the Single-processor Approach to Achieving Large-scale Computer Capabilities," *Proc. AFIPS Conf.*, 483-85, Apr 1967.

2. N.J. Gunther, "Assessing Transaction Processing Scalability in Shared Memory Multiprocessors," Pyramid Performance Notebook, Nov 1992.

3. S.S. Thakker, "Performance of Symmetry Multiprocessor Systems," in *Cache and Interconnect Architectures in Multiprocessors*, eds. M.Dubios and S.S. Thakker, Kluwer Academic, 1990.

4. M. Ajmone Marsan, G. Balbo, and G. Conte, *Performance Models of Multiprocessor Systems*, MIT Press, 1990.

5. *Quantitative Systems Performance: Computer System Analysis Using Queueing Network Models,* E.D. Lasowska, J. Zahorjan, G.S. Graham and K.C.Sevcik, Prentice-Hall, New Jersey, 1984.

6. D.L. Eager, J. Zahorjan, and E.D. Lazowska, "Speedup Versus Efficiency in Parallel Systems," IEEE Trans. Comp., vol.38, no.3, 408-423, Mar 1989.

7. *The Benchmark Handbook for Database and Transaction Processing Systems,* ed. J. Gray, Morgan Kaufmann, San Mateo, CA, 1991.

8. G. Perlman, "The |STAT Handbook: Data Analysis Programs on UNIX and MSDOS," available via anonymous ftp in pub/stat at: archive.cis.ohio-state.edu (128.146.8.52).

9. W.H. Press, B.P. Flannery, S.A. Teukolsky, and W.T. Vetterling. *Numerical Recipes in C,* Cambridge University Press, 1988.

10. *TPC Benchmark A Full Disclosure Report for the MIServer Model ES Using ORACLE7,* Pyramid Technology Corp., Oct 1992.

11. *TPC Benchmark B Full Disclosure Report for the MIServer Model S Using UNIFY 2000,* Pyramid Technology Corp., Feb 1991.

12. N.J. Gunther, "A Simple Capacity Model of Massively Parallel Transaction Systems," Proceedings of 19th International CMG Conference, vol.2, 1035-1044, San Diego, Dec 5-10, 1993.

13. N.J. Gunther, "Scaling and Shared-Nothingness," Proceedings of 5th International High Performance Transaction Processing Workshop (HPTS), Asilomar, California, Sep 26-29, 1993.

14. N.J. Gunther, "Fundamental Issues Facing Massively Parallel Commercial Database Systems," Presented at the Commercial Applications of Parallel Processing Systems (CAPPS) Conference, Austin, Texas, Oct 19-21, 1993.

15. N.J. Gunther, "Issues Facing Commercial OLTP Applications on MPP Platforms," Proceedings COMPCON'94, IEEE Computer Society Press, 1994. *To appear.*

16. *TPC Benchmark B Full Disclosure Report for the nCube 2 Scalar Supercomputer Model nCDB-1000 Using ORACLE V6.2,* Oracle Corp., 1991.

17. M. Dubois, L. Barroso, Y-S. Chen, and K. Oner, "Scalability Problems in Multiprocessors with Private Caches," *Proc. PARLE'92,* Paris, France, Jun 1992.

18. E. Gelenbe, *Multiprocessor Performance,* Wiley, 1989.

19. N.J. Gunther, "Path Integral Methods for Computer Performance Analysis," Information Processing Letters, v32, #1, 7-13, 1989.

QPN-Tool
for Qualitative and Quantitative Analysis of Queueing Petri Nets

Falko Bause, Peter Kemper

Lehrstuhl Informatik IV
Universität Dortmund
44221 Dortmund
Germany

Abstract. Synchronisation and concurrency aspects as well as sharing of resources are common features of distributed systems. Modelling the last aspect, especially the scheduling strategy amongst competing jobs, can be extremely hard using (Coloured) Generalized Stochastic Petri nets (CGSPNs). Queueing Petri nets (QPNs) provide additional elements for a convenient specification of such queueing situations. QPNs can be used for qualitative analysis employing efficient techniques from Petri net theory, and performance analysis (quantitative analysis) exploiting Markovian analysis algorithms.

QPN-Tool supports both forms of analysis and offers a convenient graphical interface enabling also unexperienced users to specify and analyse their system using the QPN model world.

1 Introduction

System analysis is often done with respect to qualitative and quantitative aspects. E.g. one feature of a fault-tolerant computer is that it will eventually recover from an error which is a qualitative property of the system. A designer of such a system is surely also interested in the time needed for recovery, a quantitative property.

Several formalisms have been developed for modelling and analysing qualitative properties. Petri nets (PNs) [17, 22] belong to these formalism. They have been proved to be suitable for representation of concurrency and synchronisation aspects in modern distributed systems. Since PNs do not involve any notion of time, temporal descriptions have been incorporated to render them suitable also for quantitative analysis leading to Timed and Stochastic Petri nets. A well-known representative of this class are Generalized Stochastic Petri nets (GSPNs) [1, 2], which have been used for modelling a variety of systems [18, 19, 20, 21].

Apart from concurrency and synchronisation another characteristic of distributed systems is sharing of common resources. Modelling this aspect, especially the scheduling rule, of a system using (GS)PN elements is quite difficult and leads to large and complex models [3].

Consider, e.g., a queue where 2 colours of tokens arrive according to exponentially distributed interarrival times and whose service times are exponentially distributed. If the scheduling strategy is FCFS, one has to encode the colour of the token in each

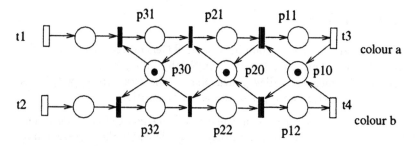

Fig. 1. GSPN model of a FCFS queue

position of the queue. Assume that an upper bound for the number of tokens in the queue is given, e.g. 3, then the GSPN in Fig. 1 would model the queue accurately. Transitions t_1 and t_2 model the arrival of a token of either colour and transitions t_3 and t_4 model the service of the token in front of the queue. The places p_{i1} and p_{i2} represent position i of the queue and the place p_{i0} ensures that this position is occupied by at most one token. Since entering a position is modelled by immediate transitions, a token entering position 3 of the queue will immediately advance to position 2 and 1, if they are free. This way of modelling a FCFS queue with GSPN elements works fine if an upper bound for the number of tokens is known *a priori*. Performing several experiments with different initial markings will necessitate a modification of the GSPN model of the queue for each experiment. If there is no upper bound known beforehand it is even more difficult. Other service times and scheduling strategies lead to very complex models. If the service time of a queue is specified by, e.g. a Coxian distribution and the scheduling strategy is Last Come First Served - Preemptive Resume, it becomes just about impossible to model such a queue with a GSPN.

A popular modelling world to represent resource sharing are queues [7, 15], where system behaviour can be modelled in a compact way. Receiving the benefits of both modelling worlds GSPNs have been enhanced by the usual descriptions of queues leading to a new model, Queueing Petri nets (QPNs) [3]. Queues can be directly integrated into Coloured GSPNs (CGSPNs) by associating them with the places of the net, since a basic property of queues is that customers entering the queue will eventually leave it. QPNs offer a specification paradigm where concurrency and synchronisation aspects are described by (CGS)PN elements and resource sharing is modelled by queues giving a structured model of a system.

Since analysis of modern systems can not be done without proper tool support, we have developed a program package (QPN-Tool) offering a convenient graphical user interface and several analysis algorithms for QPN models. Instead of specifying the transitions of a Coloured GSPN by predicates or functions, QPN-Tool automatically provides a local unfolding so that also unexperienced users can specify their QPN models. Furthermore QPN-Tool offers a variety of PN algorithms and automatically determines quantitative properties of the system employing efficient algorithms from PN and Markov theory. In Sect. 2 we introduce QPNs and the QPN-Tool is described in Sect. 3. A short comparison with other tools is given in Sect. 4.

2 The QPN world

Queueing Petri Nets (QPNs) [3, 4, 6] combine Coloured Generalized Stochastic Petri Nets (CGSPNs) [10] with Queueing Networks (QNs) by hiding stations in special places of the CGSPN which are called *timed places*. The structure of a QPN is determined by a CGSPN with two types of places and transitions:

ordinary place An ordinary place is equivalent to a place in a Coloured Petri net. Tokens fired onto such a place are immediately available for the corresponding output transitions.

timed place A timed place contains a queue and a depository. A token which is fired on a timed place is inserted into the queue according to a scheduling strategy. The scheduling strategy determines which tokens in the queue are served. Each colour has an individual service time distribution of Coxian type. After receiving service the token moves to a depository, where it is available to the place's output transitions. Figure 2 presents a timed place and its graphical shorthand notation. A timed place can be regarded as a short notation of a complex CGSPN subnet, which models a Queueing Network service station. Scheduling strategies like FCFS, which concern the order of arrival, require a ranking of token colours to handle bulk arrivals. In case of a bulk arrival all tokens are separately inserted in succession into the queue in zero time. A token of the colour with the highest rank is inserted first.

immediate transition An enabled immediate transition fires according to one of its colours without any delay in zero time. Any of its colours has a so called 'firing frequency', which allows to compute firing probabilities in case of concurrently enabled immediate transitions.

timed transition An enabled timed transition fires after a certain delay. This delay is determined by a colour-specific exponential distribution. Firing of timed transitions has a lower priority than firing of immediate transitions. Hence no timed transition can fire if an immediate transition is enabled. Like in GSPNs timed transitions obey a 'race policy' to solve conflicts. Firing of a transition is always an atomic action.

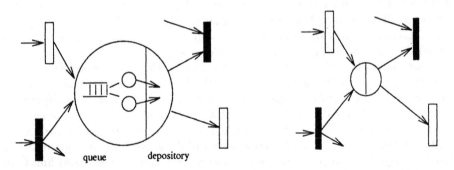

Fig. 2. Timed place in a QPN and its shorthand notation

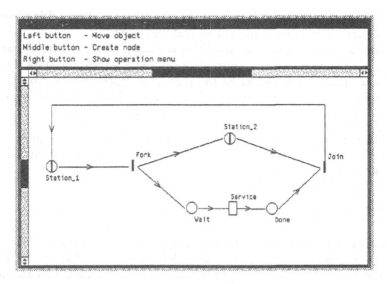

Fig. 3. Example of a QPN

The following example focuses on the illustration of the different elements a QPN can contain and not on modelling anything of practical relevance.

Example 1. Our model describes a situation such that two types of jobs - they are classified as *light* and *heavy* - require service at a first service station, then fork into two subjobs which require service from different resources before they join and then start at the first station again. Thus these jobs are never done. Figure 3 shows the corresponding QPN with timed places *Station_1* and *Station_2*, ordinary places *Wait* and *Done*, immediate transitions *Fork* and *Join* and a timed transition *Service*. It is a net with two colours at each place and transition. The scheduling strategy of *Station_1* is processor sharing (PS) and *Station_2* serves jobs according to their arrival (FCFS). For an initial marking 3 tokens of colour '*light*' and 4 tokens of colour '*heavy*' are supposed at *Station_1*, all other places are empty. Service time distribution in *Station_1* is Coxian with 2 phases for colour '*heavy*', exponential for colour '*light*' which equals a Coxian distribution with 1 phase. Actual values for rates and probabilities shall not be of further interest here.

Every QPN describes a stochastic process. A state of this process is determined by the cartesian product of the state descriptions at all timed places and the number of tokens at ordinary places with respect to their colours. If the service time within a timed place is modelled by an appropriate distribution, e.g. Cox-distribution, Markov-chain based analysis of the QPN is possible.

The state space is partitioned into two types of states similar to GSPNs (cf. [1]):

vanishing states Firing of an immediate transition has a higher priority than any other change of state. Thus the stochastic process immediately leaves a state in which an immediate transition is enabled. If several immediate transitions are enabled, the one which fires first is determined by the firing probability. Firing probabilities are deduced from firing frequencies by relating a firing frequency to the sum of firing frequencies of all enabled transitions.

tangible states If no immediate transition is enabled, firing of a timed transition or serving a token within a timed place can cause a change of state. The time for this change is determined by an exponential distribution in case of firing a timed transition or by the corresponding (exponential stage of the) service time distribution.

The initial marking of the QPN gives the initial state of its stochastic process under the assumption that initially all tokens on timed places are situated on the corresponding depository.

3 QPN-Tool

Specification and analysis of QPNs require appropriate tool support. This section contains a description of QPN-Tool, which is developed at LS Informatik IV, University of Dortmund. Furthermore a brief introduction into the qualitative and quantitative analysis of QPNs is given.

QPN-Tool is a prototype which is implemented in C and is executable on Sun3, Sun4-machines with Sunview or OpenWindows. It contains a graphical user interface and a variety of analysis algorithms for qualitative and quantitative analysis of QPNs. Figure 4 describes the modular structure of QPN-Tool. A brief description of its different modules follows.

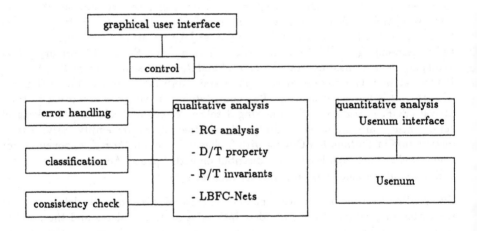

Fig. 4. Modular structure of QPN-Tool

3.1 Graphical user interface

The graphical user interface manages the complete user interaction. It supports:

- specification of QPN models,
- specification of analysis tasks, selection of analysis algorithms and required performance measures,
- presentation of results concerning classification of QPNs and results of qualitative and quantitative analysis
- aggregation of detailed information on probably undesired net properties, e.g. brief description of a firing sequence which leads to a deadlock.

The model description process consists of several main steps.

1. The net's graphical representation is clearly dominated by the Coloured Petri net part. This part is specified by creating and positioning places and transitions and establishing their connections by directed arcs. Figure 3 shows the net corresponding to Example 1.

2. Attributes of places and transitions have to be set. This includes entering the different colours in a list, choosing the type of transitions and places (timed or immediate, resp. timed or ordinary) and fixing the number of tokens for the initial marking. Especially timed places require some additional information:

1. for each colour
   (a) its rank according to bulk arrivals
   (b) its service time distribution specified by its mean and coefficient of variance. This service time distribution is approximated by a Cox distribution as described in [8].
2. scheduling strategy
   Presently available scheduling strategies are FCFS, LCFS-Pr, PS and Infinite Server (IS).
3. number of servers
4. performance figures
   to be determined in quantitative analysis (cf. Sec. 3.2.4).

Figure 5 displays attributes for *Station_1* of example 1.

3. The incidence functions of the Coloured Petri net (cf. [10]) have to be specified. This is possible in a graphical submodel at each transition. Such a submodel describes the locally unfolded net regarding a single transition and its input and output places. A special feature of QPN-Tool fills the submodel automatically with the transition's colours and all input and output places including all of their colours. Thus only arcs and their weights have to be specified manually.

Figure 6 shows the submodel of transition *Fork* in Fig. 3. A rhomb represents a coloured place. Colours of a place are represented by circles which are connected to its corresponding place by lines. Bars display colours of the transition whose submodel is regarded. Obviously transition colour '*fork_light*' takes one '*light*' token from place '*Station_1*' and puts on places '*Wait*' and '*Station_2*' one token on each

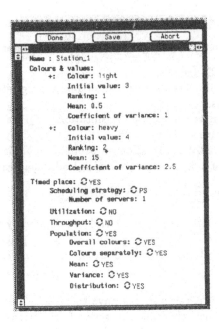

Fig. 5. Attributes of timed place *Station_1*

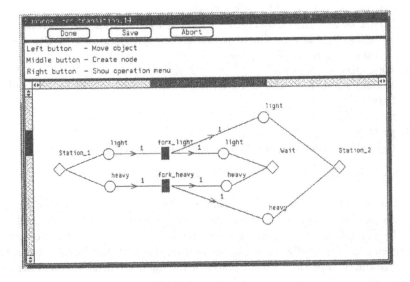

Fig. 6. Locally unfolded net at transition *Fork*

of the corresponding colour '*light*'. Transition colour '*fork_heavy*' does the same for '*heavy*' tokens.

Locally unfolding a net has certain advantages:

- It allows a detailed view on single active components with their corresponding environments.
- The Petri-net-type formalism in locally unfoldings and the whole net is homogeneous.
- Automatic generation of graphical objects in a local unfolding supports a convenient specification. '
- Furthermore generating available colours of input/output places as well as transition colours automatically avoids inconsistent specifications, because the set of colours which can be used and should be used are presented completely. Default positions reflect the common reading direction: input places and their colours to the left, transition colours in the middle, output places and their colours to the right.
- If several transitions have identical submodels, it is sufficient to specify just one. Sharing a submodel is possible as well as copying it.

Since locally unfolding is used within QPN-Tool, it combines a comfortable description technique with a clear presentation.

3.2 Analysis techniques

Typically analysis goals either refer to qualitative properties, e.g. absence of deadlocks, liveness or boundedness, or determination of performance measures. Within QPN-Tool qualitative properties are investigated by a so-called qualitative analysis which is based on Petri net theory. Performance measures are computed by analysis of the corresponding Markov-chain, which is called quantitative analysis. These analysis techniques are briefly described in the following.

Before qualitative analysis a consistency check is performed and the QPN is classified as described below.

3.2.1 Consistency check This module checks a QPN for specification inconsistencies. These could be naming inconsistencies between colour names of a place or transition and its corresponding colours in a locally unfolded transition. This type of error can be avoided by automatic generation of net components in locally unfolded transitions.

Another common error is: although the QPN appears as a connected graph, it is possible that for certain colours a place or transition happens to be a source or sink. This affects boundedness or liveness of a QPN. Thus it is checked and a user information is produced.

3.2.2 Classification Classification aims at the embedded CPN. This net is unfolded to an uncoloured Place/Transition net and classified in terms of: marked graph, state machine, free choice, extended free choice, simple or extended simple. This classification supports the choice of a suitable analysis algorithm for qualitative analysis, because for certain net classes special algorithms are available.

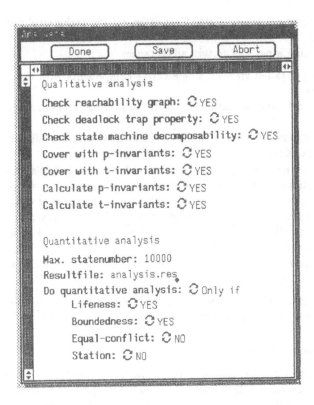

Fig. 7. Selection of analysis algorithms

3.2.3   Qualitative analysis Figure 7 presents the selection of qualitative analysis algorithms implemented in QPN-Tool. Apart from 'classical' algorithms like reachability graph analysis and calculation of P- and T-invariants, rather new algorithms for special net classes are offered. Qualitative analysis within QPN-Tool aims at liveness and boundedness. If the QPN does not have these properties, information of the employed algorithm is extracted in order to demonstrate the reason for a QPN being unbounded or not live. All implemented qualitative analysis algorithms are based on Petri net theory and ignore timing aspects as firing delays and frequencies and interpret timed places as ordinary places. Properties of this 'untimed' QPN carry over to the timed QPN under certain circumstances, if conditions Equal-conflict and Station are satisfied (cf. [3]). Condition Equal-conflict demands that only transitions of the same kind, either timed or immediate, are in conflict and condition Station states that the scheduling strategy has to be of a type like PS or IS. If the QPN is live, bounded and unveils an extended free choice net-structure these conditions are sufficient for the existence of the steady-state distribution.

The choice of algorithms contain:

reachability graph analysis The algorithm generates the reachability graph and recognises firing sequences of transitions which lead to unbounded markings. The reachability graph is checked and it is determined whether the net is bounded and live. If this is not the case a firing sequence is presented which demonstrates unboundedness or non-liveness.

computation of P- or T-invariants A system of 'base vectors' for all positive P- and T-invariants is computed.

cover of P- or T-invariants The algorithm checks whether the net can be covered by positive P- or T-invariant. If not, the set of uncovered places, resp. transitions is presented.

deadlock/trap condition For the class of simple nets it is possible to ensure liveness by checking the deadlock/trap condition. The algorithm is taken from [16] and generates the set of minimal deadlocks and checks if any minimal deadlock contains a marked trap.

check state-machine-decomposability The algorithm is taken from [13, 14]. It allows to recognise live and bounded Free-Choice nets by checking the net structure. This is highly efficient compared to other analysis algorithms.

The selection of algorithms allows to exploit the advantages of an algorithm for the particular case.

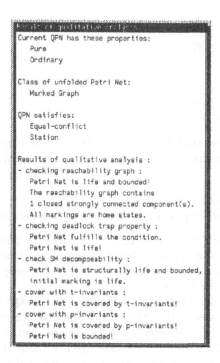

Fig. 8. Result window of classification and qualitative analysis

```
Measure POPULATION_Station_1
 mean                              8.94564e-01
 variance                         1.18166e+00
 standard deviation               1.08704e+00
 coefficient of variation         1.21517e+00
 distribution
                0        1        4.83540e-01
                1        2        2.59126e-01
                2        3        1.45815e-01
                3        4        7.71428e-02
                4        5        2.01259e-02
                5        6        3.57982e-03
                6        7        5.19505e-04
                7        8        5.22813e-05
Measure POPULATION_Station_1_COLOUR_heavy
 mean                              1.49673e-01
 variance                         1.63145e-01
 standard deviation               4.03912e-01
 coefficient of variation         2.69862e+00
 distribution
                0        1        8.66682e-01
                1        2        1.18450e-01
                2        3        1.34739e-02
                3        4        1.30068e-03
                4        5        9.34720e-05
Measure POPULATION_Station_1_COLOUR_light
 mean                              7.44890e-01
 variance                         8.57850e-01
 standard deviation               9.26202e-01
 coefficient of variation         1.24341e+00
 distribution
                0        1        5.23348e-01
                1        2        2.74085e-01
                2        3        1.35894e-01
                3        4        6.56723e-02
```

Fig. 9. Result window of quantitative analysis

3.2.4 Quantitative analysis

Quantitative Analysis is pursued with the objective of assessing performance properties for a QPN. Different performance measures are offered for

ordinary places: token population
timed places: utilisation, throughput and token population

The calculation of performance measures can result in mean value, variance and distribution. They can be computed for all colours of a place separately or aggregated overall colours. They are easily specified by setting appropriate attributes of the corresponding place, see Fig. 5.

The employed analysis technique maps the specified QPN onto a corresponding Markov chain and subsequently analyses this chain with respect to its steady state distribution. The QPN's state space is fully explored causing quantitative analysis to be restricted to QPNs with a finite state space of acceptable size. For calculating a state descriptor for FCFS stations an upper bound of the queue length has to be determined. This is performed automatically during qualitative analysis by checking the reachability graph or an appropriate P-invariant.

QPN-Tool performs quantitative analysis based on Usenum. This is a tool developed at LS Informatik IV which is specially designed for numerical analysis of finite Markov chains, see [9]. Usenum's duties include three main steps:

1. exploring state space

2. computing steady state distribution

3. calculating performance measures

It is able to handle state spaces with more than $100,000$ states and offers different algorithms for the calculation of the steady state distribution, e.g. Grassmann's algorithm, JOR, SOR.

4 Comparison

For recognising limitations of QPN-Tool a quick glance at similar and well-known existing tools might be helpful: GreatSPN [11] from the Universita di Torino, Italy, and SPNP [12] from Duke University, Durham, USA.

GreatSPN supports the specification and analysis of GSPNs and DSPNs. It provides a graphical interface which allows the modelling of only uncoloured nets but including inhibitor arcs, marking dependent rates and probabilities. Difficulties in describing non-trivial queueing situations and scheduling strategies as mentioned in Sect. 1 occur. GreatSPN offers an ample variety of algorithms for qualitative analysis which contains reachability graph analysis, computation of invariants, deadlocks, and traps, and the (inverse) token game. Net properties are nicely animated on its graphical representation. For quantitative analysis, Markov-chain based transient and steady-state analysis is provided as well as simulation. Output measures have to be textually defined. Compared to GreatSPN, future versions of QPN-Tool should be able to handle inhibitor arcs, marking dependent rates and probabilities, and deterministic times. Analysis algorithms of QPN-Tool do not contain the token game, transient analysis and simulation yet.

SPNP is based on the analysis of Markov reward models. Its textual interface is closely related to the programming language 'C', the language SPNP is implemented in. It allows marking dependent arcs, marking dependent enabling functions and general priorities. According to its descriptive power only reachability-graph based qualitative analysis is performed. Its main focus is on quantitative analysis based on Markov reward models (transient and steady state analysis). Output measures are specified by user-defined C-functions supported by a set of predefined functions. An automated sensitivity analysis is offered which derives different CTMCs from a fixed state space by variation of an independent parameter μ for firing rates and probabilities. Compared to QPN-Tool the modelling process in SPNP tends to be a programming process with a strict focus on Markov reward process analysis and few qualitative analysis features as a debugging aid. A variety of qualitative analysis algorithms like in QPN-Tool or GreatSPN is not given in SPNP. As far as quantitative analysis is concerned, SPNP differs from QPN-Tool by its transient analysis, automated sensitivity analysis and its ability to handle general reward specifications.

This comparison is not supposed to be exhaustive or to give a complete characterisation of GreatSPN and SPNP. We just wanted to demonstrate limits of QPN-Tool to draw the following conclusions.

5 Conclusions

QPNs are suitable for modelling synchronisation and concurrency situations as well as sharing of resources which appear in most distributed systems. The great benefit is that a user is not forced to model queues by ordinary (CGS)PN elements thus simplifying the description of systems. Timed places can be viewed as simple parametrisable subnets of a hierarchically specified model.

QPN-Tool offers a convenient graphical user interface. The automatic local unfolding of transitions enables also unexperienced users to get quickly acquainted with the QPN model world. The tool also offers many algorithms for efficient qualitative and quantitative analysis of QPN models.

Future developments are directed to extend this set of analysis algorithms towards transient analysis, simulation and furthermore to integrate hierarchical description and analysis techniques as proposed in [5].

References

1. M. Ajmone-Marsan, G. Balbo, G. Conti. *Performance Models of Multiprocessor Systems*. MIT Press Series in Computer Science, 1986.
2. M. Ajmone-Marsan, G. Conti, G. Balbo. A class of Generalised Stochastic Petri Nets for the performance evaluation of multiprocessor systems. *ACM Transactions on Computer Systems*, 2:93–122, 1984.
3. F. Bause. Queueing Petri Nets: a formalism for the combined qualitative and quantitative analysis of systems. In [21].
4. F. Bause, H. Beilner. Eine Modellwelt zur Integration von Warteschlangen- und Petri-Netz-Modellen. In *Proceedings of the 5th GI/ITG-Fachtagung, Messung, Modellierung und Bewertung von Rechensystemen und Netzen*, pages 190–204. Gesellschaft für Informatik (GI), Braunschweig (Germany), September 1989.
5. F. Bause, P. Buchholz, P. Kemper. Hierarchically Combined Queueing Petri Nets. 11th International Conference on Analysis and Optimizations of Systems, Discrete Event Systems, Sophia-Antipolis (France), June 1994.
6. F. Bause, P. Kemper. Queueing Petri nets. In *Proceedings of the 3rd Fachtagung Entwurf komplexer Automatisierungssysteme, Braunschweig*. Technische Universität Braunschweig (Germany), May 1993.
7. E. Gelenbe, G. Pujolle. *Introduction to Queueing Networks*. John Wiley & Sons, 1987.
8. P. Buchholz. Die strukturierte Analyse Markovscher Modelle. Informatik-Fachberichte, 282, Springer, 1991.
9. P. Buchholz, J. Dunkel, B. Müller-Clostermann, M. Sczittnick, S. Zäske. *Quantitative Systemanalyse mit Markovschen Ketten. Eine Darstellung für Informatiker und Ingenieure* Teubner-Verlag, to be published.
10. G. Chiola, G. Bruno, T. Demaria. Introducing a Color Formalism into Generalized Stochastic Petri Nets. In *Proceedings of the 9th International Workshop on Application and Theory of Petri Nets*, Venice (Italy), pp 202-215, 1988.
11. G. Chiola. GreatSPN 1.5 Software Architecture. In *Proceedings of the 5th International Conference Modeling Techniques and Tools for Computer Performance Evaluation*, Torino (Italy), Feb. 1991.
12. G. Ciardo, J. Muppala, K. Trivedi. SPNP: Stochastic Petri Net Package. In [19].

13. P. Kemper. Linear time algorithm to find a minimal deadlock in a strongly connected free-choice net. In M. Ajmone-Marsan, editor, *Application and Theory of Petri Nets 1993*, LNCS 691, pages 319–338, Berlin, 1993. Springer.

14. P. Kemper, F. Bause. An efficient polynomial-time algorithm to decide liveness and boundedness of free-choice nets. In K. Jensen, editor, *Application and Theory of Petri Nets 1992*, LNCS 616, pages 263–278, Berlin, 1992. Springer.

15. L. Kleinrock. *Queueing Systems. Volume 1: Theory.* John Wiley and Sons, 1975.

16. K. Lautenbach. Linear algebraic calculation of deadlocks and traps. In K. Voss, H.J. Genrich, and G. Rozenberg, editors, *Concurrency and Nets, Advances of Petri Nets*, Berlin, 1987. Springer.

17. J.L. Peterson. *Petri Nets and the Modelling of Systems.* MIT Press Series in Computer Science, 1981.

18. *Proceedings of the 2nd International Workshop on Petri Nets and Performance Models, Madison (USA).* IEEE Computer Society Press, 1987.

19. *Proceedings of the 3rd International Workshop on Petri Nets and Performance Models, Kyoto (Japan).* IEEE Computer Society Press, 1989.

20. *Proceedings of the 4th International Workshop on Petri Nets and Performance Models, Melbourne (Australia).* IEEE Computer Society Press, 1991.

21. *Proceedings of the 5th International Workshop on Petri Nets and Performance Models, Toulouse (France).* IEEE Computer Society Press, 1993.

22. W. Reisig. *Petri Nets. An Introduction*, volume 4. EATCS Monographs on Theoretical Computer Science, Berlin, 1985, Springer.

ACKNOWLEDGEMENTS

The authors are grateful for some valuable comments made by several anonymous referees.

Software Architecture of the EPOCA Integrated Environment

Susanna Donatelli[1], Giuliana Franceschinis[1],
Nicola Mazzocca[2] and Stefano Russo[2]

[1] Dipartimento di Informatica, Università di Torino
Corso Svizzera 185, 10149 Torino - Italy
[2] Dipartimento di Informatica e Sistemistica, Università di Napoli
Via Claudio 21, 80125 Napoli - Italy

Abstract. We describe the software architecture of EPOCA (Environment for analysis and Performance evaluation Of Concurrent Applications), a tool for the analysis of concurrent programs. The analysis is based on a formal model of the application. The class of models chosen is that of stochastic Petri nets (in particular we adopt Generalized Stochastic Petri Nets - GSPN [1]): starting from a concurrent program written in DISC (DIStributed C), an extension of C to include concurrent constructs of the CSP type [12], a GSPN model is automatically generated, and GSPN analysis tools can then be applied. EPOCA is built as an integration of the DISC environment (a graphical interface based environment that provides compiling, monitoring and profiling facilities for DISC programs) and GreatSPN [6] (a graphical interface based environment for the definition and the analysis of GSPN).

1 Introduction

The EPOCA (Environment for analysis and Performance evaluation Of Concurrent Applications) tool presented in this paper results from the integration of a distributed program development environment based on C, called DISC [11] (DIStributed C), with a modeling and performance analysis tool, called Great-SPN [6], which is based on a class of stochastic Petri nets called GSPN [1].

EPOCA supports performance oriented distributed program development, by offering the possibility of studying the behavior of an application during different stages of its implementation cycle. The analysis performed is of *static* type [14] although *dynamic analysis* is also possible in EPOCA exploiting the monitoring tool provided by DISC. Static analysis can be used since the early stages of the program development because it doesn't require a detailed implementation to be available; this analysis technique employs a probabilistic approach to data

* This work has been supported in part by the CNR project "Progetto Finalizzato Sistemi Informatici e Calcolo Parallelo, Sottoprogetto 3: Architetture Parallele" and by the Italian Ministry of University and Scientific and Technological Research within the "40%" Project.

dependency, hence it gives average results. Monitoring on the other hand, is used at an advanced implementation stage, is very data dependent, and hence may give more precise and detailed results than static analysis with respect to specific input data. It is not well–suited to obtain average results because this requires the generation of several program execution traces. Interaction between static and dynamic analysis tools is possible: traces produced by the monitor might be used to derive timing and probabilistic information to be introduced in the abstract model; on the other hand, the model could be used as a basis for visualizing monitor traces at a fairly high level of abstraction, for example to visualize a trace that led to a deadlock.

In order for static analysis to produce reliable results, the formal model used for this purpose must be a consistent representation of the program behavior. Our tool guarantees the desired consistency by automatically generating the models through translation of the process activation and interaction scheme described in the program code. The automatic translation approach has several advantages: it does not require the programmer to be an expert in Petri net modeling, it avoids the boring and error–prone manual model input phase, and finally, it avoids the risks of using a program model that does not represent the actual program behavior but rather the behavior that the programmer *thinks* the program *should have*.

The final program can also be studied by monitoring executions through the DISC monitor.

The program model is generated in two steps: the first step is embedded into the DISC compiler and consists of the translation of each process in isolation using well defined DISC statements translation rules to produce GSPN models of each process scheme. The second step is performed by the so called "Petri net linker" and generates the complete program model by properly instantiating and linking the several process models produced in the first step. The output of the PN linker is again a GSPN description file that can be read by the GreatSPN tool.

From GreatSPN it is possible to run several analysis algorithms and apply the available facilities to validate the model and study its performance. The analysis results may give hints on how to modify the program (in case some unexpected or inefficient behavior of the program is found) or they may provide useful information for a subsequent (semi)automatic mapping phase (for example as suggested in[8, 5]). The monitor provided by EPOCA, inherited from DISC, consists of a module that captures the events naturally handled by the run-time environment. Therefore measurement data are obtained without need of program instrumentation, thus keeping the degree of intrusiveness at a minimum level.

In this paper the features of EPOCA are presented and the details of the programs-to-GSPN translation module, that acts as a bridge between the DISC and GreatSPN tools, are explained.

The paper is organized as follows: Section 2 presents the characteristics of the two systems that are more relevant to the final tool, moreover it explains the structure of the tool. Special emphasis is placed in Section 3 on the software

that enables the automatic translation; the translation mechanism is explained
with the aid of an example: a Fast Fourier Transform implementation. Section 4
discusses the current state of the implementation for EPOCA. Section 5 discusses
future developments of the tool and concludes the paper.

2   The integration

Concurrent programming is a difficult task, and it calls for good programming
environments: it is difficult to write *correct* programs (due to the extra complexity of managing concurrency and interaction among processes); and it may be
even more difficult to write *efficient* programs (there is a risk of paying for the
interaction among processes more than is gained by the concurrent execution of
processes).

DISC, as well as other programming environments, supports correctness
study through debugging, and efficiency via monitoring; but correctness and
efficiency can be studied also through modeling. Given a concurrent program,
it is possible to define the corresponding model using some suitable formalism.
Having been constructed from the program code, the model can be used only for
static analysis. There is a clear trade-off in the use of modeling. Since a model
is usually at a higher level of abstraction than the modeled system, it is easier
to reason on it and to try to prove properties of the program itself; moreover if
the model also takes into consideration timing aspects, then it may be possible
to infer the efficiency of the program from the performance evaluation of the
model. On the other side, the model represents only the program code and it
may only loosely depend on input data, and therefore the conclusions drawn
from the model may not be of use for the real system. In a similar way, the low
dependency on input data provides very general results, but there is also the
drawback that they may not be significant. A classic example is that of non-
faults [14, 13], i.e., faults that are detected in the model even if there exists no
execution that exhibit a faulty behavior.

Assuming that an adequate formalism is available to model parallel programs,
there is a strong need for tools that automatically produce a model from a
program and manage relations between the program and its model; properties
to be proved on the program should be automatically translated into properties
of the model, while results obtained from the model (for example performance
results) should be automatically re-interpreted in terms of results for the program
(for example efficiency of the program).

The analysis of concurrent programs in EPOCA starts with a qualitative
model (untimed Petri net) derived from a (possibly) incomplete but consistent
version of the source code, for example a version in which only interactions
among processes are completely specified. The qualitative model is then used to
validate time independent properties, such as deadlocks and mutual exclusion
relations. The use of untimed Petri nets for the study of concurrent programs
properties has been widely studied in the literature (see for example [10, 13]),
and the particular case of the use of GSPN for studying DISC programs has

been discussed in [4]. Once a qualitative model is available, the programmer may experiment with model animation to check interactions among processes in terms of both activations and communications; model animation is not particularly exciting from a theoretical point of view, but it has proved to be very useful in many real cases (see for example [3]), for its help in program comprehension.

A quantitative model is then obtained by adding timing specifications as delays associated to timed transitions. The quantitative model is finally used for performance evaluation or to compute performance characteristics of the program, as described in [5] where it is shown, for example, how these quantities can be used to establish processes–onto–processors mappings.

The model can also provide support for dynamic analysis; in particular, monitored execution traces may be visualized using trace-driven model animation. Moreover, the results of the model can be used, possibly together with some user–defined requirements, to compute the set of objects to monitor, in order to reduce the cost of producing traces. EPOCA has been built exploiting the functionalities of the DISC and GreatSPN environments, to be discussed next.

2.1   DISC functionalities

The language DISC has been defined in [12]. Briefly, DISC programs consist of a hierarchy of processes that are statically activated through a par statement. Each process has an interface and a body: the interface defines the structure of interaction with other processes, namely channels of communication and process input and output variables, while the body implements its functionality. Input and output variables are exchanged with the parent process on the basis of a call-by-value semantic, upon process activation and termination, respectively. Communication channels are typed, many-to-one (they connect a single owner to many users), and a channel inheritance mechanism is provided. The code for the processes of a given program can be distributed onto one or more modules. Since separate compilation is supported, the environment also provides a tool (the Concurrent Linker) that performs static global consistency checks on the whole program, by analyzing process interfaces and process nesting. For instance, the Concurrent Linker checks for type compatibility of corresponding communication channels of two interacting processes. The language has been implemented on a network of UNIX systems through the Run Time Support (RTS), which represents the target virtual machine for the DISC Compiler. The RTS incorporates a monitoring system, which is able to record every relevant event during a program execution. A Profiler, that is a window-based tool for post-mortem analysis of the execution traces recorded by the monitor, is also available.

2.2   GreatSPN functionalities

GreatSPN is a tool for the (performance) analysis of GSPN models. The user can interact with GreatSPN either through its graphical interface or by running some C-shell scripts that execute the proper analysis programs.

GSPN models are characterized by the presence of immediate transitions (with no associated delay), used to model logical activities, and timed transitions (with an exponentially distributed random delay), used to model activities that take time.

GreatSPN has been extensively used for modeling parallel programs and parallel architectures [3, 8, 2, 9]. Once a GSPN model of the application under study has been devised, the GreatSPN "special purpose" graphical editor can be used to enter the model. The model description can be structured into submodels by using the "layering" facility provided by GreatSPN: several named layers can be defined for each submodel; each object (place, transition or arc) in a net can be assigned to any subset of layers. Any network portion corresponding to a subset of layers can be visualized in isolation through the graphical interface. The graphical representation of the model can be stored for later analysis; a PostScript description of the GSPN picture can be generated by clicking a button and either sent out to a laser printer or saved in a format suitable for inclusion into a LaTeXdocument (indeed all the nets included in this paper have been produced in this fashion).

The user may then experiment with the model through interactive simulation (token game) that provides model animation: all the enabled transitions in a given marking are highlighted, the user can choose the next transition to be fired by simply clicking on one of the enabled transitions. At any time during the interactive simulation the user may decide to "backtrack" to a previously reached marking to try alternative paths.

Model correctness can be checked to some extent at a relatively low computational cost by applying the structural analysis module.

The tool contains a module for the generation of the model state space called Tangible Reachability Graph [6]. Of course this analysis technique may be applied only to models with finite state space (structural boundedness can be checked first to avoid trying to build an infinite TRG). The TRG generation is followed by a graph analysis algorithm that gives some information on the presence of deadlock states, transient states and sets of absorbing states.

Another module generates the Continuous Time Markov Chain (CTMC) describing the timed behavior of the model, using the previously generated TRG. Both steady-state and transient analysis can be performed on the CTMC. The results of these analysis modules are default and user defined performance indices.

Timed behavior analysis of the model can also be performed by means of the simulation module that provides estimates on user–defined performance indices.

2.3 EPOCA functionalities

The EPOCA environment, depicted in Fig. 1, consists of several components: makefile generator, compiler, linker tools, GSPN analyzer, run time support for networks of workstations or transputer-based systems, monitor, post-mortem debugger and window-based interface. The EPOCA architecture is basically the superposition of the DISC architecture, and the GreatSPN architecture: some

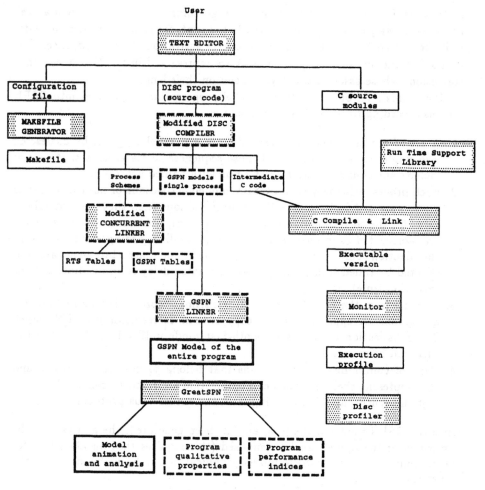

Fig. 1. Architecture of the EPOCA system.

additional software has been built and some existing software has been modified to provide the desired functionalities. New or modified parts are shown in Fig. 1 with bold dotted frames. Bold solid frames represent GreatSPN modules, while plain solid frames represent DISC modules. The new software that implements the integration is the subject of next section.

3 The translation process

In this section we shall first describe the translation process, to then explain how it has been implemented. The translation process is based on a set of translation rules, defined in [4], defined with the aim of obtaining a model including the representation of the control flow, the process activation and the communication statements. In particular we model every single communication statement as well as each **par, sequence, alt, if** and **while (repeat, for)** statement that includes in its body a communication and/or a process activation. All the sequences that do not include any of these instructions are represented in the GSPN as timed transitions whose associated delays depend on their length.

Using this level of abstraction, parallel programs are translated into GSPN models according to the following procedure:

1. A set of process schemas is derived from a parallel program by coalescing into single macrostatements all those sequences of statements that do not include any communication or any **par** with named processes.
2. Each process schema yields a GSPN model: activations of named processes are represented in the net with single transitions (i.e., they are not substituted for the corresponding process translation), communication statements are represented as immediate transitions (thus disregarding any type of synchronization connected with the rendez-vous), and macrostatements are substituted by timed transitions (whose associated delay is an estimate of their execution time).
3. Starting from the initial process, all transitions representing named process activations are expanded with a copy of their corresponding GSPN model. The substitution continues in depth-first mode until all the transitions of this kind have been replaced.
4. Pairs of communication transitions that belong to different processes and that represent their (mutual) communication are fused to concretely represent the synchronization deriving from the rendez-vous protocol.

Once the GSPN model of the program has been generated, the analysis can be performed or the model can be visualized (this requires that an algorithm for the computation of the GSPN model graphical layout be embedded into the translation procedure).

A DISC program example We shall illustrate the translation process and its implementation in EPOCA by an example. The program is a parallel version of a two-dimensional Fast Fourier Transform (2D-FFT) algorithm on complex data. The data are stored in a two-dimensional matrix (DxD) and are divided into submatrices assigned to N different processes. The root process in Fig. 2 activates the N processes in parallel. The 2D-FFT operation is obtained by first performing a 1D-FFT on the rows of the matrix, followed by a 1D-FFT on the columns. D/N rows are assigned to each of the N processes and their FTTs are evaluated in parallel. Data are then exchanged among all the processes and D/N columns of the matrix are again assigned to each process to be Fourier transformed. The DISC code implementing this operations is shown in Fig. 3. The communication protocol among the processes consists of a broadcast: each node in the network retains a portion of the data it has computed and evenly divides the remaining data among the other processes. Fig. 4 displays the DISC code of processes rx and tx used to implement the broadcast.

```
process root()
::
{ matrix a, a1,a2,a3;
   input_data(a);          /* read data from disk */
   split(a,a1,a2,a3);      /* split them into submatrices */
   par {                   /* activates the elaboration processes p */
     local c[(N-1)*N];     /*communication channels */
     process p(in a1; chan c[2],c[4],c[0],c[1]);
     process p(in a2; chan c[0],c[5],c[2],c[3]);
     process p(in a3; chan c[1],c[3],c[4],c[5]); }
} endprocess
```

Fig. 2. DISC code of the root process.

```
process p(a,c12,c13,co1,co2)
in matrix a;                  /* input submatrix */
used matrix_tx c12,c13;       /* output channels */
owned matrix_tx co1,co2;      /* input channels */
::
{   matrix_tx m_tx2, m_rx2, m_tx3, m_rx3;    /* data declaration */
    for(i = 0;i < NUMLINE;i++) four(a,i);    /* rows FFT */
    prepare_messages();
/* in order to overlap the communications the process P actives the
concurrent processes for the data input (rx) and output (tx) */
    par {process tx(in m_tx2; chan c12);
         process tx(in m_tx3; chan c13);
 /* input messages are collected in matrix a */
         process rx(in a; chan co1,co2); }
    for(i = 0;i < NUMLINE;i++) four(a,i);    /* columns FFT */
    output();
} endprocess
```

Fig. 3. DISC code of the computation process.

```
process tx(n,channel)            process rx(a,channel1,channel2)
in matrix_tx n;                  in matrix a;
used matrix_tx channel;          owned  channel1, channel2;
::                               ::
{                                { for(i = 1; i < N; i++)
    channel !! n;                    alt {
} endprocess                         channel1 ?? a => receive();
                                     channel2 ?? a => receive(); }
                                 } endprocess
```

Fig. 4. DISC code of the processes implementing the broadcast protocol.

3.1 The generation of process nets

An interesting aspect of automatic model generation in our tool is that it is obtained as a by-product of the usual program compilation phases. We shall first discuss how each process is translated in isolation. This task is accomplished by a proper module of the DISC compiler. It produces a net description for each process (referred to as *process net*), using the translation rules described in [4]. We shall then describe the second step, consisting of the integration of the process-nets.

Translation of the processes in isolation. The translation procedure has a natural recursive definition: the net representing a compound statement is obtained by properly linking the nets resulting from the translation of the component statements.

Fig. 5 illustrates the translation of the process schemas **root**, **p**, **tx** and **rx**, whose code is given in Fig. 2, Fig. 3 and Fig. 4 respectively.

Fig. 5.(a) is the net corresponding to process **root**: it is easy to recognize the *cobegin–coend* structure of the subnet comprised between the par and end-par transitions. Since the "par" activates three processes, the subnet has three branches, where each timed transition represents an activation of process **p**.

Observe now the translation of process **p** depicted in Fig. 5.(b): all the code preceding the "par" has been summarized into a single timed transition (c1). The same is true for the code that follows the "par" (transition c2).

Fig. 5.(c) depicts the very simple translation of an output statement (the input is exactly the same): since this is a translation of a process in isolation, the communication is simply modeled by an immediate transition.

Fig. 5.(d) illustrates the more complex code of process **rx**. The "for" structure is easily identified by the two transitions that represent leaving the loop (t_{exit}) and entering the loop (t_{loop}). The code inside the loop is an "alt" statement: since the two guards are communication statements, the choice between the two channels is modeled as a non deterministic choice between the two immediate transitions that represent an input from "channel1" (transition ch1) and an input from "channel2" (transition ch2).

Implementation issues Two modules have been added to the DISC compiler for the purpose of process-net generation. Having built the parse tree for a process, the compiler produces the code for the virtual machine (the run-time support), while a new module produces the process schemas as input to the Concurrent Linker. The new GSPN generation module recursively scans the parse tree: for each concurrent statement, as well as for each flow control statement which contains at least one concurrent instruction, it sets up in memory data structures for the proper set of of places, transitions and arcs that model the statement. Sequential parts are collapsed into a single timed transition.

The second module is a graphical layout generator: it performs the computation of the coordinates for each object in a subnet, i.e., places, transitions, intermediate points of the arcs and object tags.

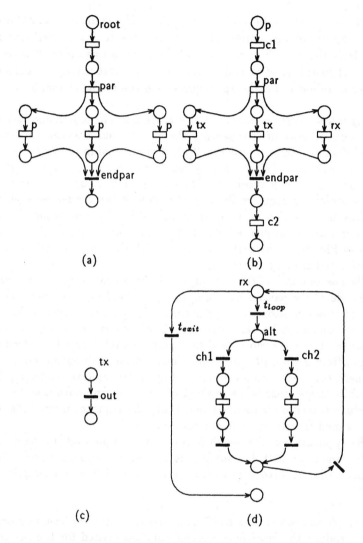

Fig. 5. GSPN models of processes **root, p, tx, rx.**

3.2 Process nets linking

Substitution of each process activation with the corresponding GSPN. This procedure is at first applied to the net of the root process obtained in the second step. In the subnet translating a parallel command, each transition which models the activation and execution of a child process is replaced by the GSPN representing the activated process. The procedure is then applied recursively to the subnet just inserted. As an example, when the subnet corresponding to the activation of process **p** is found in **root**, it is replaced by the subnet representing **p**. The procedure is then applied again to the new subnet and the three process activations contained in it (two instances of tx and one instance of rx) are replaced by the corresponding complete GSPNs. The result of this step, applied to our example, is shown in Fig. 6. Observe that since in this program three instances of p, three instances of rx and six instances of tx are activated, a corresponding number of subnets of each type is present in the model of Fig. 6.

Linking inter-process communication transitions The fourth and last step of the translation consists of representing the *rendez-vous* between processes due to communication. Referring to Fig. 7(a), if the meaning of t_A (t_B) in the translation of process A (B) is that of requiring a rendez-vous with B (A), then all we need is the superposition of t_A and t_B (Fig. 7.(b)). Moreover, the action of communication requires the logical action of the rendez-vous plus the time for transmission. This is reflected in the GSPN by expanding the immediate transition into the sequence immediate transition - place - timed transition, as shown in Fig. 7.(c); the delay associated with the timed transition depends on the estimated amount of exchanged information.

When several input (??) and output (!!) statements on the same channel exist in the code of the parallel program, they can yield many rendez-vous. Without making any assumption on the program behavior we can only say that each input statement on channel x can represent a rendez-vous with any output statement on channel x. We therefore need to superpose each transition which represents an input from x with *all* transitions which represent an output on x. In terms of Petri nets, this operation is a superposition of nets on transitions [7]; the number of resulting transitions is the product of the input and output statements. Each immediate transition is then expanded into the sequence immediate transition - place - timed transition, as in the previous case.

The application of the fourth step to our example leads to the final model of Fig. 8: transitions with two input arcs represent communications. Observe the presence of interrupted arcs, a feature of GreatSPN which simplifies the final layout.

Implementation issues The DISC Concurrent Linker has been augmented in order to produce the cross-reference information needed for the net integration. In addition to verifying channels and in/out variables consistency for the whole application by checking the process interfaces, and producing the initialization data for the run-time support, it also generates two tables, used by the GSPN linker, (Fig. 1), to build the GSPN model of the whole application.

The first table (*process nesting table*) contains the description of the process activation nesting, and a unique (link-time) identifier associated with each distinct process instance. This information is needed during net linking to substitute the proper process-net to the corresponding transition (acting as a "place-holder") in a DISC parallel command (step no. three of the above procedure). For instance, from the following table fragment for the example program, the GSPN Linker recognizes that the process-nets of process p (id=100) and p (id=104) are to be substituted for the transitions with the same name in the

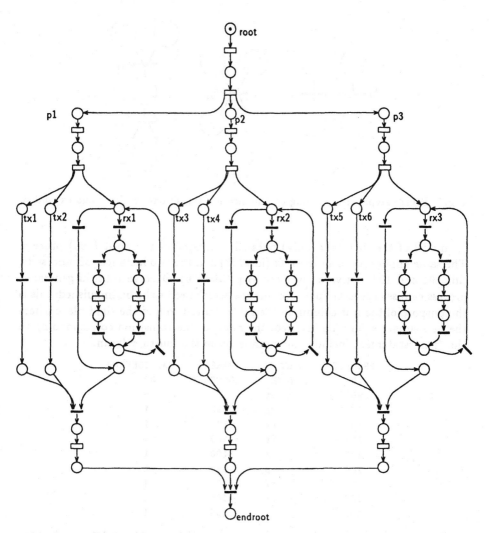

Fig. 6. Model obtained by applying the third step.

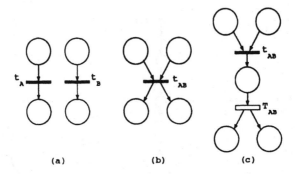

Fig. 7. Fourth step: corresponding communication transitions are fused and expanded.

model of the first (and only) parallel command of the **root** (999) process. The same applies to for process **rx** (id=103) and the two instances of process **tx** (id=101 and 102) activated by process **p** (id=100). Observe that, in general, a process can issue several "par" commands, which need to be distinguished: this is the purpose of the last column in the table. Indeed to find the right place where the new subnet should be inserted, the name of the transition corresponding to the "par" statement includes the number from the fourth column.

PROCESS ID	PROCESS NAME	PARENT PROCESS ID	PAR COMMAND NR.
999	root	–	–
100	p	999	1
101	tx	100	1
102	tx	100	1
103	rx	100	1
104	p	999	1
105	tx	105	1
106	tx	105	1
107	rx	105	1

The second table lists the actual channels in the program and, for each channel, the process instances connected to it. This table is used to establish the potential synchronization points among processes (i.e., the potentially corresponding input/output communication statements). For instance, the following table fragment reveals that actual channel "c[0]" (defined by process **root**) is connected to formal channels "c21" and "co1" of process **p** (id=104) and **p** (id=100), respectively, and has been inherited by processes **rx** (id=103) and **tx** (id=105), which name it "channel1" and "channel", respectively.

CHANNEL INTERNAL ID	FORMAL CH. NAME	PROCESS ID	ACTUAL CH. NAME	CHANNEL ATTRIBUTES	
1	c21	104	c[0]	USER	LOCAL
1	channel1	103	c[0]	OWNER	INHERITED
1	channel	105	c[0]	USER	INHERITED
1	co1	100	c[0]	OWNER	LOCAL

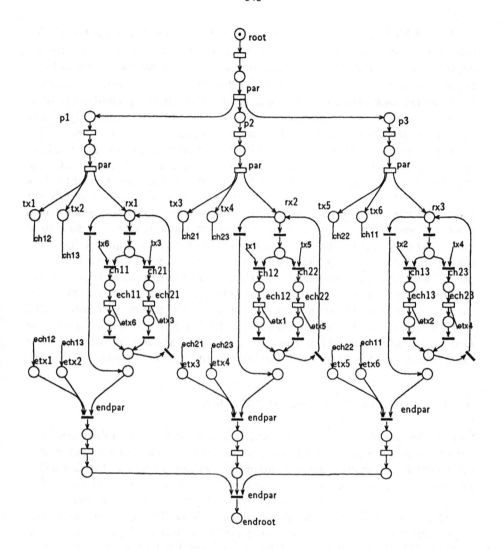

Fig. 8. Final model of the FFT program.

The GSPN Linker uses this information in the fourth step of the translation procedure, that is to fuse processes' communication transitions over corresponding channels. Communication transitions in a net representing a process instance are identified through their name, that includes the formal name of the channel.

A further task of the GSPN Linker is to determine the graphical layout of the overall program net. This has been accomplished through proper algorithms for objects' coordinates computation, as in the compilation step. By exploiting the layering feature of GreatSPN, we offer to the programmer the possibility of preserving the identity of each process subnet in the graphical representation of the overall net. In fact, during the third step of the described procedure (the first one of the linking phase) each process subnet is assigned to a different layer, and the user can then arbitrarily decide which layers to visualize in a given moment. This is useful to concentrate on the behavior of a particular process while executing the "token-game" on the net. Two graphical "policies" can be applied in this step, depending on the "regularity" of the net. Either the parallel commands are "expanded", by clearly positioning each child process net in place of the corresponding activation transition in the "par" command, or the several process nets are simply juxtaposed on the screen, with interrupted arcs accounting for the needed connections for the flow of tokens. Finally, the exploitation of the labeled interrupted arcs feature in the second link step, while fusing communication transitions, further simplifies the final layout (see Fig. 8).

4   Current state of implementation

Some of the described functionalities of EPOCA are still being implemented. The current version is based on the DISC window interface, that has been augmented incorporating the translation and linking procedures, as well as the GreatSPN package to visualize the output of the translation process: all *nets depicted in this paper have been automatically generated from the DISC code and directly included as postscript file with no additional editing.* The automatic definition of performance indices and the computation of transition delays is under development. The set of indices we intend to implement is that defined in [5], e.g., the maximal and minimal level of parallelism, the level of interference among processes, the set of phases (a phase is a set of processes simultaneously active), the probability of each phase, the probability of a given process of being active or idle, the communication throughput (general or over a given channel), and the probability of a process being blocked for a rendez-vous on a given channel.

Each index is defined in terms of basic quantities such as the probability of a state, the set of transitions that represent an I/O on a given channel, the set of places belonging to a process and the subset of places of a process that represent an active or idle state. All this information is already available from the translation process.

Concerning the computation of transitions delays, a possible idea is that of providing the compiler with a table containing the estimated duration of the basic DISC statements; moreover, the user should be allowed to insert annotations in the code, specifying the expected duration of the not yet implemented portions of code. The transition delays can be defined in a parametric form, and the analysis algorithms can be executed off-line for a set of different parameter values to produce a table describing how the performance indices change as a function of program parameters.

5 Conclusions and future developments

We have presented the software architecture of EPOCA, an environment for the development and performance analysis of concurrent programs. It was obtained by integrating two existing tools: DISC and GreatSPN. The integration has been achieved by implementing the software for the automatic translation of DISC programs into corresponding GSPN models. Using GreatSPN, the automatically built models can be analyzed to obtain both qualitative and quantitative information on the program behavior. The modeling and analysis features of the tool can be exploited from the early stages of program development, by using an incremental approach to program validation.

EPOCA has been designed for non–expert of GSPNs; in fact, in its final implementation, it will not require the user to ever deal with the automatically generated GSPN model. On the other hand, we have implemented some graphical layout computation algorithm to allow users that are familiar with the GSPN formalism to perform visual simulation of the program model or to use it as a basis for building model variations containing the representation of some new aspects (e.g., architecture and mapping) or to derive more abstract models.

Some work is still needed to complete the implementation of the mechanism that allows automatical translation of the GSPN model analysis results, in order to present them to the user in a program–oriented style. As new analysis algorithms for GSPNs are devised and implemented, they will be exploited to investigate new program properties and integrated in the tool by simply implementing an interface module for program–oriented interpretation.

A typical example of program–oriented analysis result of interest is the communication graph. Observe that since this kind of information is provided by the tool, it is conceivably possible to integrate EPOCA with an automatic mapping tool to support the user in the program placement phase.

Model–based static analysis is not the only performance analysis possibility provided by our environment. The DISC monitoring feature can also be used to obtain information on the program behavior for specific input data and for a given placement of processes on a real architecture. We think that it could be useful to integrate the dynamic and static performance analysis tools by (1) using the statements duration measured by the monitor to define a set of rate parameter values for the model, (2) using the graphical representation of the model to visualize execution traces collected by the monitor, (3) using GSPN model simulation traces to force deterministic choices during the program execution (exploiting the *trace replay* option of the monitor), (4) selecting portions of the program model of interest to users to determine the amount of information that has to be collected in a monitor trace.

We have experimented with EPOCA on several example programs. Our experience suggests that the translation module has to be extended to include the representation of control variables to reduce the non-faults found in the model. Most of these control variables are either boolean or defined on natural numbers and are modified through simple operations so that their inclusion in the program model does not pose big problems. Another aspect that becomes immediately apparent is that we do not need very complex programs to generate models that do not fit the GreatSPN graphical interface window. As far as analysis is concerned, this is not a big problem because it can be performed without necessarily visualizing the model. However in the Petri net literature there are well known approaches coping with the model size problem; for example, Colored Petri Nets (CPN) could be used to build more parametric models where similar subnets (such as those representing different instances of the same process) are folded onto one another. Since GreatSPN has some facilities for dealing with CPNs, extensions of the translation module in this direction is being considered.

In conclusion, even if the tool is not yet sophisticated enough to support the development of big and complex concurrent programs it already shows the advantages of the approach we are proposing. As new methods for dealing with Petri net model complexity are developed by the Petri net research community, they could be implemented and embedded in the tool making it usable over a broader range of applications.

References

1. M. Ajmone Marsan, G. Balbo, and G. Conte. A class of generalized stochastic Petri nets for the performance analysis of multiprocessor systems. *ACM Transactions on Computer Systems*, 2(1), May 1984.

2. M. Ajmone Marsan, G. Balbo, and G. Conte. *Performance Models of Multiprocessor Systems*. MIT Press, Cambridge, USA, 1986.

3. G. Balbo, G. Chiola, S.C. Bruell, and P. Chen. An example of modelling and evaluation of a concurrent program using coloured stochastic Petri nets: Lamport's fast mutual exclusion algorithm. *IEEE Transactions on Parallel and Distributed Systems*, 3(1), January 1992.

4. G. Balbo, S. Donatelli, and G. Franceschinis. Understanding parallel programs behaviour through Petri net models. *Journal of Parallel and Distributed Computing*, 15(3), july 1992. Special Issue on Petri Net Modelling of Parallel Computers.

5. G. Balbo, S. Donatelli, G. Franceschinis, A. Mazzeo, N. Mazzocca, and M. Ribaudo. On the Computation of Performance Characteristics of Concurrent Programs using GSPNs. Accepted for publication on Performance Evaluation, 1993.

6. Giovanni Chiola. *GreatSPN 1.5 software architecture*. In *Proc. 5th Int. Conf. Modeling Techniques and Tools for Computer Performance Evaluation*, Torino, Italy, February 1991.

7. F. De Cindio, G. De Michelis, L. Pomello, and C. Simone. Superposed automata nets. In C. Girault and W. Reisig, editors, *Application and Theory of Petri Nets*. IFB 52, New York and London, 1982.

8. F. DeCindio and O. Botti. Comparing Occam2 program placements by a GSPN model. In *Proc. 4th Intern. Workshop on Petri Nets and Performance Models*, Melbourne, Australia, December 1991.

9. A. Ferscha. Modelling mappings of parallel computations onto parallel architectures with PRM-net model. In *Proc. IFIP-WG 10.3 Working Conference on Decentralized Systems*, Lyon, December 1989.

10. U. Goltz and W. Reisig. CSP programs as nets with individual tokens. In *Proc. 5th International Conference on Application and Theory of Petri Nets*, Aarhus, Denmark, June 1984.

11. G. Iannello, A. Mazzeo, C. Savy, and G. Ventre. Parallel software development in the DISC programming environment. *Future Generation Computer Systems*, 5(4), 1990.

12. G. Iannello, A. Mazzeo, and G. Ventre. Definition of the DISC Concurrent Language. *SIGPLAN Notices*, 24(6), June 1989.

13. S.M. Shatz and W.K. Cheng. A Petri net framework for automated static analysis of Ada tasking. *The Journal of Systems and Software*, 8:343–359, October 1987.

14. R. Taylor. A general purpose algorithm for analyzing concurrent programs. *Communications of ACM*, 26, May 1983.

The PEPA Workbench:
A Tool to Support a Process Algebra-based Approach to Performance Modelling

Stephen Gilmore and Jane Hillston

Department of Computer Science, The University of Edinburgh

Abstract. In this paper we present a new technique for performance modelling and a tool supporting this approach. Performance Evaluation Process Algebra (PEPA) [1] is an algebraic language which can be used to build models of computer systems which capture information about the performance of the system. The PEPA language serves two purposes as a formal description language for computer system models. The performance-related information in the model may be used to predict the performance of the system whereas the behavioural information in the model may be exploited when reasoning about the functional behaviour of the system (e.g. when finding deadlocks or when exhibiting equivalences between sub-components). In this paper we concentrate on the performance aspects of the language.

A method of reasoning about PEPA models proceeds by considering the derivation graph obtained from the model using the underlying operational semantics of the PEPA language. The derivation graph is systematically reduced to a form where it can be treated as the state transition diagram of the underlying stochastic (in fact, Markovian) process. From this can be obtained the infinitesimal generator matrix of the Markov process. A steady state probability distribution for the system can then be obtained, if it exists.

We have implemented a prototype tool which supports this methodology from the initial checking of the well-formedness of the PEPA model through the creation of the state transition diagrams to the calculation of performance measures based on the infinitesimal generator matrix. The tool is implemented in Standard ML [2] and provides an interface to the Maple Symbolic Algebra package [3] for the solution of matrix equations.

1 Introduction

Formal descriptions of computer systems are amenable to analysis by a range of formal techniques. At the simplest level, they may be checked for conformance with the syntax, grammar and type-correctness rules of the formal language used. More advanced analysis may involve deriving properties of a system from its description: either by deduction or by calculation. For concurrent systems modelled by an algebraic description, the properties which may be checked include freedom from deadlock and algebraic equivalence under observation with a simpler description which serves as a specification of the system. When the algebraic description is enhanced with information about the system's expected performance—as in PEPA—still further properties can be calculated. These include steady-state probabilities and rewards which may be used to derive performance measures.

The goal of the work described here is to provide a workbench for a designer of a computer system who is working from an initial PEPA model. As with other languages based on algebra and processes, e.g. CCS [4], PEPA is a parsimonious language which provides the essential, simple tools for system description. The formality and succinctness of the language have enabled the authors to design and build a workbench which assists with checking and reasoning about PEPA descriptions. Use of PEPA and the workbench is illustrated by an example taken from the area of communication networks.

2 The PEPA Language

The motivation for process algebra-based techniques for the quantitative analysis of computer systems have been presented in detail elsewhere [5, 6]. Some of the advantages of such an approach are:

- The system is represented as a collection of active agents who cooperate to achieve the behaviour of the system. This *cooperator paradigm* is particularly apt for modelling many modern computer systems.
- Compositional reasoning is an integral part of the modelling language.
- The formal definition clarifies the task of providing tools for model manipulation, simplification and analysis.
- Process algebra has growing importance as a design methodology [7, 8] and so this approach offers the possibility of integrating performance analysis into the system design process.

From a performance point of view, process algebras, such as CCS, lack essential, *quantifiable* information about time and uncertainty. Timed extensions of some process algebras have been proposed [9, 10, 11, 12] but these make a distinction between time progressing and computation progressing. PEPA, and TIPP, developed at Erlangen, take an alternative approach—time is incorporated into the algebra by associating a random variable, representing *duration*, with each *activity*[1]. We assume a *race condition* between simultaneously enabled activities. Thus, as in probabilistic process algebras, we replace the nondeterministic branching by probabilistic branching, and the timing behaviour of the system is captured. This is analogous to the association of a duration with the firing of a timed transition in a generalised stochastic Petri net [13].

It was important when designing the PEPA language to retain the key features of a process algebra which had motivated the approach: compositionality, parsimony, and the existence of a formal definition. However, it was also necessary to incorporate features to make the language suitable for capturing the performance-related information about the system. This additional information can be added as an annotation to an existing model or design.

---

[1] *'Activity'* is used instead of the usual process algebra *'action'* to distinguish between timed and instantaneous behaviour respectively.

2.1 PEPA Terminology

In PEPA a system is described as an interaction of *components* and these components engage, either individually or cooperatively, in *activities*. The components will correspond to identifiable substructures in the system, or rôles in the behaviour of the system. They represent the active units within a system; the activities capture the actions of those units. For example, a queue may be considered to consist of an arrival component and a service component which interact to form the behaviour of the queue.

A component may be atomic or may itself be composed of components. Thus the queue in the above example may be considered to be a component. Each component has a behaviour which is defined by the activities in which it can engage. Actions of the queue might be *accept*, when a customer enters the queue, *service*, or *loss*, when a customer is turned away because of a full buffer.

Each activity has an *action type*. We assume that each discrete action within a system has a unique type and there is a countable set, $\mathcal{A}$, of all possible such types. The action types of a PEPA term correspond to the actions of the system being modelled. There are situations when a system is carrying out some action (or sequence of actions) the identity of which is unknown or unimportant. To capture these situations there is a distinguished action type, τ, which can be regarded as the *unknown* type. Activities of this type are private to the component in which they occur.

Every activity in PEPA has an associated duration which is a random variable with an exponential distribution. Since an exponential distribution is uniquely determined by its parameter, the duration of an activity may be represented by a single real number parameter. This parameter is referred to as the *activity rate* (or simply *rate*) of the activity; it may be any positive real number, or the distinguished symbol $\top$, which should be read as *"unspecified"*.

An $M/M/1/N/N$ queue in which the arrival process is suspended when the buffer is full, is represented as follows:

$$Arrival_0 \stackrel{def}{=} (accept, \lambda).Arrival_1$$

$$\vdots \quad \vdots$$

$$Arrival_i \stackrel{def}{=} (accept, \lambda).Arrival_{i+1} + (serve, \top).Arrival_{i-1} \qquad 1 \leq i \leq N-1$$

$$\vdots \quad \vdots$$

$$Arrival_N \stackrel{def}{=} (serve, \top).Arrival_{N-1}$$

$$Server \stackrel{def}{=} (serve, \mu).Server$$

$$Queue_0 \stackrel{def}{=} Arrival_0 \underset{\{serve\}}{\bowtie} Server$$

Each activity, a, is defined as a pair (α, r) where $\alpha \in \mathcal{A}$ is the action type and r is the activity rate. It follows that there is a set of activities, $Act \subseteq \mathcal{A} \times \mathbb{R}^+$, where $\mathbb{R}^+$ is the set of positive real numbers together with the symbol $\top$.

When enabled, an activity $a = (\alpha, r)$, will delay for a period determined by its associated distribution, denoted $F_a(t)$ ($= 1 - e^{-rt}$). We can think of this as the activity setting a timer whenever it becomes enabled. The time allocated to the timer is determined by the rate of the activity. If several activities are enabled at the same time each will have its own associated timer. When the first timer finishes that activity takes place—the activity is said to *complete* or *succeed*. This means that the activity is considered to "happen": an external observer will witness the event of an activity of type α. An activity may be *preempted*, or *aborted*, if another one completes first.

2.2 The Syntax and Semantics of PEPA

Components and activities are the primitives of the language PEPA; the language also provides a small set of combinators. As explained in the previous section the behaviour of a component is characterised by its activities. However, this behaviour may be influenced by the environment in which the component is placed. The combinators of the language allow expressions, or terms, to be constructed defining the activities which components may undertake and the interactions between them.

The syntax for terms in PEPA is defined as follows:

$$P ::= (\alpha, r).P \mid P \bowtie_L Q \mid P + Q \mid P/L \mid X \mid A$$

Prefix: $(\alpha, r).P$ Prefix is the basic mechanism by which the behaviours of components are constructed. The component $(\alpha, r).P$ carries out activity (α, r), which has action type α and a duration which is exponentially distributed with parameter r (mean $1/r$). The time taken for the activity to complete will be some Δt, drawn from the distribution. The component subsequently behaves as component P. When $a = (\alpha, r)$ the component $(\alpha, r).P$ may be written as $a.P$.

It is assumed that there is always an implicit resource, some underlying resource facilitating the activities of the component which is not modelled explicitly. Thus the time elapsed before activity completion represents use of the resource by the component enabling the activity. For example, this resource might be processor time or CPU cycles within a processor, depending on the system and the level at which the modelling takes place.

Choice: $P + Q$ The component $P + Q$ represents a system which may behave either as P or as Q. $P + Q$ enables all the current activities of P and all the current activities of Q. Whichever enabled activity completes it must clearly belong to either P or Q. In this way the first activity to complete distinguishes one of the components. The other component of the choice is discarded. The continuous nature of the probability distributions ensures that the probability of P and Q both completing an activity at the same time is zero. The system will subsequently behave as P' or Q' respectively, where P' is the component which results from P completing the activity, and similarly Q'.

There is an underlying assumption that P and Q are competing for the same implicit resource. Thus the choice combinator represents competition between components.

Cooperation: $P \bowtie_L Q$ The cooperation combinator is in fact an indexed family of combinators, one for each possible set L of action types. The set L, the *coopera-tion set*, defines the action types on which the components P and Q must synchronise or *cooperate*, i.e. it determines the interaction between the components.

All activities of P and Q which have types which do not occur in L will proceed unaffected. These are termed *individual* activities. In contrast *shared* activities, activities whose type does occur in L, will only be enabled in $P \bowtie_L Q$ when they are enabled in both P and Q. Thus one component may become blocked, waiting for the other component to be ready to participate. These activities represent situations in the system when the components need to work together to achieve an action. In general both components will need to complete some work, corresponding to their own representation of the action. Thus a new *shared* activity is formed by the cooperation $P \bowtie_L Q$, replacing the individual activities of P and Q. This activity will have the same action type as the two contributing activities and a rate reflecting the rate of the slower participant.

If an activity has an unspecified rate in a component, the component is *passive* with respect to that action type. This means that although the cooperation of the component may be required to achieve an activity of that type the component does not contribute to the work involved. An example might be the rôle of a channel in a message passing system: the *cooperation* of the channel is essential if a transfer is to take place but the transfer involves no work on the part of the channel. This may be regarded as one component *coopting* another.

In contrast to choice, it is assumed that P and Q each have their own implicit resource. Activities with action types in the set L are assumed to require the simultaneous involvement of both components, both resources. The unknown action type, τ, may not appear in any cooperation set.

Hiding: P/L The component behaves as P except that any activities of types within the set L are *hidden*, meaning that their type is not witnessed upon completion. Instead they appear as the unknown type τ and can be regarded as an internal delay by the component.

Hiding does not have any effect upon the activities a component may engage in individually, but a hidden activity is witnessed only as a delay of the unknown type, τ. The duration of an activity is unaffected if it is hidden. However, a hidden activity cannot be carried out in cooperation with any other component. In effect the action type of a hidden activity is no longer externally accessible, to an observer or to another component.

Variable: X If E is a component expression which contains a variable X, then $E\{P/X\}$ denotes the component formed when every occurrence of X in E is replaced by the component P. More generally, an indexed set of variables, $\tilde{X}$, may be replaced by an indexed set of components $\tilde{P}$, as in $E\{\tilde{P}/\tilde{X}\}$.

Constant: $A \stackrel{\text{def}}{=} P$ We assume that there is a countable set of *constants*. Constants are components whose meaning is given by a defining equation such as $A \stackrel{\text{def}}{=} P$ which gives the constant A the behaviour of the component P. This is how we assign names to components (behaviours).

The semantics of the language, presented in structured operational semantics style, are shown in Figure 1. The transitional semantics over PEPA is then given by the least multi-relation $\longrightarrow \subseteq PEPA \times Act \times PEPA$ satisfying the rules.

358

Prefix

$$(\alpha,r).E \xrightarrow{(\alpha,r)} E$$

Cooperation

$$\frac{E \xrightarrow{(\alpha,r)} E'}{E \underset{L}{\bowtie} F \xrightarrow{(\alpha,r)} E' \underset{L}{\bowtie} F} \quad (\alpha \notin L)$$

$$\frac{F \xrightarrow{(\alpha,r)} F'}{E \underset{L}{\bowtie} F \xrightarrow{(\alpha,r)} E \underset{L}{\bowtie} F'} \quad (\alpha \notin L)$$

$$\frac{E \xrightarrow{(\alpha,r_1)} E' \quad F \xrightarrow{(\alpha,r_2)} F'}{E \underset{L}{\bowtie} F \xrightarrow{(\alpha,R)} E' \underset{L}{\bowtie} F'} \quad (\alpha \in L)$$

where $R = \dfrac{r_1}{r_\alpha(E)} \dfrac{r_2}{r_\alpha(F)} \min(r_\alpha(E), r_\alpha(F))$
and $r_\alpha(E)$ is the apparent rate of α in E

Choice

$$\frac{E \xrightarrow{(\alpha,r)} E'}{E + F \xrightarrow{(\alpha,r)} E'}$$

$$\frac{F \xrightarrow{(\alpha,r)} F'}{E + F \xrightarrow{(\alpha,r)} F'}$$

Hiding

$$\frac{E \xrightarrow{(\alpha,r)} E'}{E/L \xrightarrow{(\alpha,r)} E'/L} \quad (\alpha \notin L)$$

$$\frac{E \xrightarrow{(\alpha,r)} E'}{E/L \xrightarrow{(\tau,r)} E'/L} \quad (\alpha \in L)$$

Constant

$$\frac{E \xrightarrow{(\alpha,r)} E'}{A \xrightarrow{(\alpha,r)} E'} \quad (A \stackrel{def}{=} E)$$

Fig. 1. Operational Semantics of PEPA

When the set L is empty, $\bowtie_L$ has the effect of parallel composition, allowing components to proceed concurrently without any interaction between them. We use the more concise notation $P \parallel Q$ (the *parallel* combinator) to represent $P \bowtie_\emptyset Q$.

Execution Strategies and the Exponential Distribution The *race condition* governs the dynamic behaviour of a model whenever more than one activity is enabled. This means that we may think of all the activities attempting to proceed but only the 'fastest' succeeding. Of course, which activity is 'fastest' on successive occasions will vary due to the nature of the random variables determining the durations of activities. The probability that a particular activity completes will be given by the ratio of the activity rate of that activity to the sum of the activity rates of all the enabled activities.

We assume that the introduction of cooperation between two components implies that in general they are independent and running on separate resources. Thus we can think of their individual activities as interleaving. On the other hand, when there is a choice between components we assume that they are competing for the same underlying resource and that in fact only one of them gains the use of that resource. Thus we have two different preemption scenarios: *preemptive-resume* for cooperation and *preemptive-restart with resampling* for choice. However, we take advantage of the memoryless property of the exponential distribution which makes the two equivalent and always assume a preemptive-restart policy (with resampling). This allows us to formulate Expansion Laws of the form shown below: this would not be possible if another distribution were associated with activity durations as in some versions of TIPP [14].

Expansion Law Let $P \equiv P_1 \bowtie_L P_2$. Then

$$P = \sum \{(\alpha, r).(P_1' \bowtie_L P_2) : P_1 \xrightarrow{(\alpha, r)} P_1'; \ \alpha \notin L\}$$
$$+ \sum \{(\alpha, r).(P_1 \bowtie_L P_2') : P_2 \xrightarrow{(\alpha, r)} P_2'; \ \alpha \notin L\}$$
$$+ \sum \{(\alpha, r).(P_1' \bowtie_L P_2') : P_1 \xrightarrow{(\alpha, r_1)} P_1'; \ P_2 \xrightarrow{(\alpha, r_2)} P_2'; \ \alpha \in L\}$$

Recent work on TIPP [5] has concentrated on the subset of the language in which all activity durations are exponentially distributed. The major differences between PEPA and this subset are in the definition of the cooperation or parallel composition, and more importantly, the choice of a multi-relation, rather than a relation, to capture the operational semantics of the language.

2.3 Generating and Solving the Underlying Markov Process

For any PEPA model we can define a multigraph—the *derivation graph*—based on the operational semantics. This is a graph in which language terms form the nodes and the arcs represent the possible transitions (activities) between them; it is a multigraph since we distinguish between different instances of the same activity. This derivation graph provides a useful way to reason about the behaviour of a model. Moreover it is used to generate the stochastic process underlying any PEPA model. Each node of the derivation graph is taken to be a state in the stochastic process and the transition rate between states is the sum of the rates shown on arcs connecting the nodes in the multigraph. This is analogous to the use of the reachability graph in stochastic extensions of Petri nets such as GSPN [13]. For the $M/M/1/N/N$ queue considered earlier the derivation graph is shown in Figure 2.

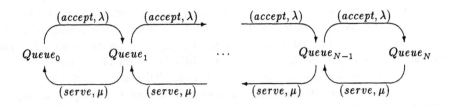

Fig. 2. The derivation graph for an $M/M/1/N/N$ queue

Theorem 2.1 *In a PEPA model if we define the stochastic process $X(t)$, such that $X(t) = C_i$ indicates that the system behaves as component C_i at time t, then $X(t)$ is a Markov process.*

We can construct transition rates $q(C_i, C_j)$ between components of the system as follows:

$$q(C_i, C_j) = \sum_{a \in \mathcal{A}ct(C_i|C_j)} r_a \qquad \text{where } \mathcal{A}ct(C_i \mid C_j) = \{\!|\, a \in \mathcal{A}ct(C_i) \mid C_i \xrightarrow{a} C_j \,|\!\}$$

Typically this multiset will only contain one element. The $q(C_i, C_j)$, or q_{ij}, are the off-diagonal elements of the infinitesimal generator matrix of the Markov process, Q. Diagonal elements are formed as the negative sum of the non-diagonal elements of each row, i.e. $q_{ii} = -\sum_{j \neq i} q_{ij}$.

The conditions which must be satisfied in order to guarantee the existence of an equilibrium distribution for a Markov process, and for this to be the same as the limiting distribution, are well-known—a stationary *or* equilibrium *probability distribution,* Π, *exists for every time homogeneous irreducible Markov process whose states are all positive-recurrent.*

All PEPA models are time-homogeneous since all activities are time-homogeneous: the rate and type of activities enabled by a component are independent of time. The other conditions, irreducibility and positive-recurrent states, are easily expressed in terms of the derivation graph of the PEPA model. We only consider PEPA models with a finite number of states so if the model is irreducible then all states must be positive-recurrent i.e. the derivation graph is strongly connected. In terms of the PEPA model this means that all behaviours of the system must be recurrent; in particular, for every choice, whichever path is chosen it must eventually return to the point where the choice can be made again, possibly with a different outcome.

It is interesting to note that *deadlock* and *livelock* in the process algebra model will correspond to an absorbing state, or set of states respectively, in the underlying Markov process. However, in this paper, we restrict ourselves to models without such features; for more details see [1].

Solving the Markov Process For finite state PEPA models whose derivation graph is strongly connected, (ergodic Markov process) the equilibrium distribution of the model, Π, is found by solving the matrix equation

$$\Pi Q = 0 \tag{2.1}$$

subject to the normalisation condition

$$\sum \Pi(C_i) = 1 \tag{2.2}$$

The computer algebra package Maple[2] [3] is used to find Π. The equations 2.1 and 2.2 are combined by replacing a column of Q by a column of 1s and placing a 1 in the corresponding row of 0. Moreover, since Maple deals with row vectors instead of column vectors, this modified Q is transposed.

---

[2] Maple is a registered trademark of Waterloo Maple Software.

Reward Structures and Derivation of Performance Measures Performance measures are derived by defining a *reward structure* over a model in a similar way to the use of reward structures in [15]. Reward structures have generally been explicitly treated only in the context of performability modelling, where reliability and performance aspects of a system are considered together. However, such structures may also be used to define performance measures.

As the emphasis in a PEPA model is on the behaviour of the system in terms of activities, rather than states, we associate rewards with certain activities within the system. The reward associated with a derivative (and underlying state), is then the sum of the rewards attached to activities enabled by the derivative. The performance measure is then defined as the total reward based on the steady state probability distribution, i.e. if ρ_i is the reward associated with derivative C_i, and $\Pi(\cdot)$ is the steady state probability distribution of the underlying Markov process, then the total reward R is

$$R = \sum_i \rho_i \, \Pi(C_i)$$

In this way, as in Stochastic Reward Networks [15], the rewards can be defined at the level of the PEPA model, rather than at the level of the underlying Markov process.

3 The PEPA Workbench

The design philosophy behind the PEPA workbench was to provide a set of simple tools to allow a skilled user of the PEPA language to delegate to machine assistance some of the routine tasks in checking PEPA descriptions and performing calculations of transition graphs and rewards. The Standard ML language was chosen as the implementation language for the workbench because it had previously been successfully used for the implementation of the Concurrency Workbench (for CCS and TCCS) [16] and choosing the same language may allow us to re-use some of the Concurrency Workbench code. Standard ML has also been used for theorem provers and other software tools locally since it provides high-level functionality via higher-order polymorphic functions. However, these functional language features are smoothly integrated with imperative assignment which allows the convenient construction of efficient programs. Standard ML is a strongly-typed, secure programming language and its use gives us confidence in the correctness of the workbench.

3.1 The Workbench Implementation

The workbench takes the form of a Standard ML image with the functionality implemented as Standard ML functions which have been pre-compiled. This provides a convenient and secure mechanism for exporting the PEPA workbench while also conveniently providing a powerful command line interface in the Standard ML language itself. A screen dump showing the workbench being accessed via the Lemacs editor is given in Figure 3. Some simple Emacs Lisp routines provide pull-down menus with sub-menus for issuing workbench commands. The benefits of the design of Standard ML are inherited by this process. For example, PEPA descriptions can easily be stored as

Fig. 3. The PEPA workbench

Standard ML values in the Standard ML environment and moribund values will then be taken away by the built-in garbage collector of the system, freeing the user of the workbench from the problem of managing and conserving space while generating large graphs. As a further example, it is easy to interrupt a PEPA workbench session at any time and still be able to return to it later simply by exporting the Standard ML image. No re-compilation of the PEPA description will be necessary upon returning to the session.

3.2 The PEPA Parser

PEPA is a mathematical notation and in designing a parser for the notation it was necessary to decide whether to use an extended character set for input or to decide to devise a replacement concrete syntax for the mathematical symbols. The second option was chosen. Distinct precedences were assigned to the connectives: the hiding operator was given highest precedence with prefix next, followed by co-operation. The choice operator was given lowest precedence. Parentheses were provided to allow the user to enforce the alternative parsing. The language does not have a local block construct so the processing of names is simplified. Separate name spaces are maintained for activities and components. Rates may either be entered as symbolic values or as numeric literals.

Notationally, even with the above additions, PEPA is certainly not a large language. For this reason, we decided not to use the Standard ML versions of the well-known

Lex and Yacc tools to generate a lexical analyser and a parser. This decision has a favourable consequence since using these tools would mean that they would be added to the exported image of the workbench, making it larger than really necessary. Instead, a Burge-style parser [17] has been produced for the PEPA language. This is a compact, elegant functional program which uses infix function symbols to encode the operators which combine productions in a formal description notation such as BNF. This provides a simple correspondence with the grammar for the language which makes the parser both easy to construct and easy to modify. Coding a Burge-style parser elegantly requires the language to provide polymorphic functions as first-class objects, which Standard ML does. In general, these parsers are not as efficient as Yacc-generated parsers but the efficiency of the PEPA parser is perfectly acceptable.

3.3   Computing the Transition Graph

The possible transitions of a PEPA component are obtained by following the transitions given in the operational semantic rules in Figure 1. The built-in exhaustiveness checking of the pattern-matching process deployed in this function checks that all program forms are handled by the function. Initially, the semantic rules were encoded in a naïve functional prototype implementation. This had the virtue of being obviously faithful to the language definition as given by the operational semantics but, as expected, this implementation was intolerably inefficient. Even when a sophisticated optimizing compiler was used to compile the workbench, small PEPA descriptions executed on a SparcStation 10/52 with 160Mb of memory had a running time of several hours. For some mid-sized PEPA descriptions, this prototype would exhaust the machine's memory and fail without delivering the transition graph.

After some study and analysis, a minor modification was made to produce the next version of the workbench. This used the imperative features of Standard ML to avoid some redundant re-computation which was being performed by the functional prototype. This modification was modest enough that we may be sure it did not alter the program's output from the results which would have been obtained from the prototype, thus maintaining our confidence in its correctness. However, now the workbench will calculate the transition graphs of mid-sized PEPA descriptions in a few seconds when running on a more modest SparcStation ELC with only 16Mb of memory!

This decrease in run time makes possible the interactive form of experimentation which we hoped that the workbench would provide, making it a considerably more useful tool. In addition, a decrease in memory utilization was achieved, facilitating the analysis of models greater than the largest which could have been handled by the functional prototype of the workbench.

3.4   Interfacing with Maple

The matrix manipulation routines which are required to solve the generator matrix either symbolically or numerically are provided by the Maple computer algebra package. It was judged to be simpler to use the existing Maple routines rather than re-implement these in Standard ML. Thus we have implemented the functionality to allow a workbench user to call Maple from the workbench. This enables a workbench user to pass PEPA values between Standard ML and Maple, manipulating them using whichever system is more useful for the processing task at hand.

Using the derivation graph the workbench specifies entries for the generator matrix in Maple syntax. Thus, results from the workbench can be written as Maple files and loaded into Maple. These files contain the results of the workbench analysis of the PEPA model and it is important that the PEPA user should be able to read these files in order to be able to check that the PEPA model has the behaviour which was expected. For this reason, the Maple input file is annotated with PEPA transition notation explaining the significance of the transition in the user's terms. These are written using Maple's comment notation and are therefore ignored by Maple.

4 Investigating a Simple MSMQ System

We illustrate the use of PEPA as a modelling paradigm, and the workbench, in an example taken from the study of communication systems. Polling systems have been used extensively over the last twenty years to investigate many computer and communication systems [18]. In these systems a single server circulates amongst a number of queues providing service according to a predetermined discipline. Extracting performance measures for these systems is non-trivial since the congestion at any one queue is dependent on the congestion at the other queues in the system. Recently these systems have been extended by the introduction of one or more additional servers to form *multi-server multi-queue* (MSMQ) systems [19]. MSMQ systems have been used to model applications in which multiple resources are shared among several users, possibly with differing requirements. In particular these models have been applied to local area network architectures, with ring topologies and scheduled access, in which more than one node may transmit simultaneously. For example, slotted rings and rings with multiple tokens are modelled as MSMQ systems by Yang et al. in [20].

Exact solutions for MSMQ systems have only recently been provided by Ajmone Marsan et al., [19]. In this paper we extend the class of asymmetric models considered by those authors. In [19] they consider a system of N nodes in which one node has capacity K and arrival rate $K\lambda$ while all other nodes have capacity 1 and arrival rate λ. This represents a network in which one node has high traffic and the other nodes have light traffic, such as a LAN connecting several diskless workstations and one file server. It was shown that the presence of the heavily loaded node did not greatly affect the mean waiting time of customers at lightly loaded nodes. Here we consider a system of N nodes each with capacity 1 and arrival rate λ but with customers at one node placing a larger service requirement on the server. We investigate the effect of this on the average waiting time of customers at the other nodes.

4.1 Model

We consider an MSMQ system in which there are four nodes, and two servers. Service is *limited*, meaning that each server serves at most one customer at each visit to each node. This corresponds to the *release-by-source* access mechanism for slotted rings. Moreover, only one server may service a node at any given time. Buffering is *restricted*:

a customer occupies a place in the buffer until its service is complete, and the arrival process is suspended whenever the buffer is full. We assume that the arrival process at each node is Poisson with parameter λ, and that normal service, heavy service and walk times in the system are exponentially distributed with rates μ, $m\mu$ and ω respectively.

The PEPA model of this system is shown in Figure 4. The components of the model of the system are the servers, and the nodes. Since the structure of the system is simple we model each node as a single entity.

$$Node_{j0} \stackrel{def}{=} (in, \lambda).Node_{j1} + (walk\_E_j, \top).Node_{j0} \qquad\qquad \text{for } 1 \le j \le N$$
$$Node_{j1} \stackrel{def}{=} (walk\_F_j, r_N).Node_{j2}$$
$$Node_{j2} \stackrel{def}{=} (serve_j, \mu_j).Node_{j0} + (walk\_E_j, \top).Node_{j2}$$
$$\text{where } \mu_j = \begin{cases} \mu & \text{if } j = 1 \\ m\mu & \text{if } 1 < j \le N \end{cases}$$

$$S_j \stackrel{def}{=} (walk\_F_j, \omega).(serve_j, \top).S_{j\oplus1} + (walk\_E_j, \omega).S_{j\oplus1}$$
$$\text{where } j \oplus 1 = 1 \text{ when } j = N$$

$$MSMQ \stackrel{def}{=} (Node_{10} \parallel Node_{20} \parallel Node_{30} \parallel Node_{40}) \underset{\substack{\{walk\_F_j, \\ walk\_E_j, serve_j\}}}{\bowtie} (S_1 \parallel S_1) \qquad \text{for } 1 \le j \le 4$$

Fig. 4. PEPA model of an asymmetric MSMQ system with restricted buffering

S_j denotes a server ready to approach the jth node in the system. There are two possibilities: either it walks to the node and finds it empty or occupied, or it walks to the node and finds a customer requiring service and no other server currently present. These two possibilities are represented by the two activities $walk\_E_j$ and $walk\_F_j$ respectively. After the former activity the server is ready to approach the next node, but after the latter it must remain at $Node_j$ until the service is complete. The rate at which service occurs is determined by the node. All the nodes appear alike to the server but we must distinguish between them in order to maintain the cyclic scheduling. Similarly, each of the servers appear alike to the nodes. The two servers do not directly interact with each other so they may be represented as $S_j \| S_j$ or $S_j \| S_k$.

Each node, $Node_j$ has three distinguishable states depending on whether the buffer is empty or full, and whether a full buffer is occupied by a server. These are represented by the three derivatives of the node component, $Node_{j0}$, $Node_{j1}$ and $Node_{j2}$. An arrival may occur only when the node is empty and this is represented by an in activity with rate λ. The node will enable a walk to the node without engaging the server, $walk\_E$, when it is empty ($Node_{j0}$) or when it is already occupied by a server ($Node_{j2}$). It will enable a walk and engage the server, $walk\_F$, whenever the buffer is full but there is no server currently present ($Node_{j1}$). In each case the rate of the walk activity is determined by the server. Although the nodes are not passive with respect to the $walk\_F$ action type, we assume that the corresponding activity rate r_N is greater than ω. When the buffer of the node is full and a server is present a *serve* activity will be enabled with a rate determined by the node. This activity must be completed before arrivals are resumed at the node.

The system has four nodes, so that when the server leaves $Node_4$ it walks on to $Node_1$. The nodes are independent, but must cooperate with a server to complete a *walk_E*, *walk_F* or *serve* activity.

4.2 Solution

The values which were assigned to the parameters are shown in Table 1. The effect of varying the service rate of customers at $Node_1$ was investigated with respect to the mean customer waiting time at the other nodes. The model has 560 states and 2064 transitions.

in	$serve_j$ $(j = 2, 3, 4)$	$serve_j$ $(j = 1)$	walk_E	walk_F
λ	μ	$m\mu$	ω	ω
0.1	1	$1 \leq 1/m \leq 5$	10	10

Table 1. Parameter values assigned to the PEPA MSMQ model

For each node we calculate the mean customer waiting time, W_j, by applying Little's Law to the node. The mean number of customers present at the node, N_j, is found by noting that there is exactly one customer present whenever the activity *in* is *not* enabled. Thus if we associate a reward of 1 with the activity *in* we can calculate the reward R_{in_j}. This has the effect of associating a reward of 1 with all states in which $Node_j$ is unoccupied. Then

$$N_j = 1 - R_{in_j}.$$

The throughput at the node, X_j, is found as the throughput of the activity $serve_j$, calculated by associating a reward of μ_j with the activity. Little's Law calculates the mean time spent in the node by a customer so the mean customer waiting time, W_j is:

$$W_j = \frac{N_j}{X_j} - \frac{1}{\mu_j} \tag{4.3}$$

The mean customer waiting time at each of the nodes, as the service demand at $Node_1$ increases, is shown in the graph shown in Figure 5. The expected waiting time for customers at $Node_1$ increases only slightly as the service demand at that node is increases. However at the other nodes the expected customer waiting time grows as the service demand at the $Node_1$ increases. It is interesting to note that this rate of growth is slightly slower at the node immediately downstream from the distinguished node ($Node_2$) as it is able to take advantage of the second server overtaking the server occupied at $Node_1$.

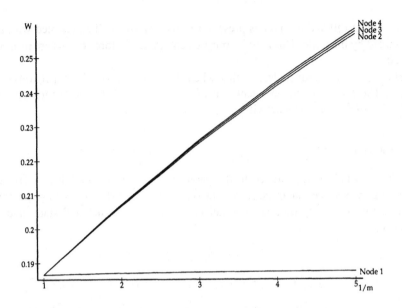

Fig. 5. A plot of mean customer waiting times

Note that using this approach asymmetric systems are handled as easily as symmetric ones. As with GSPNs the major problem of the approach is state space explosion. However, unlike GSPNs, the formal nature of the language makes it easy to detect symmetries within the system and to take advantage of these to simplify the model. A full description of these simplification techniques is beyond the scope of this paper but details can be found in [6].

5 Future Extensions

Although it would be possible to extend the PEPA language to include more combinators, we are not tempted to do this. The economy of PEPA makes reasoning about PEPA descriptions easier and made it straightforward for us to implement the workbench.

Some features could be added to the input language of the workbench to make the concrete syntax version of PEPA models shorter. These would include providing an array mechanism to allow the convenient description of families of related components. This feature is already present in the PEPA mathematical notation in the use of subscripting to denote component families.

More interesting planned extensions to the workbench include the addition of increased support for the experimentation process, allowing the workbench to take advantage of Maple's ability to solve global balance equations symbolically. In part this will rely on implementing an equivalence checker for PEPA components. This algebraic equivalence (known as *bisimulation*) will enable the user of the workbench to solve more complex models by replacing complex components with simpler ones which are algebraically equivalent. Finally, we intend to investigate the use of alternative algorithms for the balance equations to replace the Gaussian elimination with partial pivoting currently used.

References

1. J. Hillston. PEPA - Performance Enhanced Process Algebra. Technical report, Dept. of Computer Science, University of Edinburgh, March 1993.
2. R. Milner, M. Tofte, and R. Harper. *The Definition of Standard ML*. The MIT Press, 1990.
3. B.W. Char, K.O. Geddes, G.H. Gonnet, M.B. Monagan, and S.M.Watt. *Maple Reference Manual*. 1988.
4. R. Milner. *Communication and Concurrency*. Prentice-Hall, 1989.
5. N. Götz, U. Herzog, and M. Rettelbach. Multiprocessor and Distributed System Design: The Integration of Functional Specification and Performance Analysis using Stochastic Process Algebras. In *Performance'93*, 1993.
6. J. Hillston. *A Compositional Approach to Performance Modelling*. PhD thesis, Department of Computer Science, University of Edinburgh, 1994. to appear.
7. I.S.O. LOTOS : A Formal Description Technique Based on the Temporal Ordering of Observational Behaviour. IS 8807, TC97/SC21, 1989.
8. C.J. Koomen. *The Design of Communicating Systems: A System Engineering Approach*. Kluwer, 1991.
9. F. Moller and C. Tofts. A Temporal Calculus for Communicating Systems. In J.C.M. Baeten and J.W. Klop, editors, *CONCUR'90*, volume 458 of *LNCS*, pages 401–415. August 1989.
10. X. Nicollin and J. Stifakis. An Overview and Synthesis on Timed Process Algebras. In *Real-Time: Theory in Practice*, pages 526–548. Springer LNCS 600, 1991.
11. J. Davies and S. Schneider. A Brief History of Timed CSP. Technical report, Programming Research Group, Oxford University, Oxford OX1 3QD, September 1992.
12. B. Strulo. *Process Algebra for Discrete Event Simulation*. PhD thesis, Imperial College, 1993. to appear.
13. M. Ajmone Marsan, G. Conte, and G. Balbo. A Class of Generalised Stochastic Petri Nets for the Performance Evaluation of Multiprocessor Systems. *ACM Transactions on Computer Systems*, 2(2):93–122, May 1984.
14. N. Götz, U. Herzog, and M. Rettelbach. TIPP—a language for timed processes and performance evaluation. Technical Report 4/92, IMMD7, University of Erlangen-Nürnberg, Germany, November 1992.
15. J.K. Muppala and K.S. Trivedi. Composite Performance and Availability Analysis Using a Hierarchy of Stochastic Reward Nets. In G. Balbo and G. Serazzi, editors, *Computer Performance Evaluation: Modelling Techniques and Tools*, pages 335– 349. Elsevier, February 1991.
16. R. Cleaveland, J. Parrow, and B. Steffen. The concurrency workbench: A semantics-based tool for the verification of concurrent systems. *ACM Transcations on Programming Languages and Systems*, 15(1):36–72, January 1993.
17. W.H. Burge. *Recursive Programming Techniques*. Addison-Wesley, 1975.
18. H. Takagi. Queueing Analysis of Polling Models: An Update. In H. Takagi, editor, *Stochastic Analysis of Computer and Communication Systems*, pages 267 – 318. IFIP, 1990.
19. M. Ajmone Marsan, S. Donatelli, and F. Neri. GSPN Models of Markovian Multiserver Multiqueue Systems. *Performance Evaluation*, 11:227–240, 1990.
20. Q. Yang, D. Ghosal, and L. Bhuyan. Performance Analysis of Multiple Token Ring and Multiple Slotted Ring Networks. In *Proceedings of Computer Network Symposium*, pages 79–86, Washington DC, 1986. IEEE.

On the Accuracy of Memory Reference Models[1]

K. Grimsrud[a], J. Archibald[b], R. Frost[b], and B. Nelson[b]

[a]Intel Corporation, JF2-53, 5200 N.E. Elam Young Parkway, Hillsboro, OR
[b]Department of Electrical & Computer Engineering, Brigham Young University, Provo, UT

Abstract. This paper presents a new method of quantifying and visualizing the locality characteristics of any reference stream. After deriving a locality function, we show the correspondence between features of the locality function and common low-level program structures. We then apply the method to determine the locality characteristics of reference streams generated by a variety of synthetic models. These characteristics are shown to be substantially different from those of the reference trace used to determine the parameters of the models. We conclude that these synthetic models have serious inadequacies for evaluating the performance of memory hierarchies.

1   Introduction

Synthetic workload models have not been widely adopted in the industrial computer performance community due to concerns about the accuracy of those models in representing real reference streams. Many commercial computer designers rely on trace-driven simulation to evaluate machine performance, despite the expense and inconvenience of capturing, storing, and processing records of real machine execution. To be widely adopted, synthetic techniques must generate traces that are closely matched to actual trace characteristics.

This paper is concerned about quantifying the accuracy of synthetic models in representing reference locality, a characteristic essential in modelling memory hierarchies. It has two principle contributions. First, it describes a new approach for directly evaluating and visualizing the locality present in any reference stream. Second, it applies this technique to compare the locality characteristics of an actual workload with those of reference streams generated using several synthetic models. The locality visualization technique is shown to be an effective means of identifying the strengths and weaknesses of each model.

In Section 2 we present the new method of quantifying the locality in a reference stream, and we explore properties and characteristics of this notion of locality in Section 3. In Section 4, we evaluate the strengths and weaknesses of five synthetic memory reference models on the basis of their locality. Section 5 contains conclusions and a summary.

---

[1]This work funded in part by Amdahl Corporation and Intel Corporation.

2   Locality: a Quantitative Approach

It is well-known that computer programs exhibit statistical dependencies in their patterns of memory access. If a particular memory location has just been addressed, there is an increased probability that both that address and also nearby addresses will be accessed during the next few memory accesses of the program. These tendencies have been called *temporal locality* and *spatial locality*, respectively. Modern computer systems exploit locality in a variety of ways.

Despite the intuitive appeal of the concepts of temporal and spatial locality, they have not been previously formalized. In this work we present a quantitative definition that unifies both concepts, and that provides a convenient means of summarizing and visualizing essential locality characteristics for any reference sequence.

Let a trace, $\vec{T}$, be a vector (time-ordered list) of address references such that $\vec{T}[t_0]$ is the t_0^{th} address referenced. Temporal locality suggests that references tend to be repeated, that is,

$$\vec{T}[t_0] = \vec{T}[t_0+d]$$

for some small $d>0$. We refer to d as the *distance* or *delay* in making the repeated access which has $d-1$ intervening references. Temporal locality therefore refers to the *probability* that the address of the current reference will occur again d accesses in the future:

$$Pr\left(\vec{T}[t_0] = \vec{T}[t_0+d]\right) .$$

Similarly, spatial locality suggests that address locations close to the currently referenced address are more likely to be accessed in the next few references than locations far away. Defining s as the *stride* or *offset* in address space to the nearby memory location, we can extend our previous equation by considering

$$Pr\left(\vec{T}[t_0]+s = \vec{T}[t_0+d]\right) ,$$

the probability that an address s away from the currently referenced address will be accessed d references in the future. In this formulation temporal locality is simply a special case of spatial locality where $s=0$.

Of course, for any given value of s, there can exist multiple d such that $T[t_0]+s = T[t_0+d]$. In practice, only the first occurrence (minimum d) is of interest when analyzing locality since any subsequent occurrence is considered a repeated reference (temporal locality) at a later time index. Including the requirement of first occurrence, we define the locality of trace T as

$$L(\vec{T},s,d) = Pr\left(\vec{T}[t_0]+s = \vec{T}[t_0+d] \wedge \vec{T}[t_0]+s \notin \{\vec{T}[t_0+1] \cdots \vec{T}[t_0+d-1]\}\right) .$$

For numerical calculation and display, it is convenient to define a quantized version of L by defining bins in both delay d and stride s and by doubling the size of successive bins. This quantized locality function L^* is given as

Definition #1:

$$L^*(\vec{T},s,d)=Pr\left(\vec{T}[t_0]+s \ \epsilon \ \left\{\vec{T}\left[t_0+\frac{d}{2}+1\right]\cdots\vec{T}[t_0+d]\right\} \bigwedge \ \vec{T}[t_0]+s \ \epsilon \ \left\{\vec{T}[t_0+1]\cdots\vec{T}\left[t_0+\frac{d}{2}\right]\right\}\right)$$

Figure 1 A sample trace with the locality elements shown for $t_0=2$.

Figure 1 illustrates the calculation of locality. For $t_0=2$, we see the following locality characteristics:

Delay 1 -	Stride 1
Delay 2 -	Stride -1
Delay 3 -	Stride 0
Delay 4 -	Stride 3
Delay 5 -	Stride 4

Dist	Stride								
	-2	-1	0	1	2	3	4	5	6
1	1/7			5/7		1/7			
2		2/7			2/7		2/7		
3-4			2/7		1/7	3/7		1/7	
5-8							2/7	2/7	1/7

Table 1 The overall locality of the sample trace of Figure 1.

To obtain the locality for an entire trace, the locality characteristics are computed separately for all trace elements, quantized into the appropriate bins, and averaged across all trace elements for which it was computed -- in this case one less than the number of trace elements. (The last trace element has no successors with which to compute its locality.) For the trace above, the overall locality is presented in Table 1.

It is convenient to represent locality as a three dimensional surface with s and d as the base axes and $L*(T,s,d)$ as the magnitude. To make the surface more concise, the stride axis is quantized in $\log_2$ bins in the same way as the distance axis, with the resultant values computed as the average across the interval. Figure 2 shows the locality surface for instruction references from a 462 million reference trace of the KENBUS benchmark executing on an i486 system running SVR4 UNIX [6]. Since some parts of the surface are behind other features, a top view of the locality surface is also presented.

Much insight about the reference stream can be gained by analyzing the features of the associated surface. In the next section, we discuss typical surface features and describe the basic reference patterns responsible for producing each of them. (A thorough study of these features can be found in [8].) For the figures in this paper, the stride axis is limited to ±256. This is sufficient to capture most features of interest.

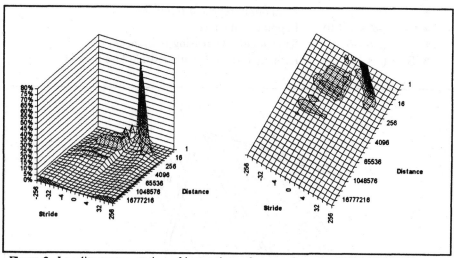

Figure 2 Locality representation of instruction references from the SVR4 KENBUS trace.

3   Properties of the Locality Surface

Although the above figure is limited in stride and distance, the locality surface actually stretches to infinity in both dimensions. The complete locality surface has two properties worth noting. First, the cross section of the surface for $d=1$ sums to one. This is equivalent to noting that the next address is within ±∞ of the current address. Other cross sections for constant d do not necessarily sum to one because they reflect only first occurrences; some addresses may have already been referenced with a smaller d. The second property is that, ignoring address boundary conditions (e.g., negative strides taken from address zero), cross sections of the surface for constant s sum to one. This is equivalent to saying that, for an infinitely long trace, each memory location in the machine is used and then reused, or that at most a finite number of references are not reissued. In practice, locality plots are truncated and traces are finite, so the area of any cross section can be less than one. However, for those parts of the graph where major locality features are located, it is a reasonable approximation and provides valuable insights.

3.1   Surface Features

This section describes the relationship between features of the locality surface and trace structures that cause them. The locality surface can be partitioned into the seven regions indicated in Figure 3 and listed below. The characteristic features of each region and sample traces that generate these features are discussed below.

- B+ $(s > d)$ Sequential references with average stride greater than 1.
- S+ $(s = d)$ Sequential references with a stride of 1.
- L+ $(0 < s < d)$ Sequential references with average stride less than 1.
- T $(s = 0)$ Temporally repeated references.
- L- $(-d < s < 0)$ Looping structures.
- S- $(s = -d)$ Sequentially decreasing references.
- B- $(s < -d)$ Reference streams with a stride less than -1.

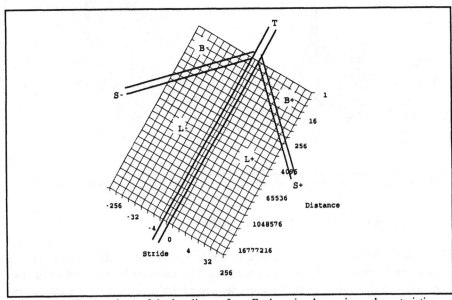

Figure 3 The 7 regions of the locality surface. Each region has unique characteristics.

Sequentiality.   Perhaps the easiest trace structure to analyze lies in region S+. Sequentiality is manifest as a ridge along the positive diagonal (S+ region) of the locality surface. A sequential run of references has the property that a stride of one, distance of one is followed by a stride of two, distance of two, and so on. The sequential ridge typically decays away from the origin as a function of the likelihood of a branch. Its length reflects the distribution of sequential run lengths in the reference stream, while its amplitude reflects sequential run frequency. The ridge may curve slightly into the L+ region as will be discussed in a later section.

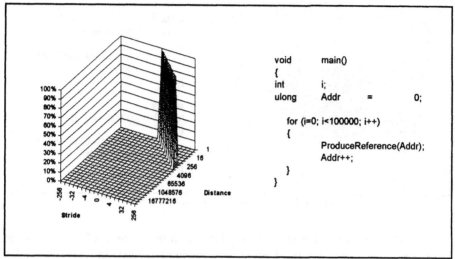

Figure 4 The locality of a sequential reference stream.

Figure 4 is the locality of the purely sequential reference stream generated by the algorithm of the associated listing. As can be seen in the figure, the ridge does not decay away from the origin since the sequential run length is essentially infinite. Also, the magnitude of the sequential ridge is 100% since all references are sequential.

Features in the S+ region are typical of sequential instructions being fetched and executed in the traditional Von Neumann approach; these features can be accentuated by instruction prefetch units. Some data accesses may exhibit this locality feature -- string operations for example -- although it is typically less pronounced than for instructions.

The S- region is similar to the S+ region except that consecutive references are to sequentially lower addresses instead of sequentially higher addresses. (Features on the locality surface can be reflected across the s=0 axis by reversing the order of the associated references.) Such access patterns are typical of some stack accesses and reverse order block transfers, but they are rare for instruction fetches.

Striding. The B+ region is similar to the S+ region except the sequential references have stride greater than one. Figure 5 displays the locality surface of a reference stream consisting solely of striding references; the algorithm used to generate it is also included. The reference stream consists of a series of references with a fixed stride of two followed by a series of references with stride eight.

Striding reference streams are typical of numerical algorithms such as matrix operations where the elements are accessed in row order instead of in column order. Striding is rare for instruction fetches.

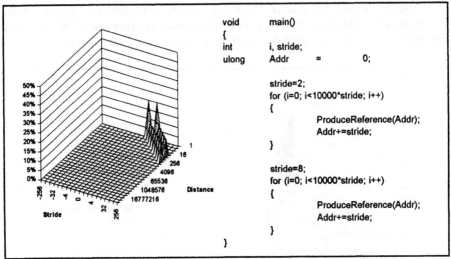

Figure 5 The locality of striding references.

Corresponding features can be produced in the B- region by a decreasing sequence of addresses with fixed strides -- accessing array data in reverse order for example -- but they are much less common.

Temporality.   The T region of the locality surface corresponds with the traditional notion of temporal locality. The shape of this region indicates the distribution of distances between repeated accesses. Figure 6 presents the locality surface of a reference stream consisting of repeated accesses to the same location separated by

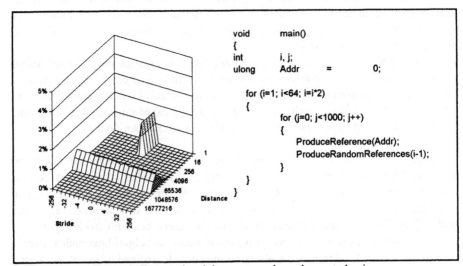

Figure 6 Temporal activity appears along the central spine.

376

strings of random references. The algorithm used to generate the references is also included in the figure. The elevated region near *d*=65k is an artifact of the random references used to separate the successive temporal accesses. The distance of temporal features from the origin is determined by the number of intervening references -- controlled by variable *i* in the listing.

Loops. The primary features in the L- region are ridges parallel to the stride axis that correspond to loops in the reference stream. Figure 7 shows the locality surface of a reference stream consisting of repeated loops of various lengths, together with the associated algorithm. Looping ridges are usually accompanied by a pronounced sequential ridge (due to the sequential references within the loop body) which decays as a function of the loop lengths.

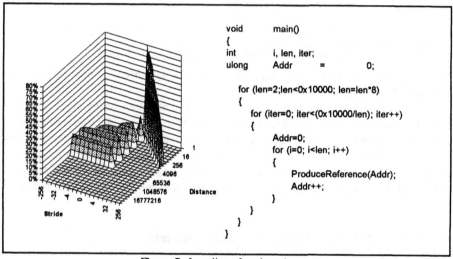

Figure 7 Locality of various loops.

The distance from the origin to the looping ridges is a function of the length of the loops, and the amplitude of the ridges is a function of the loop frequency. The looping ridges decay as the stride increases because only those few references at the end of each loop have strides large enough to form the ridge at its outermost point. The maximum stride is obtained at the very end of each loop where the distance is one -- the next reference is a large backward stride. For the same reason, the looping ridges also curve slightly toward the *d*=0 axis as the stride decreases. (On a linear scale the looping ridges would actually be diagonal.) The curvature is de-emphasized by the logarithmic scale of the axes and is most noticeable for short loops.

Fractional Striding. The most common features in the L+ region are created by *fractional strides* in which an otherwise sequential reference stream has intervening references between each sequential access. Figure 8 shows the locality surface of a reference stream with this characteristic. The associated algorithm produces a series of sequential addresses with a number of random intervening references between each.

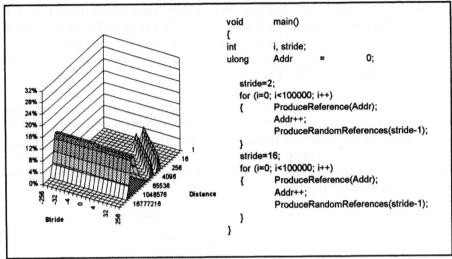

```
void        main()
{
int         i, stride;
ulong       Addr        =          0;

            stride=2;
            for (i=0; i<100000; i++)
            {      ProduceReference(Addr);
                   Addr++;
                   ProduceRandomReferences(stride-1);
            }
            stride=16;
            for (i=0; i<100000; i++)
            {      ProduceReference(Addr);
                   Addr++;
                   ProduceRandomReferences(stride-1);
            }
}
```

Figure 8 A locality surface illustrating fractional strides.

The ramp at the outer distance regions of the locality surface is again due to the random intervening references.

Fractional striding is quite common for data accesses. Many algorithms that appear sequential actually produce fractional striding features. A program that sums a vector of numbers, for example, may result in sequential data references to the vector with intervening references to the summing variable. Fractional striding is rare for instruction streams, but is common when analyzing merged instruction and data streams since data references typically fall between sequential instruction fetches.

4   Locality of Synthetic Reference Models

This section evaluates the ability of five synthetic reference models to reproduce the locality characteristics of a reference trace. The model parameters were chosen to fit the statistics of the instruction references of the SVR4 KENBUS trace (see Figure 2).

4.1   Independent Reference Model

The *independent reference model* (IRM) [2, 4, 5, 10] assigns a fixed probability of access to each address. References are selected independently without regard to past history. Because of the independence assumption, the IRM cannot be expected to produce a reference sequence with accurate temporal locality.

Figure 9 presents the cumulative probability distribution of accessing physical memory locations in the reference trace. It appears that the operating system resides in the lower part of the memory space; about 35% of the references are to locations in the first 100 KB of the address space. Applications are loaded throughout the remaining space.

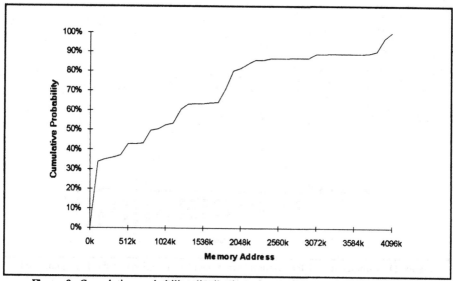

Figure 9 Cumulative probability distribution of accessing each word in memory.

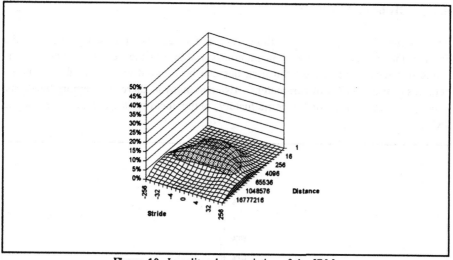

Figure 10 Locality characteristics of the IRM.

Figure 10 presents the locality surface of the reference stream generated by the IRM. The locality surface does not reflect the locality characteristics of the actual reference stream. Indeed, it has few if any of the features introduced in the previous section.

Figure 11 contrasts the temporal locality of the IRM generated references with that of the original trace. As expected, the two are quite dissimilar. We conclude that the IRM has limited value in the performance evaluation of locality dependent components.

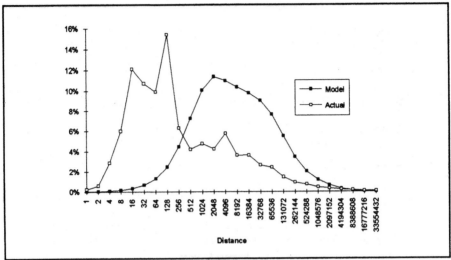

Figure 11 Comparison of the temporal cross section (T region) of the IRM generated references and the original trace.

4.2   Distance Model

The distance model [11] can be thought of as the relative addressing analog of the IRM; it models as an independent random variable, not the next reference, but the stride to the next reference. Since strides can be forward or backward, this model requires a large number of parameters -- twice as many as there are memory locations. Since the model incorporates some history, better locality is expected than for the IRM.

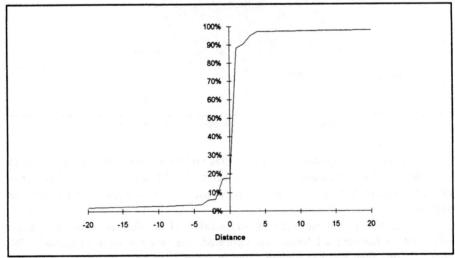

Figure 12 Cumulative probability of inter-reference spatial distances.

Figure 12 presents the probability distribution for strides in the KENBUS trace. Clearly, small strides dominate the reference stream. A distance of one is the most common, but a distance of zero is rare. While negative strides typically indicate looping structures, we believe that many of the short negative strides are due to the prefetch unit of the i486 processor fetching 4-word cache lines in different word orderings.

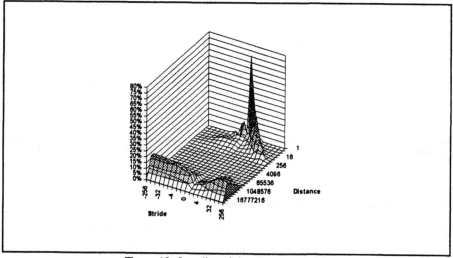

Figure 13 Locality of the distance model.

Figure 13 presents the locality characteristics of the references produced with the distance model. As expected, the sequential ridge is very prominent and a short spur exists on the negative diagonal (the S- region). However, the surface lacks the looping and other low-amplitude structures present in the original trace. These structures, which depend on higher-order statistics for their representation, are considered by the (approximately first-order) distance model to be random references, and are incorrectly accounted for by the prominent ramp at d = 4M. Recall that the notch on this ramp is due to the sequential ridge -- the sum along a locality cross section parallel to the $s = 0$ axis approaches 1.0.

Figure 14 presents the cross sections of the original and modeled locality surfaces along the S+ diagonal. The curves are qualitatively similar, but there are some important differences. For example, at a distance of 4, the actual curve is greater than the model curve by 50%. The performance of prefetch units and caches are particularly sensitive to such differences. We conclude that the distance model, although an improvement over the IRM, is probably not sufficiently accurate to evaluate the performance of memory hierarchies.

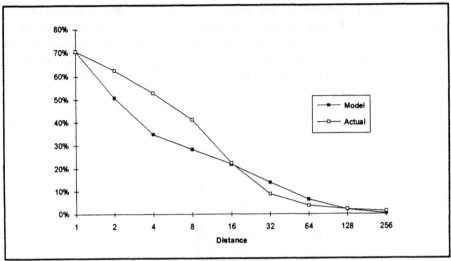

Figure 14 Comparison of sequential cross section for the distance model and actual trace.

4.3   Partial Markov Reference Model

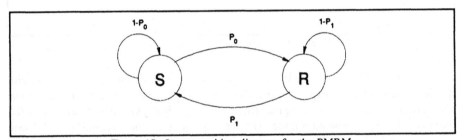

Figure 15 State transition diagram for the PMRM.

The partial Markov reference model (PMRM) [1] uses a simple two-state Markov chain to generate reference streams. Figure 15 illustrates the model. State S generates sequential references, while state R generates independent random references. The model produces runs of sequential references starting at random locations and with lengths whose distribution is determined by the arc weights.

As seen in Figure 16, the locality characteristics of the PMRM are similar to those of the distance model. In particular, both surfaces are dominated by two features: the sequential ridge and the ramp at $d = 4M$ caused by random references. However, the PMRM surface lacks the small-amplitude features found in the surface produced by the distance model.

Figure 17 compares the sequential ridge of the actual instruction stream and that of the PMRM. The general shape of the PMRM ridge is somewhat better than that of the distance model, which is encouraging, given that the PMRM has many fewer parameters than the distance model. However, the PMRM underestimates the

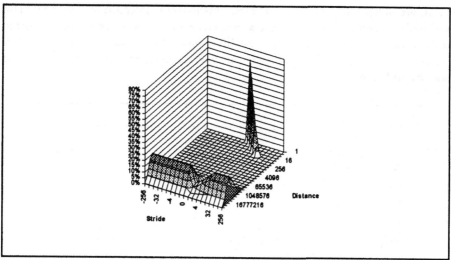

Figure 16 Locality characteristics of the PMRM.

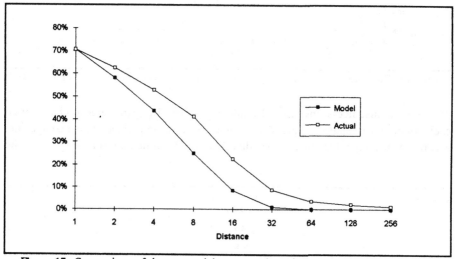

Figure 17 Comparison of the sequential cross section of the PMRM and actual trace.

sequential locality by as much as 60%. As with the distance model, we conclude that the PMRM is probably not suitable for memory hierarchy studies.

4.4   Distance-Strings Model

The distance-strings model [11] is an extension of the distance model that constructs a trace as concatenated bursts of sequential references. Both the length of the next burst, and the distance from the current address to the starting address of the next burst, are independent random variables. This model is intended to reflect program sequentiality better than either the distance model or PMRM by providing a realistic distribution of burst lengths. It is a simplified form of the model proposed by [7] in which a mixture of local deterministic behavior is combined with randomness in the starting addresses.

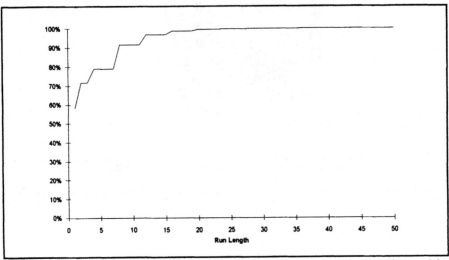

Figure 18 Cumulative probability of generating various length strings in the string model.

The distance-strings model requires a probability parameter to characterize each possible run length. In practice the number of these parameters is manageable. Figure 18 presents the cumulative distribution of sequential run lengths in the

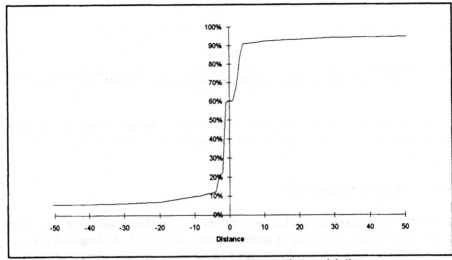

Figure 19 Cumulative probability of inter-string spatial distances.

reference trace. The majority of sequential runs are short; only 10% are longer than 10 references and only 1% exceed 20. This is in agreement with other studies [9] that have indicated that branch instructions occur every 5-6 instructions. Based on these statistics, we have limited the maximum run length in our implementation of the model to 255.

Figure 19 presents the cumulative distribution of distances between sequential runs. About 60% of all branches are backward and about 90% of all branches are to a location within ±50 of the current address. These statistics are consistent with other studies [9].

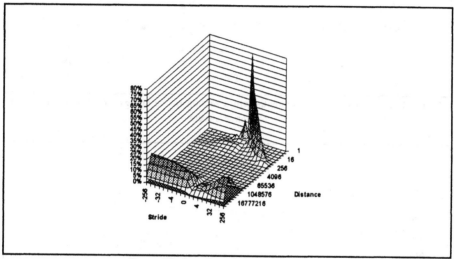

Figure 20 Locality of the strings model.

Figure 20 presents the locality characteristics of the references generated with the distance-strings model. Again, the only recognizable feature is along the sequential diagonal. Figure 21 compares the sequential locality characteristics of the generated and actual reference streams. It is disappointing that this model does not match this feature much better than the other models. It appears that the extra complexity of this model is not warranted.

4.5   Modified Stack Model

The stack model [3] assumes that with high probability the next reference will be to an address accessed in the recent past. This is accounted for by maintaining a stack of past references. The stack is reordered after every reference placing the most recently referenced address on top. The probability of accessing the element at each stack depth is fixed and computed before the model is used; the probabilities generally decrease with distance from the top of the stack.

In this paper, we have modified the stack algorithm to limit the size of the

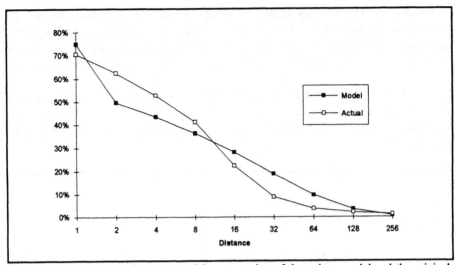

Figure 21 Comparison of the sequential cross section of the strings model and the original reference stream.

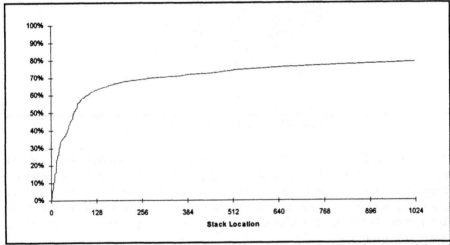

Figure 22 Cumulative probability of accessing each element in the stack for the stack-based model.

LRU stack to 1024 entries. Since the stack is smaller than the address space, there is a non-zero probability that the next reference will be to an address not currently in the stack. In this case we either generate a sequential access by incrementing the address by one, or we generate a random address. In either case the stack reordering puts the new element on top and pushes the bottom entry off the stack. Figure 22 presents the cumulative probability distribution for accessing each element of the stack model. (The top of the stack has index 0.)

Figure 23 presents the locality of the references generated with the stack model. Note the lack of sequential access and looping structures, and the presence of

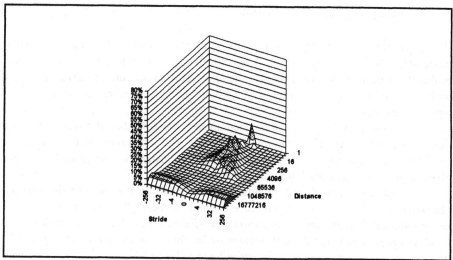

Figure 23 Locality characteristics of the stack model.

strong features along the temporal spine.

Figure 24 compares the temporal cross section (the T region) of the locality surface of the KENBUS trace with that of the model. The model references have temporal characteristics remarkably close to those of the original trace. The poor fit at distances greater than 2048 reflects the small stack used in the model. We conclude that the model accurately reflects temporal but not spatial locality.

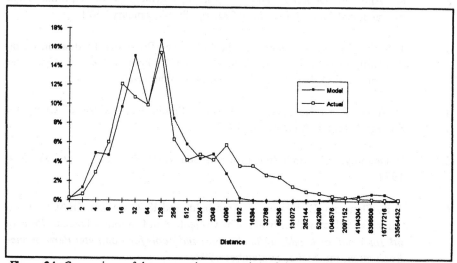

Figure 24 Comparison of the temporal cross section of the stack model and original trace.

5   Summary and Conclusions

This paper introduces a new technique for quantifying and visualizing the locality characteristics of reference streams. A new locality function was derived based on the probability that an address at a fixed stride or offset from the current reference occurs within a given number of references. This function unifies the traditional notions of temporal and spatial locality.

The locality function can be decomposed into a number of different regions that correspond to typical programming constructs. The position and shape of features in the various locality regions allow substantial insight to be gained into the characteristics of the underlying reference stream.

The various workload models were evaluated on the basis of the locality characteristics of the resultant reference streams. None of the examined workload models accurately represent the locality characteristics of actual reference streams. We believe that this new approach to measuring locality provides the means by which new, more accurate models can be devised and validated.

References

1.   A. Agarwal, M. Horowitz, and J. Hennessy, "An Analytical Cache Model," *ACM Transactions on Computer Systems*, Vol. 7, No. 2, pp 184-215, May 1989.

2.   A. Aho, P. Denning, and J. Ullman, "Principles of Optimal Page Replacement," *Journal of the ACM*, pp. 80-93, January 1971.

3.   J. Archibald and J-L. Baer, "Cache Coherence Protocols: Evaluation Using A Multiprocessor Simulation Model," *ACM Transactions on Computer Systems*, Vol. 4, No. 4, pp. 273-298, November 1986.

4.   O. Aven, E. Coffman, and Y. Kogan, "Stochastic Analysis of Computer Storage," Reidel, Amsterdam, 1987.

5.   P. Denning, "On Modelling Program Behavior," *Proc. SJCC*, pp. 937-944, 1972.

6.   J. K. Flanagan, B. Nelson, J. Archibald, and K. Grimsrud, "BACH: BYU Address Collection Hardware; The Collection of Complete Traces," *Proc. of 6th Int. Conf. on Modeling Techniques and Tools for Computer Performance Evaluation*, pp. 51-65, September 1992.

7.   C. Fricker and P. Robert, "A Memory Reference Model for the Analysis of Cache Memories," *Performance '90*, pp. 255-269.

8.   K. Grimsrud, "Quantifying Locality," PhD Dissertation, Dept. of Electrical & Ccomputer Engineering, Brigham Young University, December 1993.

9.   J. Hennessy and D. Patterson, "Computer Architecture, A Quantitative Approach," Morgan Kaufman, San Mateo, CA, 1990.

10.   G. Rao, "Performance Analysis of Cache Memories," *Journal of the ACM*, No. 25, pp. 378-395, 1978.

11.   J. Spirn, "Program Behavior: Models and Measurements," Elsevier, New York, 1977.

Automatically Estimating Network Contention of Parallel Programs

Thomas Fahringer

Dept. of Software Technology and Parallel Systems
University of Vienna
Bruennerstr. 72, A-1210 Vienna, Austria
e-mail: tf@par.univie.ac.at

Abstract

Efficient parallelization and optimization of programs on distributed memory parallel systems demands a considerable amount of knowledge about the underlying target machine and parallelizing compiler. Clearly, an automatic performance estimator would significantly alleviate the parallelization effort by relieving the programmer from the burden of studying machine and compiler details.

This paper describes how to automatically estimate the critical network contention behavior as induced by a parallel program on a parallel architecture. The approach assumes a hypercube network topology and a fixed and static routing policy. Careful modeling of the virtual to physical processor mapping, communication implementation, network topology and the underlying routing policy allows to prevent expensive simulation of the network contention behavior. The method is therefore low in computational complexity and memory requirements. A proof is presented which demonstrates the absence of network contention for an important class of communication statements.

The network contention cost function is fully implemented as part of the P^3T, which is a static parameter based performance prediction tool under the *Vienna Fortran Compilation System (VFCS)*.

1 Introduction

A key measure of the performance of a parallel program on a parallel machine is the communication overhead. On most state-of-the-art parallel systems sending data from a local to a remote processor still takes one or two orders of magnitude longer than the time to access the data on a local processor. It is not only the number of transfers[1] and the amount of data transferred which significantly influences the communication overhead. It has been shown ([6, 5, 17, 15]) that in particular network contention, which refers to the sharing of a communication channel by two or more transfers, may severely degrade all network traffic. For example, Shahid Bokhari reports in [5, 4] that on the iPSC/860 Hypercube, which uses circuit-switched communications with e-cube routing, channel contention can increase the time to communicate by a factor of more than seven.

It is very difficult for a programmer to be aware of the network contention induced by a parallel program. This requires solid knowledge of the underlying architecture and parallelizing compiler. Therefore, providing the programmer with feedback on the network contention behavior of a parallel program represents a major help in the parallelization effort.

This paper describes a novel static approach to compute the network contention of parallel programs for distributed memory systems. A single profile run ([8]) of the original program derives concrete values for program unknowns such as statement execution counts (frequencies). This is done by the *Weight Finder* ([9]), which is an advanced profiler for Fortran programs. The number of transfers of each processor are statically computed by the P^3T. Assuming e-cube routing allows one to obtain the exact path of each transfer. All channels are marked as occupied if traversed by a transfer. If a channel is already occupied then a channel contention counter is incremented by one. The source code is examined whether communication occurs before or inside of loop nests. In the first case, the probability of network contention is much higher, because all processors are likely to communicate in concert according to the loosely synchronous communication model ([11]). Therefore, the network contention counter is multiplied by a high probability value. The second case represents the asynchronous communication model, where it is unlikely that all processors communicate at the same time. For this case, the network contention counter is multiplied by a rather low probability value. This modeling technique is extended to procedures and the overall program.

Moreover, a formal proof for the absence of network contention for an important class of communication statements is presented.

The rest of this paper is organized as follows: The next section describes the underlying compilation and programming model. The third section describes the algorithm to compute the network contention of a communication statement, loop nest, procedure and the overall program; furthermore the network contention with respect to individual processors can be examined. The fourth section presents experiments on the iPSC/860 hypercube to validate the usefulness of the described network contention parameter. In the fifth section related work is discussed. Finally some concluding remarks are made.

---

[1] A transfer refers to a single transmission of data between two different processors.

2 Model

The described estimation method for the network contention of a parallel program is implemented as part of the P^3T ([10]), which is a static parameter based performance prediction tool. The P^3T computes a set of parallel program parameters which report on the communication overhead, data locality and the work distribution of a parallel program. It is integrated in the *VFCS* ([7]), which is an automatic parallelization system for massively parallel distributed-memory systems (DMS) translating Fortran programs into explicitly parallel message passing programs.

The parallelization strategy of the *VFCS* is based on domain decomposition in conjunction with the Single-Program-Multiple-Data (SPMD) programming model. This model implies that each processor is executing the same program based on a different data domain. The input to *VFCS* can either be a Fortran77 program and a separate data distribution specification – which defines the partitioning of the program's arrays and the mapping of the corresponding array segments to a set of parallel processors – or a program written in Vienna Fortran ([21]). Vienna Fortran is a machine-independent language extension to Fortran77, which provides annotations for the specification of data distributions. The performance prediction techniques described in this paper are applicable to both strategies; for the sake of simplicity only the first alternative is discussed in this paper. The output of *VFCS* is a parallel Fortran program with explicit message passing.

Let P denote the set of processors available for the execution of the parallel program, and A an array with *index domain* I^A, as determined by the declaration of A. A *data distribution* for A with respect to P is a total function $\delta^A : I^A \to \mathcal{P}(P)$, which associates one or more processors with every element of A. If $p \in \delta^A(i)$ for some $i \in I^A$, then a processor p *owns* $A(i)$, or $A(i)$ is local to p. The set of all local variables of a processor p is denoted by $\lambda(p)$; the corresponding projection containing only elements of A is specified by $\lambda^A(p)$. In this paper, only arbitrary *block* distributions, and total replication of arrays are considered. For each p and each A, $\lambda^A(p)$ is called the *segment* of A associated with p. For an m-dimensional array A, $\lambda^A(p)$ can then be characterized by a Cartesian product $X_{i=1}^m D_i^p$, where each $D_i^p = [l_i^p : u_i^p]$ describes the range of index values in dimension i of A that is associated with processor p. P^A refers to the set of all processors to which A is distributed.

The *work distribution* of the parallel program is determined – based on the data distribution – according to the *owner computes rule*[2].

Processors can only access local data. Non-local data referenced by a processor

---

[2] There are no other work distribution concepts considered such as those associated with FORALL loops in Vienna Fortran.

are buffered in so-called *overlap areas* that extend the memory allocated for the local segment. If $A \in \mathcal{A}^+$, the set of distributed arrays, and $p \in P$, then the overlap area, $OA(A,p)$, is defined as the smallest rectilinear contiguous area around the local segment of a processor p, containing all non-local variables accessed. The updating of the overlap areas is automatically organized by the system via explicit message passing. This is the result of a series of optimization steps that determine the communication pattern, extract single communication statements from a loop, fuse different communication statements, and remove redundant communication. These optimizations are applied after the first step in the parallelization of the program has been performed. This first step results in a *defining parallel program* as described below: for each potentially non-local access to a variable, a communication statement *EXSR* (EXchange Send Receive) is automatically inserted. An *EXSR* statement ([12]) is syntactically described as $EXSR\ A(I_1, ..., I_n)\ [l_1/u_1, ..., l_n/u_n]$, where $v = A(I_1, ..., I_n)$ is the array element inducing communication and $ovp = [l_1/u_1, ..., l_n/u_n]$ is the overlap description for array A. For each i, l_i and u_i respectively specify the left and right extension of dimension i in the local segment. The dynamic behavior of an *EXSR* statement can be described as follows:

```
IF (executing processor p owns v) THEN
        send v to all processors p', such that
        (1) p' reads v, and
        (2) p' does not own v, and
        (3) v is in the overlap area of p'
ELSE IF (v is in the overlap area of p) THEN
            receive v
        ENDIF
ENDIF
```

If C is a communication statement which implies the transfer of data messages due to a reference to a distributed array A in a parallel program Q, and $OA(A,p)$ is the overlap area of a processor $p \in P^A$ with respect to C, then the set of *neighboring segments* of p with respect to C is defined as follows:

$$NS(C,p) := \{K | K = OA(A,p) \cap \lambda^A(q) \text{ for some } q \in P^A \text{ and } p \neq q; K \neq \phi\}$$

Processor p can only receive data from one of its neighboring segments with respect to C. Neighboring segments are inherently owned by some neighboring processor $q \in P^A$ $(p \neq q)$. This is of major importance in order to analyze the network contention behavior induced by C. It is a prerequisite - provided a fixed and static routing policy - to obtain the exact message transfer path traversed between two processors on a network topology.

In Vienna Fortran ([21]) the user may declare processor arrays by means of the *PROCESSORS* statement. For instance, *PROCESSORS P(4,8)* declares a two-dimensional processor array with 4x8 processors. Distribution annotations may be appended to array declarations to specify distributions of arrays to processors. For instance, *REAL A(N,N) DIST(BLOCK,BLOCK) TO P* distributes array A two dimensional block-wise to the processor array P.

The loops of a loop nest L are consecutively numbered from 1 to n, where the loop at level 1 is the outermost loop. The loop at level i is denoted by L_i. Loop L_i consists of a loop body and a loop header statement. Loops are normalized, so that the loop increment is equal to 1. I_i is the loop variable of L_i.

3 Network Contention of a Parallel Program

A fixed and static routing policy such as the e-cube routing (cf. [8]) in a hypercube topology is assumed. This routing mechanism works as follows: each processor is identified by a unique binary label; starting with the right hand side of the binary label of the current processor, a message moves to the processor whose label most closely matches the label of the destination processor.

Furthermore, it is assumed that network contention occurs, iff two or more transfers occupy at least one specific network (communication) channel[3] at the same time in the same direction (*channel contention*). This can occur for the iP-SC/860 hypercube ([5]) which uses e-cube routing.

The *traversal direction* of a channel can be easily defined by the binary representation of the adjacent processors of a channel. For example, traversing a channel from the processor with the smaller binary label to the one with the higher binary label can be defined as the *positive traversal direction*; the opposite direction as the negative one.

The following definition specifies a network contention parameter in terms of number of channel contentions as implied by a specific communication statement.

Definition 3.1 Network contention for a communication statement

> *The network contention induced by $C \in C_L$ for a single instantiation of L is defined by a total function $nc(C)$.*

$nc(C)$ is the count of all channel contentions incurred by C. This means, if n transfers use the same channel at the same time in the same direction, then $nc(C)$ is incremented by $n - 1$. If network contention for C occurs on a set of channels CH, and for each $c \in CH$ there is a set of transfers TR_c – originating from C – inducing contention on c, then $nc(C) = \sum_{c \in CH} MAX\left(\left(\sum_{t \in TR_c} 1\right) - 1, 0\right)$.

---

[3] A channel refers to a physical link between to processors on a network topology.

3.1 Computing the Network Contention for a Communication Statement

The following describes how to derive $\overline{nc}(C)$, an estimated value for $nc(C)$, where A is the distributed array referenced by C; $ct^+(c)$ and $ct^-(c)$ the number of transfers respectively traversing a channel c of the target architecture in the positive and negative channel direction. The transfers induced for every $p \in P^A$ with respect to C are supposed to be evenly distributed across all neighboring segments of p, namely $NSa(C, p)$. For a precise analysis the exact number of transfers and associated amount of data transferred would have to be determined for each neighboring segment at the cost of additional computational complexity (cf. [8]). $nt^p(C)$ denotes the number of transfers induced by a processor p with respect to a C. This figure is equal to the number of receive operations processed by p. [8, 10] explains in detail how to statically estimate $nt^p(C)$ with high accuracy.

1. Initialize NC, a network contention counter, and $ct^+(c)$, $ct^-(c)$ for all channels c of the target architecture with zero

2. For every $p \in P^A$ do the following:

   (a) Compute $NSa(C, p)$, the set of all actually accessed neighboring segments of p, $nt^p(C)$, the number of transfers for p induced by C, and $w = nt^p(C)/NSa(C, p)$.

   (b) For each $h \in NSa(C, p)$ – which refers to a data portion to be received by p – find the owning processor p' (responsible to send h to p), and derive the exact message transfer path between p' and p according to the e-cube routing scheme (cf. [8]); for each network channel c traversed, the traversal direction is evaluated; For a positive channel traversal direction: if $ct^+(c) = 0$ which means that the channel is assumed to be free, then $ct^+(c) := w$ (w messages are supposed to traverse c consecutively in the positive direction) without changing NC; otherwise c is assumed to be already occupied due to previous transfers. Therefore NC is incremented by $\text{MIN}(w, ct^+(c))$ – number of channel conflicts – and $ct^+(c) := |ct^+(c) - w|$ in this order; For a negative channel traversal direction this step is done for $ct^-(c)$ only;

3. NC is multiplied by a probability value φ ($0 \leq \varphi \leq 1$).

φ in the above algorithm specifies a probability value between 0 and 1, which depends on C. For communication statements outside of a loop nest the loosely synchronous communication model is assumed, which means that all processors are supposed to interact simultaneously. For those inside of a loop nest, which represent the asynchronous communication case, there is a lower probability that all processors communicate in concert. Therefore, if $C \in C_o$, φ is supposed to be rather large (for instance, 0.8), otherwise it is assumed to have a small value (for instance, 0.2); The actual value for φ strongly depends on the application.

It might well be the case that the processors executing a parallel program do not reach a loop L at the same time during execution. In such a case φ should be obviously smaller than 0.8 with respect to communication statements outside of L. Only accurate and expensive simulation techniques or actually executing the program would make this analysis more precise.

$\overline{nc}(C)$ is defined by the value of NC at the end of the above algorithm.

3.2 Computing the Network Contention for a Loop Nest

In the following it is described how to extend the above algorithm for a loop nest L with communication statements. In order to take procedure calls into account, the parameter outcome for a single procedure call instantiation is supposed to be independent of the call site. This means that the parameter outcome at a particular call site is the same as the parameter outcome of the procedure over all call sites. This assumption is commonly made in performance estimators ([18]) to prevent more expensive analysis techniques such as simulation. All call graphs[4] are supposed to be acyclic. Let $\mathcal{F}_L$ be the set of procedure call statements contained in L. $proc(q)$ is the associated procedure of a procedure call $q \in \mathcal{F}_L$. $call(K)$ is the set of all associated procedure calls to a procedure K. $ncE(K)$ refers to the accumulated network contention implied by K.

1. for every $C \in \mathcal{C}_o$ do the following:

    (a) same as item 2 in the network contention algorithm for a single communication statement, replacing NC by NC1.

2. $NC1$ is multiplied by a rather large probability value φ (for instance, 0.8)

3. initialize $NC2$, a network contention counter with zero

4. for every $C \in \mathcal{C}_L \setminus \mathcal{C}_o$ do the following:

    (a) same as item 2 in the network contention algorithm for a single communication statement, replacing NC by NC2.

5. $NC2$ is multiplied by a rather small probability value φ (for instance, 0.2)

6. $NC3 = NC1 + NC2 + \sum\limits_{q \in \mathcal{F}_L} \dfrac{ncE(proc(q)) * freq(q)}{\sum\limits_{g \in call(proc(q))} freq(g)}$

$freq(q)$ refers to a characteristic number of times q is executed during a single program run. This figure is derived by the *Weight Finder* as outlined in Section 1. The rightmost term in 6. models the assumption that the parameter outcome at a particular call site is the same as the parameter outcome of the procedure over all call sites. $ncE(proc(q)$ is distributed across all procedure calls to $proc(q)$ based on the corresponding frequencies.

---

[4] A call graph describes calling relationships among procedures.

The value of $NC3$ at the end of the above algorithm specifies $\overline{ncL}(L)$, an approximated value for $ncL(L)$, the network contention implied by a loop nest L.

In the following it is shown how to extend the network contention parameter to procedures and an entire program.

Definition 3.2 Network contention for a procedure or an entire program

Let E be a procedure or an entire program, E contains a set of nested loops $\mathcal{L}_E$, $freq(L)$ is the frequency of $L \in \mathcal{L}_E$, and $\mathcal{F}_E$ is the set of procedure calls – outside of loops – in E, then the network contention induced by all communication statements in E is defined as follows:

$$ncE(L) := \sum_{L \in \mathcal{L}_E} ncL(L) * freq(L) + \sum_{q \in \mathcal{F}_E} \frac{ncE(proc(q)) * freq(q)}{\sum\limits_{g \in call(proc(q))} freq(g)}$$

The estimated values for $ncL(L)$ and $ncE(E)$ are referred to as $\overline{ncL}(L)$ and $\overline{ncE}(E)$, respectively. $\overline{ncE}(E)$ is computed by incorporating $\overline{ncL}(L)$ instead of $ncL(L)$ in Def.3.2.

Note that the estimated number of channel contentions is an upper bound. This is because time constraints are ignored. If two transfers ever occupy the same channel in the same direction based on the static e-cube routing then they are supposed to induce a single channel contention. This ignores the fact that channel contention only occurs, if both transfers traverse the same channel at the same time. Only precise and therefore expensive simulation techniques or actually executing – supposing that a channel contention measurement tool is available – the parallel program would allow to determine the exact channel contention behavior.

In the following a necessary condition for the existence of a channel contention between two different transfers on a hypercube topology is presented assuming e-cube routing. The binary representation of a processor identification on the target architecture is considered, which is specified by an array of bits. If the dimensionality of the underlying hypercube is $d \geq 1$, then a processor identification for a processor s is defined by $s[1 : d]$, where $s[1]$ and $s[d]$ is the least and highest significant bit, respectively. A transfer t from a sending processor s to a receiving processor r is described by $t : s \rightarrow r$.

Definition 3.3 Channel contention based on e-cube routing

Let $t1, t2$ be two transfers induced by a communication statement C on a hypercube topology with dimension d assuming e-cube routing, then a channel contention occurs, iff $t1$ and $t2$ traverse a channel in the same direction at the same time.

Lemma 3.1

Let $t1 : s1 \rightarrow r1$ and $t2 : s2 \rightarrow r2$ be two transfers implied by a communication statement C on a hypercube topology with dimension d assuming e-cube routing, then for a necessary channel contention condition between $t1$ and $t2$, there exists a k $(1 \leq k \leq d)$ such that

1. $s1[k : d] = s2[k : d]$, and

2. $s1[k] \neq r1[k]$ and $s2[k] \neq r2[k]$, and

3. $r1[1 : k] = r2[1 : k]$

Intuitively the above lemma can be explained as follows: The path between a source processor s and the target processor r is uniquely defined by the static and fixed e-cube routing. The $k-1$ lowest significant bits are already switched[5] from the source processor to the target processor binary representation. In the k-th bit both transfers have the same source processor and target processor binary representation, where the source and target binary representations are different. The $d - k$ highest significant bits are equal for both source processors in $t1$ and $t2$. When switching the k-th bit from the source to the target processor a channel contention may occur with respect to both transfers. Note that there might be several channel contentions because k is arbitrary.

For example, Fig.1 illustrates a four dimensional hypercube with the binary labels of all 16 processors. If a message is sent from processor 0 to processor 7 according to the e-cube routing policy, then the following path has to be traversed: $0 \rightarrow 1 \rightarrow 3 \rightarrow 7$. This path implies 3 network hops. Let $t1 : 0001 \rightarrow 0011 \rightarrow 0111$ and $t2 : 0000 \rightarrow 0001 \rightarrow 0011 \rightarrow 1011$. For this example $k = 2$ and the channel conflict between $t1$ and $t2$ occurs in $0001 \rightarrow 0011$.

Proof 3.1

Let the path for $t1$ be described by $a_1 \rightarrow a_2 \rightarrow ... \rightarrow a_r$ and for $t2$ by $b_1 \rightarrow b_2 \rightarrow ... \rightarrow b_s$ where $a_1 = s1$, $a_r = r1$, $b_1 = s2$, $b_s = r2$, and $1 \leq s, r \leq d$. If there is a channel contention induced by $t1$ and $t2$ based on e-cube routing, then there exists a triple (i, j, k) with $1 \leq i \leq r, 1 \leq j \leq s$ and $1 \leq k \leq d$ such that:

1. $a_i[k] = b_j[k]$, and

---

[5] A switch of a single processor binary representation bit yields the binary representation of an immediate neighboring processor according to the e-cube routing mechanism.

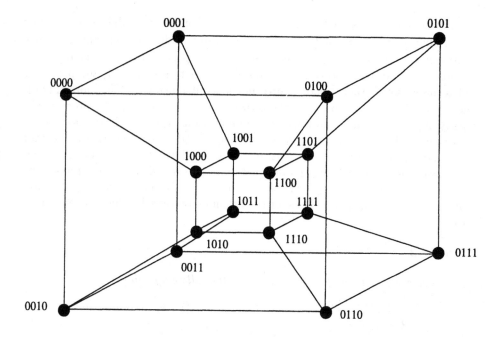

Figure 1: 4-dimensional hypercube topology

2. $a_{i+1}[k] = \neg a_i[k]$, and

3. $b_{j+1}[k] = \neg b_j[k]$, and

4. all bit positions besides the k-th one in a_i, b_i, a_{i+1} and b_{i+1} are identical, and

5. $a_i[k:d] = b_j[k:d] = a_1[k:d] = b_1[k:d]$, and

6. $a_i[k] = a_1[k]$ and $b_j[k] = b_1[k]$ and $a_r[k] \neq a_1[k]$ and $b_s[k] \neq b_1[k]$, and

7. $a_i[1:k] = b_j[1:k] = a_r[1:k] = b_s[1:k]$.

In the following an important class of communication statements will be discussed for which a proof – showing the absence of network contention on the iPSC/860 hypercube assuming e-cube routing – is presented. This proof is independent of the mapping strategy because it is based on the target processor's binary representation.

Theorem 3.1

Let $t1 : s1 \rightarrow r1$ and $t2 : s2 \rightarrow r2$ be two transfers, which are implied by a communication statement C, then if the trivial channel contention case ($s1 = s2$ and $r1 = r2$) is excluded there is no channel contention induced between $t1$ and $t2$, iff $r1 - s1 = r2 - s2$.

This theorem means that there is no channel contention induced by a single communication statement C, if all of its transfers induce a constant difference between sending and receiving processor, with respect to their binary representation. For example if processor 7 sends a message to processor 4, and processor 11 to processor 8, their binary representation difference is 3. Therefore there is no network contention between these two transfers. The above theorem covers many important stencil and wavefront computations, where each processor receives a single data message from a specific neighboring processor with a constant processor binary representation distance. However, the trivial network contention case where both source and destination of two transfers are equal has to be excluded. For this case – by definition of the e-cube routing – both transfers imply exactly the same path. Thus for each channel traversed a channel contention occurs.

Proof 3.2

For a non-trivial channel contention the following holds:

1. *$s1[k : d] = s2[k : d]$ and $r1[1 : k] = r2[1 : k]$ according to Def. 3.1, and*

2. *$s1 \neq s2$ or $r1 \neq r2$ excluding the trivial channel contention case*

If $s1 \neq s2 \Longrightarrow s1[1 : k-1] \neq s2[1 : k-1] \Longrightarrow r1[1 : k-1] - s1[1 : k-1] \neq r2[1 : k-1] - s2[1 : k-1] \Longrightarrow r1 - s1 \neq r2 - s2$.

If $r1 \neq r2 \Longrightarrow r1[k+1 : d] \neq r2[k+1 : d] \Longrightarrow if\, r1 - s1 = r2 - s2$ has to be true, then $s1[1 : k-1] \neq s2[1 : k-1] \Longrightarrow r1[1 : k-1] - s1[1 : k-1] \neq r2[1 : k-1] - s2[1 : k-1] \Longrightarrow r1 - s1 \neq r2 - s2$.

Corollary 3.1

Let C be a communication statement, A the array referenced by C and distributed according to the block distribution as outlined in Section 2. If

- *the overlap area of A extends the local area in exactly one dimension side in the same direction such that the overlap area does not extend the local area across more than one neighboring segment, and*

- *assuming a virtual to physical processor mapping strategy such that the absolute physical processor identification difference of all pairs of communicating processors with respect to C is a constant value,*

then C does not induce any network contention.

Proof 3.3

The conditions of the above corollary guarantee that every processor $p \in P^A$ communicates with at most one neighboring processor and the distance between source and target for each transfer is constant. Theorem 3.1 states the absence of network contention for such a communication statement.

For example, every specific communication statement in the Gauss/Seidel, Jacobi and Red-black relaxation kernel ([16]) do not imply any network contention based on the previous corollary. Note that this corollary refers only to single communication statements. The absence of network contention is not guaranteed for more than one of such a communication statement in consecutive order.

Moreover, a good network contention figure does not necessarily induce a good overall performance of a parallel program. Together with the network contention also other performance aspects such as number of transfers, amount of data transferred, transfer time, work distribution and data locality ([8, 10]) play an important role to obtain a well optimized program. P^3T fully adheres to this need by computing parallel program parameters for all of these performance aspects.

4 Experiments

The author has not yet found a reasonable way to measure the actual network contention for real programs on the iPSC/860 hypercube. This appears to be rather difficult at the program statement level using the limited measurement features available for the iPSC/860; on the other hand simulation might distort the real network behavior. However, as the network contention parameter is an upper bound, it will be probably good enough to rank program sections (e.g. communication statements and loops) with respect to their actual network contention outcome. In the following experiments the communication overhead of several parallel programs has been measured and related to the associated network contention figures. The following shows the Livermore Fortran Kernel-3 ([14]), which computes the inner product of two vectors.

Example 4.1 *Livermore Fortran Kernel 3 (LFK-3)*

```
PARAMETER (N=1000)
DOUBLE PRECISION Z(N), X(N)
...
Q= 0.000d0
DO K = 1,N
   Q= Q + Z(K)*X(K)
ENDDO
```

A sequential program $v1$ and four different parallel programs $(v2, \ldots, v5)$ have been created by respectively distributing array Z and X to 1, 2, 4, 8 and 16 processors. The following shows the parallel program version of $v4$ as created by the *VFCS*:

Example 4.2 *Parallel Livermore Fortran Kernel 3*

```
        PARAMETER (N=1000)
        PROCESSORS P(8)
        DOUBLE PRECISION Z(N), X(N) DIST(BLOCK)
        ...
        Q= 0.000d0
C1:     EXSR X(:) {{875/875}}
C2:     EXSR Z(:) {{875/875}}
L:      DO K = 1,N
            Q= Q + Z(K)*X(K)
        ENDDO
```

The "all" row in Table 1 displays the sum of the measured execution time for the loop nest L and the communication time for C1 and C2 in LFK-3. The "loop" and the "com" row respectively stand for the execution and communication time of L only. They are tabulated in percentages proportionally to the associated "all" values. The last row in this table displays $\overline{ncL}$, the estimated number of channel contentions for L.

Table 1: Measured execution and communication times and predicted network contention

program section	$v1$	$v2$	$v3$	$v4$	$v5$
all (secs)	0.000342	0.02013	0.051162	0.356298	1.607763
loop (% of all)	100	3	1	0.2	0.04
com (% of all)	0	97	99	99.8	99.96
$\overline{ncL}$	0	1.6	19.2	134.4	768

From Table 1 it can be clearly seen that for all program versions except $v1$, which inherently does not contain any communication, the communication overhead dominates the program's runtime. This effect is increasing for larger number of processors utilized. Furthermore the communication overhead is increasing with larger number of processors incorporated for a program. This is because the communication statements induce communication across the entire processor array P. Thus the more processors are involved in the computation of the kernel the more number of transfers are implied. The communication overhead is clearly increasing from $v2, \ldots, v5$ in this order. This order is also indicated by the predicted network contention figure. Based on these and other experiments the author believes that the described network contention parameter is able to rank different program versions with respect to the actual network contention outcome.

Network contention usually depends on the data volume transferred over the network. So far $\overline{ncL}$ is independent of this factor. Future work will be devoted to eliminate this modeling drawback. The current version of P^3T provides a separate parameter for the amount of data transferred, which can put the described network contention parameter into the right perspective. A low amount of data transferred might indicate that the estimated number of channel contentions is to pessimistic. While a high amount of data transferred probably makes the network contention parameter a realistic figure.

5 Related Work

G. Pfister and V. Norton ([15]) compute the asymptotically maximum value of the network throughput per processor and the asymptotic limit of the total communication bandwidth available as a function of the number of processors and a hot spot[6] parameter. These parameters are derived for shared memory multiprocessors with multi-stage networks.

V. Sarkar ([18]) uses a queuing theory model to approximate the effect of communication load, when the communication bandwidth cannot support the peak communication load. Even though some of his parameters are difficult to obtain at the program level during compile time, the overall method is usually sufficient to prevent large values of average waiting time for a communication request.

Ko-Yang Wang ([20]) estimates the influence of hot spots by analyzing patterns of a data dependence graph[7] and outlines its usability to generate network traffic in simulations.

V. Balasundaram et al. ([2]) describe a performance estimator which evaluates data partitioning schemes for distributed memory systems. Their tool indirectly relates to the network contention of collective communication statements by pre-measuring the associated runtime on a variety of target machines. They do not handle asynchronous communication.

In [1, 3, 13, 19] other interesting research done in this area is described.

---

[6] A hot spot refers to a module in the multi-stage blocking network that has sufficient concentration of network traffic.

[7] This graph describes the input/output dependences among program variables.

6 Conclusion

The communication overhead and in particular the network contention incurred by a parallel program may severely degrade the program's performance. It is therefore vital to provide the programmer with feedback on this performance aspect.

This paper describes a parallel program performance parameter, which reports on the number of channel contentions of a parallel program. This figure is statically estimated by the P^3T at the program level assuming a hypercube topology with e-cube routing. Moreover, for an important class of communication statements the absence of network contention was proven.

The complexity to compute the described network contention goodness function is independent of problem size, loop iteration and statement execution counts. As a consequence, the described method is faster than simulating or actually executing the parallel program.

Much of the described analysis is done for the iPSC/860 hypercube. In particular, the underlying e-cube routing is important for this modeling approach. However, the outlined techniques can be partially applied to other parallel machines as well, provided they are based on a hypercube topology with a static and fixed routing mechanism.

The lack of accurate measurement tools has prevented so far a precise evaluation of this method. However, as the network contention parameter is an upper bound of the number of channel conflicts, it most likely can be used to rank different program versions with respect to the actual network contention behavior.

References

[1] D. Atapattu and D. Gannon. Building Analytical Models into an Interactive Performance Prediction Tool. In *Proc. Supercomputing 89*, pages 521–530, Reno, Nevada, 1989. ACM Press.

[2] V. Balasundaram, G. Fox, K. Kennedy, and U. Kremer. A Static Performance Estimator to Guide Data Partitioning Decisions. In *3rd ACM Sigplan Symposium on Principles and Practice of Parallel Programming (PPoPP)*, Williamsburg, VA, April 21-24 1991.

[3] L. Bhuyan, Q. Yang, and D. Argawal. Performance of multiprocessor interconnection networks. *IEEE Computer*, pages 25–37, Februar 1988.

[4] S. Bokhari. A network flow model for load balancing in circuit-switched multicomputers. NASA Contractor Report 182049, Institute for Computer Applications in Science and Engineering, NASA Langley Research Center, Hampton, VA 23665, May 1990.

[5] S. Bokhari. Communication overhead on the intel iPSC-860 Hypercube. ICASE Interim Report 10, NASA Contractor Report 182055, Institute for Computer Applications in Science and Engineering, NASA Langley Research Center, Hampton, VA 23665, May 1990.

[6] S. Bokhari. Complete Exchange on the iPSC-860. ICASE Report No. 91-4, NASA Contractor Report 187498, Institute for Computer Applications in Science and Engineering, NASA Langley Research Center, Hampton, VA 23665, January 1991.

[7] B. Chapman, S. Benkner, R. Blasko, P. Brezany, M. Egg, T. Fahringer, H.M. Gerndt, J. Hulman, B. Knaus, P. Kutschera, H. Moritsch, A. Schwald, V. Sipkova, and H.P. Zima. *VIENNA FORTRAN Compilation System - Version 1.0 - User's Guide*, January 1993.

[8] T. Fahringer. *Automatic Performance Prediction for Parallel Programs on Massively Parallel Computers*. PhD thesis, University of Vienna, Department of Software Technology and Parallel Systems, October 1993.

[9] T. Fahringer. The Weight Finder, An Advanced Profiler for Fortran Programs. In *Automatic Parallelization, New Approaches to Code Generation, Data Distribution, and Performance Prediction*. Vieweg Advanced Studies in Computer Science, ISBN 3-528-05401-8, Verlag Vieweg, Wiesbaden, Germany, March 1993.

[10] T. Fahringer and H. Zima. A Static Parameter based Performance Prediction Tool for Parallel Programs. In *Invited Paper, In Proc. of the 7th ACM International Conference on Supercomputing 1993*, Tokyo, Japan, July 1993.

[11] G. Fox, M. Johnson, G. Lyzenga, S. Otto, J. Salmon, and D. Walker. *Solving Problems on Concurrent Processors*, volume 1. and 2. Prentice Hall, Englewood Cliffs, NY, 1988.

[12] H.M. Gerndt. *Parallelization for Distributed-Memory Multiprocessing Systems*. PhD thesis, University of Bonn, December 1989.

[13] C. Kruskal and M. Snir. The performance of multistage interconnection networks for multiprocessors. *IEEE Transactions on Computers*, C-32:1091–1098, Dec. 1983.

[14] F.H. McMahon. *The Livermore Fortran Kernels Test of the Numerical Performance Range*, pages 143–186. Elsevier Science B.V.,North Holland, Amsterdam, 1988.

[15] G. Pfister and V.A. Norton. Hot Spot Contention and Combining in Multistage Interconnection Networks. In *Proc. of the 1985 International Conference on Parallel Processing*, pages 790–797, 1985.

[16] W.H. Press, B.P. Flannery, S.A. Teukolsky, and W.T. Vetterling. *Numerical Recipes in C; The Art of Scientific Computing.* Cambridge University Press, 1988.

[17] D. Russel. *The Principles of Computer Networking.* Cambridge University Press, Cambridge, Great Britain, 1991.

[18] V. Sarkar. *Partitioning and Scheduling Parallel Programs for Multiprocessor.* The MIT Press, Cambridge, Massachusetts, 1989.

[19] A. Sethi and N. Deo. Interference in multiprocessor systems with localized memory access probabilities. *IEEE Transactions on Computers,* C-28(2), Feb. 1979.

[20] K.Y. Wang. A Performance Prediction Model for Parallel Compilers . Technical report, Computer Science Dept., Purdue University, November 1990. Technical Report CSD-TR-1041, CAPO Report CER-90-43.

[21] H. Zima, P. Brezany, B. Chapman, P. Mehrotra, and A. Schwald. Vienna Fortran - a language specification. Technical report, ICASE, Hampton,VA, 1992. ICASE Internal Report 21.

Modelling of Communication Contention in Multiprocessors

Cécile Tron and Brigitte Plateau

Laboratoire de Génie Informatique, IMAG-INPG,
46, avenue Félix Viallet,
38031 Grenoble cedex, FRANCE,
(emails: Cecile.Tron@imag.fr, Brigitte.Plateau@imag.fr)

Abstract. This paper deals with performance evaluation and modelling of point-to-point communications in parallel machines with distributed memory. Existing models of point-to-point communications in contention-free networks are presented. A major drawback of these models is that they do not reflect the behavior of the network during the execution of a real parallel application. A general methodology for studying the effect of contention on point-to-point communications under "reasonable" load is presented here. This methodology has been applied on the software router VCR on a Meganode.

Keywords: point-to-point communications, performance evaluation, modelling, contention, VCR.

1 Introduction

Parallel machines are used to run faster, and from this point of view performance is essential. In parallel machines with distributed memory the speed-up is limited by the cost of communication and synchronization. The ratio communication speed/computation speed determines the best possible level of granularity in order to use efficiently the architecture. Thus performance evaluation of communication is a key problem.

Communication delay is generally characterized by two parameters: *bandwidth* and *start-up delay*. These two parameters have different values depending on whether the evaluated communication primitive is a system call or a message passing library call, and depend on the performance of the hardware. Such parameters are taken into account in existing models of communication delay, but these models characterize only point-to-point communication in a *contention-free* network.

A general methodology to measure performance of communications under contention does not exist. Benchmarks dedicated to specific machines [7, 13, 4] generally fill in this gap. Results of these benchmarks are difficult to use in order to compare different machines. In this paper a general methodology is proposed to evaluate performance under light to average contention of point-to-point communications. Very high contention should not appear frequently and can be a sign of bad implementation choices or it can mean that the machine is not appropriate for the application. There are two types of communications in a parallel machine: *point-to-point communications* (one process communicates with another one), and *global communications* (groups of processes communicate). Global communications make an intensive use of the communication network, but are implemented in a synchronous mode in order to avoid unexpected delay. This paper focuses on modelling point-to-point communications, and the proposed methodology is planned to be extended to global communications.

In Section 2, we present existing models of point-to-point communication and methodologies generally used in order to evaluate communication performance. Section 3 describes the proposed methodology, and experiments on the software router VCR are presented in section 4 and section 5.

2 Performance Evaluation of Communication in Parallel Machines

The main parameters used to characterize communication are described in this section. These parameters do not generally take into account the contention of the network. The second part of this section describes the method used to measure these parameters.

2.1 Modelling of Point-to-Point Communications

In most of the parallel machines available on the market, communication delay can be characterized by a linear model. This means that communicating a message of size l between two physically connected processors takes:

$$T(l) = \beta + \tau l \tag{1}$$

where β is the *start-up delay*, the time required for the machine to handle the message at both source and destination node [15] ; τ is a characteristic of the link, its inverse $\frac{1}{\tau}$ is generally called the *bandwidth*. If the communication primitive is a system call or a message passing library call, these two parameters may have different values for the same hardware.

If the processors are not physically connected, messages must be routed. Such *routing* capabilities are integrated in the hardware of recent machines. In this case, communication delay depends on the size l of the message and on the distance d of the two processors which communicate. Modelling distant communication generally assumes that the linear model given by (1) is verified for two physically connected processors.

Popular routing modes are *message switching* and *packet switching*, and have been replaced in the "new" parallel machines by the *wormhole routing mode* [11, 17]. In the packet-switching routing mode the unit of routed data is the *packet*. Each packet includes a header composed of the identifier of the destination and the useful data (the message or the last packet can be smaller than a unit). Packets are independent, and at each step the packet is stored in the memory of the intermediate node before being sent to the next node. Packets are pipelined and the communication time depends on whether the message to be routed is lower or greater than a packet. The router has to split the message and this operation takes a time proportional to the number of packets included in the message. Given l the size of the message to be routed, d the distance between source node and destination node, p the size of a packet, h the size of the header, if l is smaller than p, communication delay can be modelled by the following equation:

If $l \leq p$ then

$$T(l,d) = \alpha_0 + \alpha_1 + d(\beta + \tau(l+h)) \tag{2}$$

where β and τ have the same meaning as in (1), α_0 is the time needed to initialize the packetization and α_1 the time needed to prepare a packet. If l is greater than or equal to p communication delay can be characterized by the following equation:

If $l > p$, $l = pq + r$ then

$$T(l,d) = \alpha_0 + \alpha_1(q+1) + d(\beta + \tau(p+h)) + (q-1)(\beta + \tau(p+h)) + \beta + \tau(r+h)(3)$$

where α_0 is the initialization time of the packetization, α_1 the preparation time of packet, β and τ are the same as in (1). Generally r is assumed to be equal to 0 or 1, h is assumed negligible in comparison to l, and α_1 negligible in comparison to $\beta + \tau p$. With these assumptions the approximated equations are :

If $l \leq p$ then

$$T(l,d) = \alpha_0 + \alpha_1 + d\beta + \tau dl \tag{4}$$

If $l > p$ then

$$T(l,d) = (\alpha_0 - (\beta + \tau p)) + d(\beta + \tau p) + \frac{l}{p}(\beta + \tau p) \tag{5}$$

The *wormhole routing mode* may be characterized by a linear model too [15]. In this mode, the unit of routed data is the packet (called *flit*). A flit (flow control digit) has the size of the queues of the channels. Unlike in the packet switching routing mode, a communication at distance d is not equal to d communications between neighbors. One packet, the *header*, is added, and only this includes the address of the destination, so packets are not independent. The header is sent first, and commutes routers on each intermediate node, the other packets following in a pipelined mode. The last packet frees the queue of the channel it is leaving.

First the router has to split the message in flit (the size of one flit is noted p); this operation is done within the following delay:

$$T_0(l) = \alpha_0 + \alpha_1 \frac{l}{p} \tag{6}$$

As in the packet-switching mode the message is generally assumed to include an exact number of flits. In (6) α_0 is the time needed by the router to handle the message, plus the initialization time of the packetization, α_1 is the time needed to prepare one flit.

The delay needed to forward the header at a distance d can be formulated as:

$$T_1(d) = \delta d + \tau p \tag{7}$$

where δ is the time needed to commute on each intermediate node, $\frac{1}{\tau}$ the bandwidth and p the size of a flit.

The rest of the message reaches the destination in a pipelined mode within :

$$T_2(l) = \tau l \tag{8}$$

where l is the size of the message and $\frac{1}{\tau}$ the bandwidth. The sum of (6), (7) and (8) gives the delay formula of the wormhole routing mode ; this model is generally approximated assuming $l \gg f$ and $\frac{\alpha_1}{p}$ negligible in comparison to τ, by the following equation:

$$T(l,d) = \alpha_0 + \delta d + \tau l \tag{9}$$

Linear model ($T(l,d) = a_1 + a_2 d + a_3 l$) is verified for a lot of routing modes (*packet switching, wormhole, circuit switching* ...) [18, 2].

Further, models described above depend on the topology of the network because they include the distance as a parameter. There are mainly three types of recent machines: those with a network (grid, hypercube, etc) including the computing nodes (Telmat's Meganode, Intel's Paragon, Fujitsu's AP1000), networks where computing nodes are connected to ports with non-homogeneous distances among them (like the fat-tree of Thinking Machine's CM-5) and networks where computing node are connected to ports at the same distance of each other (like multi-stage topology of PCI's CS2 or IBM's SP1). In grid-like and in the fat-tree topologies, the distance is relevant, but in multi-stage networks all nodes are at the same distance and the distance does not appear in the model.

Finally the evaluation of these model parameters is useful to assess the communication efficiency of a machine. The method used for such an evaluation is described in the following section.

2.2 Measurement of Communication Parameters

Manufacturers give values of the parameters β and τ of (1), but these values are generally hardware characteristics, and take into account neither software management nor resource contention. So β and τ have to be evaluated more accurately during real execution.

The aim of *benchmark* applications is to evaluate performance of machines during real execution, but most of them are developed on monoprocessor machines, and are not specific to parallel machines. For instance, the Livermore Loops [14], and LINPACK/LAPACK [7] use automatic parallelizers to evaluate parallel computers. The first benchmark specific to distributed memory parallel machines is GENESIS, which gives performance metrics, and includes specific codes for problems like scheduling, load balancing, communications and synchronisation. For instance, GENESIS [9] includes an implementation of the *ping-pong* method generally used to evaluate communications in MIMD (Multiple Instructions Multiple Data) parallel machines [10].

However, the ping-pong application executes only one communication at a time. This application is very useful to evaluate parameters described in paragraph 2.1, but it does not evaluate communications in the case of contention or of global communications. For these two cases, specific benchmarks have been developed for specific machine (for instance, evaluation of communications on the CM5 [1] [13, 4, 16]).

Boyd et al. developed a benchmark in the which user can control some communication parameters like the average number of point-to-point data communications per processor, the degree of sharing (the number of variables read but not owned by a processor), the computation-to-communication ratio [3]. However all these parameters are averages, and if the user controls the average traffic in the network, the distribution is not controlled. This benchmark uses synthetic sparse matrix multiplication.

2.3 Conclusion

If "classical" benchmarks are useful to evaluate the efficiency of a single processor or the overall power of a parallel machine, they are difficult to use for evaluation of communication in a parallel machine. GENESIS tries to fill this gap, but it includes a generic method only for point-to-point communications. One initiative to introduce contention in evaluation of communication is synthetic sparse matrix multiplication [3], but it gives results only for one "scheme of communication".

3 Methodology for Studying Communication in a Network Under Contention

In our experiment, the *load of a network* is defined as the average load of each physical link in the network. The *load of a physical link* is defined as the number of bytes placed on its input during a time unit. If a link is bidirectional it has an input load per direction.

An application devoted to performance evaluation of communications should not be specific to an architecture, because it should be executed on different machines in order to compare performance of communication networks. Such an application should allow users to know communications contentions (averages and variances) of each physical links.

We propose an application composed of a ping-pong application (called *tagged ping-pong*) and an application which generates the contention of the network. The application generating the contention is composed of a set of pairs *(master, slave)* of tasks. The slave task is always waiting for a message from the master.

---

[1] Trademark Thinking Machine Corporation

To implement this application on a machine, only communication primitives like *send* and *receive* and multithreading is required.

The characteristics of the contention are to be chosen and the possibilities are numerous. In parallel applications there are generally two types of messages: *control messages* and *data messages*. Control messages are relatively small and data messages may be large. Depending on the type of application, the proportion of small messages and of big messages may be different. For instance, in *divide and conquer* algorithms a lot of data messages are exchanged, and in some fault tolerant applications a lot of control messages are regularly exchanged. In order to mimic these behaviors, the size of messages sent by masters of the contention application is uniformly chosen in $[A_{small}, B_{small}]$ with probability p_1 or in $[A_{big}, B_{big}]$ with probability $p_2 = 1 - p_1$. In order to obtain different average load, the average time between each message varies.

The two tasks, master and slave, are placed in such a way that the load is the same on each physical link. The load is statistically homogeneous in the network. Even if it is not generally verified in real applications, it will be sufficient for our purpose which is to assess performance of a communication device from an applicative level. The goal is to derive a simple model from a portable experiment.

The mapping of the couples (master,slave) may be different according to the physical topology of the studied network. On grid-like topologies these two tasks are placed on each pair of physically connected processors, if the link is bidirectional a master *and* a slave are placed on each processor of the pair. On multi-stage topologies like omega networks, the simplest way to obtain the same average load on each physical link, is to place a couple on each pair of processors (the omega network is fully-connected). In order to compare the two implementations and measurements, the load generated per node has to be kept constant. On fat-tree topology, load may be tuned such that all intermediate routing nodes support the same load (mapping described in [13] verifies this condition).

The communication delay between the two tasks of the tagged ping-pong depends now on the size of the message exchanged by the tasks of the tagged ping-pong, the distance between the two tasks of the tagged ping-pong and statistical characteristics of the load of the network. As in a contention-free network, the *average* and the *variance* of the communication time between the two ping-pong tasks are measured. A heavy contention surely increases this variance, so the number of samples has to be chosen in order to obtain relevant values. Variation of the size of the message exchanged by the ping-pong tasks, and distance between these two tasks have to be chosen according to the domain of interest.

As described in section 2.1 the communication time on a contention-free network is generally a linear function of the size of the message and of the distance between processors which communicate. Messages using the same physical links (let us call them *background messages*) add extra delays, which means that the β parameter in (4) and (5) or parameter δ in (9) should be a function of the load. The experiments will identify the effect of contention on communication delay.

An experiment where the ping-pong tasks is replaced by a set of tasks performing a group communication could be easily done. Thus the methodology allows to evaluate the performance of global communication primitives in a loaded network, the load being generated by point-to-point communication. The definition of a load composed of point-to-point and global communication is more complex and would require more modelling work.

4 Experiments on Point-to-Point Communications in Absence of Network Traffic

The experiments were done on a Meganode [2], which is a MIMD machine with distributed memory, including 128 processors (transputers). This machine is reconfigurable, all physical topologies with a degree of four being possible. It does not have hardware routing capabilities, so the software router VCR (Virtual Channel Router) is used [5, 6]. VCR implements a packet-switching routing mode (see section 2.1). In our configuration of VCR the size of a packet is 160 *bytes* and the size of the header is 4 *bytes*. For all experiments presented in this paper the Meganode was configured as a grid with 32 nodes (a 4x8 grid). The grid topology is chosen because it has the maximal possible degree and it is easy to extend.

4.1 Ping-Pong Experiment

In this experiment only the two tasks of the tagged ping-pong were executed. A description of this application can be found in [10]. Figure 1 shows the average communication time as a function of the size of the message for different distances. The hops are due to the packetization.

In Figures 2 and 3, average communication time is a function of the two parameter sizes and distance. A change in the curve appears clearly when the size of ping-pong messages is lower or greater than the size of a packet (160 bytes).

The linear model given by (4) and (5) is verified. Figures 4 and 5 show the standard deviation of the average delay for two distances, this standard deviation (σ) is smaller than 5% for distance 2 and smaller than 2% for distance 10. According to this model, a linear regression on data gives:

If $l \leq 160$ then

$$T(l, d) = 168 + 111d + 1.2ld \quad (\mu s) \quad \sigma \leq 5\% \tag{10}$$

If $l > 160$ then

$$T(l, d) = -48 + 321d + 2l \quad (\mu s) \quad \sigma \leq 5\% \tag{11}$$

---

[2] Trademark Telmat

For these two regressions the sample squared multiple correlation coefficients are respectively 0.998 and 0.997. This confirms that these linear models are a good approximation of the experimental data. The identification of (4) and (10) with (5) and (11) gives results shown in Table 1.

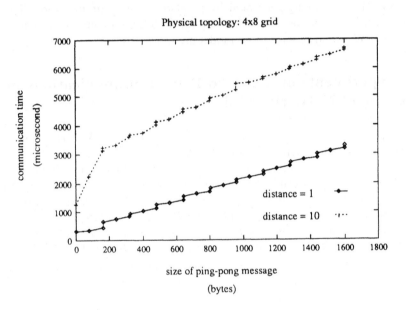

Fig. 1. Average communication time as a function of the size of the ping-pong message. Each value is the mean obtained with 1000 samples.

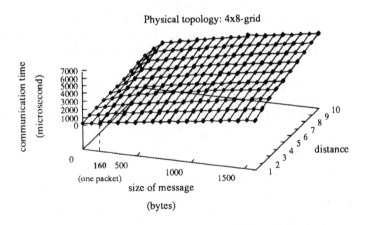

Fig. 2. Average communication time as a function of the size of the ping-pong message and the distance. Each value is the mean obtained with 1000 samples.

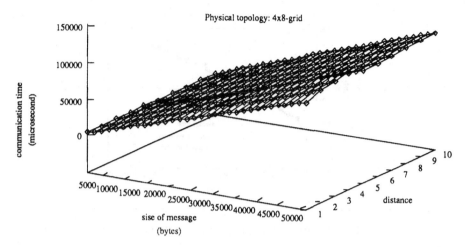

Fig. 3. Average communication time as a function of the size of the ping-pong message and the distance, for big sizes.

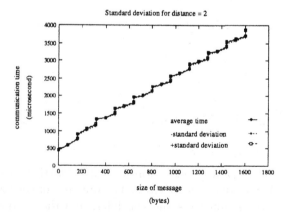

Fig. 4. Standard deviation of average communication time, distance = 2.

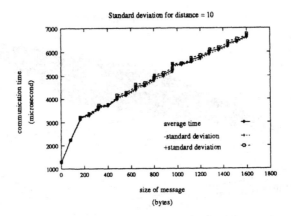

Fig. 5. Standard deviation of average communication time, distance = 10.

The measured value of τ is also equal to 1.2 $\mu second/bytes$, which means that the measured bandwidth is equal to 0.8 $Mbytes/second$. The bandwidth given by the manufacturer is 1.25 $Mbytes/second$, so the measured bandwidth is equal to 70% of the hardware bandwidth [19]. The difference between the measured values when l is smaller or greater than p is equal to 6%, and is due to the standard deviation of the measured communication time; this standard deviation is small but for the computation of $\beta + \tau p$ the errors are added.

Table 1. Results of identification

	α_0 (μs)	β (μs)	τ ($\mu s/bytes$)	$\beta + \tau p$ (μs)
$l \le p$	168	111	1.2	303
$l > p$	273			321

4.2 Influence of other Activities

Parameters of model (1) are "pure", which means that they characterize communications in absence of other activities. The presence of concurrent processes on the same processor influences the scheduler, and the routing process executes less frequently even if it has a high priority in a scheduling system without preemption.

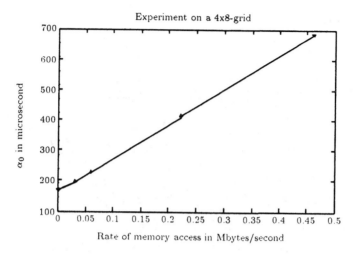

Fig. 6. Influence of other activities, α_0 as a function of the memory access rate

On the Meganode, communications are handled by a DMA on each node and the scheduler does not implement preemption. In order to observe the influence of other activities on communication parameters, a task accessing the memory and performing computations has been placed on each processor of the network. The memory access rate of this task varies, and a ping-pong is executed concurrently. Figure 6 shows the parameter α_0 as a function of the memory access rate; α_0 is time needed by the router to handle the message, and the router has to wait for the descheduling of the current activity before executing. The parameters β and τ are not influenced.

5 Experiments with the Loaded Network

This section presents an experiment of the methodology described in section 3. Conditions of the experiment are first detailed, and the results are presented in order to observe the influence of contention. A linear model using the size of the tagged message, the distance between ping-pong tasks and the average load as parameters is proposed and experimented. This model is refined in order to take into account the variance of the load. Finally the interpretation of the model is given.

5.1 Conditions of the Experiment

For these experiments the Meganode was configured as a 4x8-grid. The tagged ping-pong was executed concurrently with the application generating the contention as described in section 3. Because the Meganode has little memory on each node (1Mbyte/node), only light load was studied.

5.2 Results

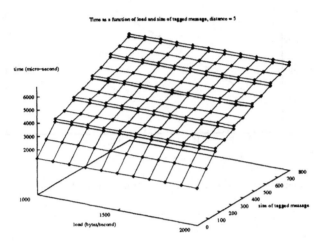

Fig. 7. Average communication time between tagged tasks when the distance is fixed, $p_1 = 0.5$.

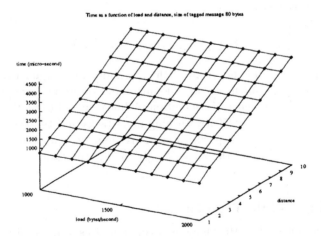

Fig. 8. Average communication time between tagged tasks when the size of the tagged message is equal to 80 *bytes*, $p_1 = 0.5$.

Results for p_1 equal to 0.5 are presented in the following figures. Figures 7, 8, 10, and 9 show the average communication time between the two tasks of the tagged ping-pong when one of the three parameters (size of tagged ping-pong message, distance between tagged ping-pong tasks and average contention) is fixed. In all these three cases, neglecting the effect of packetization, curves are approximatively planes. Change of slope when the size of the tagged message is lower or greater than the size of one packet (160) appears also on these curves.

Given the general aspect of the curves we assume that the influence of the load on the average communication time between tagged tasks is linear, and that this influence modifies the parameter β of (4) and (5). The load adds a delay and parameter β could be replaced by $\beta_0 + \beta_1 \bar{b}$ where $\bar{b}$ is the average load of the network, measured in bytes per second. So the average communication time in a network under light contention, in the case of the packet-switching mode, may be modelled as:

If $l \leq p$ then

$$\overline{T}(l, d, \bar{b}) = \alpha + \beta_0 d + \beta_1 d\bar{b} + \tau l d \tag{12}$$

If $l > p$ then

$$\overline{T}(l, d, \bar{b}) = (\alpha - (\beta_0 + \tau p)) + (\beta_0 + \tau p)\frac{l}{p} + (\beta_0 + \tau p)d - \beta_1\bar{b} + \beta_1\frac{l}{p}\bar{b} + \beta_1 d\bar{b} \tag{13}$$

where l is the size of the tagged message, d is the distance between the source and the destination of the tagged message, $\bar{b}$ the average load of the network and p the size of a packet.

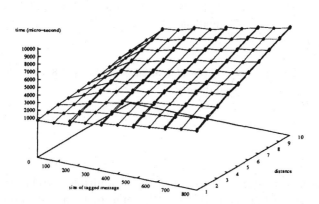

Fig. 9. Average communication time between tagged tasks when the average load is equal to 1500 $bytes/s$, $p_1 = 0.5$.

For these experiments, a regression on the data gives:

If $l \leq 160$ then

for $p_1 = 0.8$

$$\overline{T}(l, d, \overline{b}) = 216 + 101d + 0.012d\overline{b} + 1.16ld$$
$$R^2 = 0.9986$$

for $p_1 = 0.5$

$$\overline{T}(l, d, \overline{b}) = 182 + 106d + 0.005d\overline{b} + 1.17ld$$
$$R^2 = 0.9985$$

for $p_1 = 0.2$

$$\overline{T}(l, d, \overline{b}) = 174 + 107d + 0.004d\overline{b} + 1.17ld$$
$$R^2 = 0.9985$$

for $p_1 = 0.0$

$$\overline{T}(l, d, \overline{b}) = 170 + 107d + 0.003d\overline{b} + 1.18ld$$
$$R^2 = 0.9985$$

Where R^2 is the sample squared multiple correlation coefficient.

If $l > 160$ then

for $p_1 = 0.8$

$$\overline{T}(l, d, \overline{b}) = -74 + 1.9l + 318d + 0.016\overline{b} + 0.0001l\overline{b}$$
$$+ 0.016d\overline{b}$$
$$R^2 = 0.988$$

for $p_1 = 0.5$

$$\overline{T}(l, d, \overline{b}) = -107 + 2l + 321d + 0.008\overline{b} + 5 * 10^{-5}l\overline{b}$$
$$+ 0.008d\overline{b}$$
$$R^2 = 0.989$$

for $p_1 = 0.2$

$$\overline{T}(l, d, \overline{b}) = -114 + 2l + 322d + 0.005\overline{b} + 3.3 * 10^{-5}l\overline{b}$$
$$+ 0.005d\overline{b}$$
$$R^2 = 0.989$$

for $p_1 = 0.0$

$$\overline{T}(l, d, \overline{b}) = -117 + 2l + 322d + 0.005\overline{b} + 2.8 * 10^{-5}l\overline{b}$$
$$+ 0.005d\overline{b}$$
$$R^2 = 0.989$$

For all these regressions the sample squared multiple correlation coefficient (R^2) is greater than 0.9. Thus this model seems to be a good approximation of the data. However some of the regression coefficients depend on the value of p_1: all except the coefficients of l and d. This may imply that these coefficients depend on the variance of the load (the variance decreases when p_1 decreases). We study this in the next section.

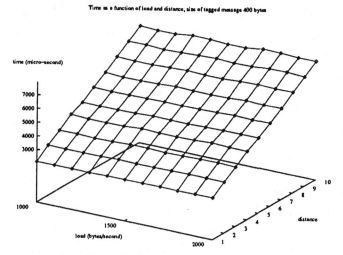

Fig. 10. Average communication time between tagged tasks when the size of the tagged message is equal to 400 *bytes*, $p_1 = 0.5$.

5.3 Refinement of the Model

In order to get an intuition of how this variance might interfere, consider that a link can be modelled by a single server. In this case, the extra delay added by the contention can be modelled by the average waiting time in the queue (generally noted $E[W]$). For one server with general service and input rate distribution, an approximation of $E[W]$ of a FIFO GI/GI/1 may be [8]:

$$E[W] = \frac{\rho(K_a + K_s)}{2\mu(1 - \rho)} \tag{14}$$

where $\rho = \frac{\lambda}{\mu}$, λ is the queue input rate, μ is the service rate, K_a the coefficient of variation of the input rate and K_s the coefficient of variation of the service rate. In our case λ corresponds to what is defined as *average load of a link* in section 3, measured in *packet/second*. The service time of our queue is equal to $(\beta + \tau p)$ *second/packets*, so the service rate is equal to $\mu = \frac{1}{\beta + \tau p}$ *packets/second*. K_s is also very low and may be approximated by zero. Let us note $\overline{b}'$ the input traffic expressed in *packet/second* ($\overline{b}' = \frac{\overline{b}}{p}$). With our notations, formula (14) is also equivalent to:

$$E[W] = \overline{b}' \frac{(\beta + \tau p)^2 K_{\overline{b}'}}{2(1 - ((\beta + \tau p)\overline{b}'))} \tag{15}$$

where $\overline{b}'$ is the average load expressed in packets/second, β and τ are the parameters defined in (1), $K_{\overline{b}'}$ is the coefficient of variation of $\overline{b}'$.

The first model used in section 5.2 approximates $\frac{(\beta+\tau p)^2 K_{\overline{b}'}}{2(1-((\beta+\tau p)\overline{b}'))}$ by a constant, and we have seen that the influence of $K_{\overline{b}'}$ cannot be neglected. We assume now that $(\beta + \tau p)\overline{b}'$ is small (light load), so

$$\frac{(\beta + \tau p)^2}{2(1 - ((\beta + \tau p)\overline{b}'))} = \frac{(\beta + \tau p)^2}{2}(1 + O((\beta + \tau p)\overline{b}')) \tag{16}$$

Given 15 and 16 we approximate $E[W]$ by:

$$E[W] = \overline{b}'\frac{(\beta + \tau p)}{2}K_{\overline{b}'}$$

$$= \frac{\overline{b}}{p}\frac{(\beta + \tau p)}{2}K_{\overline{b}}$$

$$= \beta_1\overline{b}K_{\overline{b}} \tag{17}$$

For each packet the communication time is now equal to $\beta + \tau p$ plus the waiting time ($E[W]$). Replacing β in (4) and (5) by $\beta_0 + \beta_1\overline{b}K_{\overline{b}}$ gives the following model:
If $l \le p$ then

$$\overline{T}(l, d, \overline{b}, K_{\overline{b}}) = \alpha + \beta_0 d + \beta_1 d\overline{b}K_{\overline{b}} + \tau l d \tag{18}$$

If $l > p$ then

$$\overline{T}(l, d, \overline{b}, K_{\overline{b}}) = (\alpha - (\beta_0 + \tau p)) + \frac{\beta_0 + \tau p}{p}l + (\beta_0 + \tau p)d - \beta_1\overline{b}K_{\overline{b}}$$

$$+ \frac{\beta_1}{p}l\overline{b}K_{\overline{b}} + \beta_1 d\overline{b}K_{\overline{b}} \tag{19}$$

where l is the size of the tagged message, d is the distance between the source and the destination of the tagged message, $\overline{b}$ the average contention of the network, $K_{\overline{b}}$ is coefficient of variation of $\overline{b}$ ($\frac{Variance(\overline{b})}{\overline{b}^2}$), and p the size of a packet.

A regression on all data of all the four experiments gives the following result:
If $l \le p$ then

$$\overline{T}(l, d, \overline{b}, K_{\overline{b}}) = 186 + 108d + 0.17d\overline{b}K_{\overline{b}} + 1.2ld \tag{20}$$

If $l > p$ then

$$\overline{T}(l, d, \overline{b}, K_{\overline{b}}) = -106 + 2l + 320d - 0.22\overline{b}K_{\overline{b}} + 0.0014l\overline{b}K_{\overline{b}} + 0.22d\overline{b}K_{\overline{b}} \tag{21}$$

The sample squared multiple correlation coefficient of these two regressions is respectively 0.99 and 0.98. The identification of (18) and (20) on one hand, and (19) and (21) on the other hand give results reported in table 2.

In presence of contention, the difference between the values of $\beta + \tau p$ is equal to 7%. In Table 2 we can see that the parameters β and τ are not modified by the contention: the parameter α_0 should increase because of the presence of other activities, however it is modified but not increased because its variation may be included in the parameter β_1 (the generated load ($\bar{b}$) and the memory access rate in the application are bound).

Given these values of β and τ, $\frac{(\beta+\tau p)^2}{2(1-((\beta+\tau p)*\bar{b}))}$ is plotted in figure 11. In this figure we can see that the approximation by a constant is valid for values of $\bar{b}$ smaller than 50% of the asymptotic value ($\frac{1}{\beta+\tau p}$). Our experiments clearly fall into this range.

Table 2. Results of identifications for experiments without contention and with contention

	α_0 (μs)	β or β_0 (μs)	τ ($\mu s/bytes$)	$\beta+\tau p$ (μs)	β_1 (μs^2)
Without contention					
$l \leq p$	168	111	1.2	303	
$l > p$	273			321	
With contention					
$l \leq p$	186	108	1.2	300	0.17
$l > p$	220			320	0.22

$$f(\bar{b}') = \frac{(\beta+\tau p)^2}{2(1-((\beta+\tau p)\bar{b}'))} \quad \text{in } s^2/packet$$

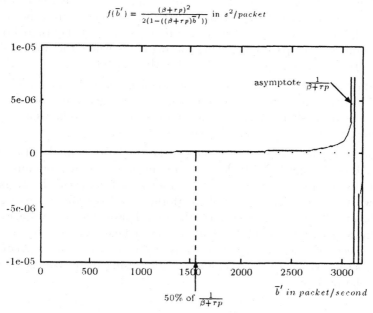

Fig. 11. Variation of the function $\frac{(\beta+\tau p)^2}{2(1-((\beta+\tau p)*\bar{b}'))}$ with $\beta+\tau p = 320\mu s$.

6 Conclusion

A benchmark has been proposed here, which allows user to evaluate point-to-point communication delay in a network under contention. This methodology is based on a synthetic application generating a traffic on the network in such a way that the effective traffic is statistically homogeneous. This benchmark can be implemented on different architectures as it requires only message passing capabilities and multithreading.

The proposed methodology has been applied on the Meganode with VCR, and a model of the point-to-point communications in this network under contention is proposed. This model is based on a linear model of the packet-switching routing mode without contention and is extended to take into account contention effects. The extension is still linear as the network is used under 50% of its capacity. We advocate that this should cover cases of normal usage. Heavy load is generated by group communications which should be studied as such. Modelling group communications is an extension of this work which we plan to do in the next future.

A similar model may exist for other routing modes. The methodology has to be adapted slightly. For instance, the circuit-switching routing mode is modelled by a linear equation in absence of other traffic [11]. In this routing mode, a header including the address of the destination is sent first to reserve the route. If it is successful, the message is transmitted and the route is freed. If there is contention, the message has to wait until the entire route is freed before being sent. So the waiting delay is added to the router start-up delay (noted α in the paper), unlike in the packet-switching routing mode where the waiting delay is added to the link start-up delay (noted β in this paper).

Further work includes testing the methodology on different machines such as CM5[3], SP1[4] and Paragon[5] in order to assess the generality of the method presented here. It should allow to the compare these networks using the same benchmarks and the same methodology.

Another application of these models is the emulation a model of a machine on a real one. Within the ALPES [12] environment, synthetic programs are used to model applications and are run on a parallel platform (here the Meganode). If one wants to experiment what the performance of the application would be if the bandwidth (τ) is twice bigger, the model says that all messages of the synthetic application should be shortened by a factor of two. By measuring relative speed-up, one can play with relative values of the computing power and network speed within the synthetic application using these models identifying simple relationships.

---

[3] Trademark Thinking Machine Corporation
[4] Trademark IBM
[5] Trademark Intel

References

1. *Proceedings of the fourth symposium on the frontiers of massively parallel computation, Frontiers'92,* 1992.
2. I. Bomans and D. Roose. Benchmarking the iPSC/2 hypercube multiprocessor. *Concurrency: Practise and Experience,* September 1989.
3. E. L. Boyd, J. D. Wellman, S. G. Abraham, and E. S. Davidson. Evaluating the communication Performance of MPPs Using Synthetic Sparce Matrix Multiplication Workloads. In *ICS93,* 1993.
4. Z. Bozkus, S. Ranka, and G. Fox. Benchmarking the CM-5 Multicomputer. [1], pages 100–107.
5. Mark Debbage, Mark Hill, and denis Nicole. *Virtual Channel Router Version 2.0 User Guide.* Department of Electronics & Computer Science, University of Southampton, Southampton, SO9 5NH,United Kingdom, June 1991.
6. Mark Debbage, Mark B. Hill, and Denis A. Nicole. Global communications on locally-connected message-passing parallel computers. University of Southampton, October 1992.
7. J.J. Dongarra. Performance of various computers using standard linear equations software in a Fortran environment. Technical Report MCS-TM-23, Argonne National Laboratory, April 1987.
8. E. Gelenbe and G. Pujolle. *Introduction aux réseaux de files d'attente.* Eyrolles, 1982.
9. A.J.G. Hey. The Genesis benchmarks. *Parallel Computing,* 17, 1991.
10. R. Hockney. Performance Parameters and Benchmarking of Supercomputers. *Parallel Computing,* 17(10-11):1111–1130, December 1991.
11. P. Kermani and L. Kleinrock. Virtual Cut-Through: a New Computer Communication Switching Technique. *Computers Networks,* 3:267–286, 1979.
12. João Paulo Kitajima, Cécile Tron, and Brigitte Plateau. ALPES: a tool for the performance evaluation of parallel programs. In J. J. Dongarra and B. Tourancheau, editors, *Environments and Tools for Parallel Scientific Computing,* pages 213–228, Amsterdam, The Netherlands, 1993. North-Holland.
13. M. Lin, R. Tsang, D. H. C. Du, A. E. Klietz, and S. Saroff. Performance Evaluation of the CM-5 Interconnection Network. Technical Report AHPCRC Preprint 92-111. University of Minnesota AHPCRC, October 1992.
14. F.H. McMahon. The Livermore Fortran Kernels: A computer test of the numerical performance range. Technical Report UCRL-53745, Lawrence Livermore Laboratory, December 1986.
15. L.M. Ni and P.K. McKinley. A Survey of Wormhole Routing Techniques in Direct Networks. *Computer,* pages 62–76, February 1993.
16. R. Ponnusamy, A. Choudhary, and G.Fox. Communication overhead on CM-5 : An Experimental Performance Evaluation. [1], pages 108–115.
17. C. L. Seitz. *VLSI and Parallel Computation,* pages 1–84. Morgan Kaufmann, 1990.
18. M. Syska. *Communication dans les architectures à mémoire distribuée.* PhD thesis, Université de Nice - Sophia Antipolis, France, December 1992.
19. TELMAT INFORMATIQUE, Soultz 68360 France. *Tnode Hardware Manual,* September 1990.

Using Interval Timed Coloured Petri Nets to Calculate Performance Bounds

W.M.P. van der Aalst

Eindhoven University of Technology
Dept. of Mathematics and Computing Science
P.O. Box 513, 5600 MB Eindhoven, The Netherlands
e-mail: wsinwa@win.tue.nl

Abstract. Nearly all existing techniques in the field of performance evaluation provide estimates (i.e. average and variation) for performance measures like: response times, throughput times, occupations rates, etc. However, when evaluating the design of a time-critical system, we are particularly interested in accurate bounds for these performance measures. We are concerned with the maximal response time of a controller in a nuclear power plant, the average response time of this controller is of less importance.

This is the reason we propose an alternative approach based on the *Interval Timed Coloured Petri Net* (ITCPN) model. The ITCPN model allows for the modelling of the dynamic behaviour of large and complex systems, without losing the possibility of formal analysis. In addition to the existing analysis techniques for coloured Petri nets, we provide a new analysis method to analyse the temporal behaviour of the net. This method can be used to calculate bounds for all kinds of performance measures.

In this paper we will show that this approach can be used to analyse complex time-critical systems.

1 Introduction

When designing a system (e.g. a computer system) we are often concerned with two things: (1) correctness and (2) performance. To verify the correctness of the system we have to prove that it performs its intended functions correctly. The performance is often characterised by performance measures like: response times, waiting times, maximum capacity, etc. To evaluate the correctness and performance of a complex system we need powerful analysis methods and tools.

In this paper we focus on the analysis of *time-critical* systems, i.e. *real-time* systems with 'hard' deadlines. These hard (real-time) deadlines have to be met for a safe operation of the system. An acceptable behaviour of the system depends not only on the logical correctness of the results, but also on the time at which the results are produced. Examples of such systems are: real-time computer systems, process controllers, communication systems, flexible manufacturing systems and just-in-time manufacturing systems. Note that for these systems correctness and performance issues are tightly coupled. An approach based on Petri nets is used to support the modelling and analysis of these time-critical systems.

Petri nets have been widely used for the modelling and analysis of concurrent systems (Reisig [21]). There are several factors which contribute to their success: the graphical nature, the ability to model parallel and distributed processes in a natural manner, the simplicity of the model and the firm mathematical foundation. Nevertheless, the basic Petri net model is not suitable for the modelling of many processes encountered in logistic, production, communication, flexible manufacturing and information processing systems. Petri nets describing real systems tend to be complex and extremely large. Sometimes it is even impossible to model the behaviour of the system accurately. To solve these problems many authors propose extensions of the basic Petri net model.

Several authors have extended the basic Petri net model with *coloured* or *typed tokens* ([8, 11, 12, 14]). In these models tokens have a value, often referred to as 'colour'. There are several reasons for such an extension. One of these reasons is the fact that (uncoloured) Petri nets tend to become too large to handle. Another reason is the fact that tokens often represent objects or resources in the modelled system. As such, these objects may have attributes, which are not easily represented by a simple Petri net token. However, these 'coloured' Petri nets allow the modeller to make much more succinct and manageable descriptions, therefore they are often called *high-level nets*.

Other authors have proposed a Petri net model with explicit quantitative time (e.g. [9, 16, 17, 20]). We call these models *timed Petri net* models.

In our opinion, only timed *and* coloured Petri nets are suitable for the modelling of large and complex real-time systems. There seems to be a consensus of opinion on this matter and several timed coloured Petri net models have been proposed in literature (cf. Van Hee et al. [8, 9], Morasca [18], Jensen and Rozenberg [14]). However, only a few methods have been developed for the analysis of the temporal behaviour of these nets. This is one of the reasons we propose the *Interval Timed Coloured Petri Net* (ITCPN) model and an analysis method, called *MTSRT*, based on this model.

The ITCPN model ([2, 3]) uses a rather new timing mechanism where time is associated with tokens. This timing concept has been adopted from Van Hee et al. [9]. In the ITCPN model we attach a *timestamp* to every token. This timestamp indicates the time a token becomes available. Associating time with tokens seems to be the natural choice for high-level Petri nets, since the colour is also associated with tokens. The *enabling time* of a transition is the maximum timestamp of the tokens to be consumed. Transitions are *eager* to fire (i.e. they fire as soon as possible), therefore the transition with the smallest enabling time will fire first. Firing is an atomic action, thereby producing tokens with a timestamp of at least the firing time. The difference between the firing time and the timestamp of such a produced token is called the *firing delay*. The (firing) delay of a produced token is specified by an *upper* and *lower bound*, i.e. an *interval*.

We have developed an analysis method, called the Modified Transition System Reduction Technique (MTSRT), which can be used to analyse ITCPNs (see Van der Aalst [2, 3]). The MTSRT method can be used to calculate *bounds* for all kinds of performance measures.

Nearly all other approaches aiming at performance analysis by means of Petri nets are based on a *stochastic Petri net model*. A stochastic Petri net is a timed Petri net with delays described by probability distributions ([7, 15, 16]). Analysis of stochastic Petri nets is possible (in theory), since the reachability graph can be regarded, under certain conditions, as a Markov chain or a semi-Markov process. However, these conditions are quite restrictive: all firing delays have to be sampled from an exponential distribution or the topology of the net has to be of a special form (Ajmone Marsan et al. [15]). Moreover, these Markov chains tend to be extremely large and extensions with colour are impracticable. Since there are no general applicable analysis methods, several authors resorted to using simulation to study the behaviour of the net (see section 3).

To avoid these problems, we propose delays described by an *interval* specifying an upper and lower bound for the duration of the corresponding activity. On the one hand, interval delays allow for the modelling of variable delays, on the other hand, it is not necessary to determine some artificial delay distribution (as opposed to stochastic delays). Instead, we have to specify bounds. These bounds can be used to verify time constraints.

This does not mean that we that we advise against the use of stochastic Petri nets! Moreover, 'stochastic analysis' and 'interval analysis' can be used in conjunction. In fact, we often use simulation along with the analysis method described in this paper.

The main purpose of this paper is to show that the ITCPN model can be used to model and analyse complex time-critical systems. For this purpose we model two non-trivial systems: a computer system (section 4) and a manufacturing system (section 5). Introductions to the ITCPN model and the MTSRT method are given in section 2 and section 3 respectively.

2 Interval Timed Coloured Petri Nets

In this section we give an informal introduction to the ITCPN model. The formal definition is given in appendix A. For the formal semantics of the ITCPN model the reader is referred to [2] or [3]. We use an example to introduce the notion of interval timed coloured Petri nets. Figure 1 shows an ITCPN composed of four **places** ($p_{in}, p_{busy}, p_{free}$ and p_{out}) and two **transitions** (t_1 and t_2). At any moment, a place contains zero or more **tokens**, drawn as black dots. In the ITCPN model, a token has three attributes: a position, a value and a timestamp, i.e. we can use the tuple $\langle\langle p, v\rangle, x\rangle$ to denote a token in place p with value v and timestamp x. The value of a token is often referred to as the **token colour**. Each place has a **colour set** attached to it which specifies the set of allowed values, i.e. each token residing in place p must have a colour (value) which is a member of the colour set of p.

The ITCPN shown in figure 1 represents an abstract computer system, jobs arrive via place p_{in} and leave the system via place p_{out}. The computer is composed of a number of parallel processors. Each processor is represented by a token which is either in place p_{free} or in place p_{busy}. The beginning and termination of the execution of a job on a processor are represented by the transitions t_1 and t_2 respectively.

There are three colour sets $\mathcal{P} = \{P1, P2, ..\}$ and $\mathcal{J} = \{J1, J2, ..\}$ and $\mathcal{P} \times \mathcal{J}$. Colour set $\mathcal{J}$ (job types) is attached to place p_{in} and place p_{out}, colour set $\mathcal{P}$ (processor types) is attached to place p_{free}. Colour set $\mathcal{P} \times \mathcal{J}$ is attached to place p_{busy}.

Places and transitions are interconnected by **arcs**. Each arc connects a place and

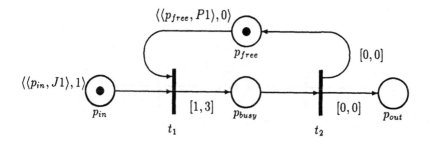

Figure 1: An interval timed coloured Petri net

a transition in precisely one direction. Transition t_1 has two input places (p_{in} and p_{free}) and one output place (p_{busy}). Transition t_2 has one input place (p_{busy}) and two output places (p_{free} and p_{out}).

Places are the passive components, while transitions are the active components. Transitions cause state changes. A transition is called **enabled** if there are 'enough' tokens on each of its input places. In other words, a transition is enabled if all input places contain (at least) the specified number of tokens (indicated by the number of interconnecting input arcs). An enabled transition may **occur (fire)** at time x if all the tokens to be consumed have a timestamp not later than time x. The **enabling time** of a transition is the maximum timestamp of the tokens to be consumed. Because transitions are eager to fire, a transition with the smallest enabling time will fire first.

Consider the state shown in figure 1, place p_{in} contains a token with value $J1$ and timestamp 1 and place p_{free} contains a token with value $P1$ and timestamp 0. Transition t_2 is not enabled because input place p_{busy} is empty. Transition t_1 is enabled and the enabling time of t_1 is equal to 1 (the maximum of 1 and 0).

Firing a transition means consuming tokens from the input places and producing tokens on the output places. If, at any time, more than one transition is enabled, then any of the several enabled transitions may be 'the next' to fire. This leads to a non-deterministic choice if several transitions have the same enabling time.

Firing is an atomic action, thereby producing tokens with a timestamp of at least the firing time. The difference between the firing time and the timestamp of such a produced token is called the **firing delay**. This delay is specified by an **interval**, i.e. only delays between a given upper bound and a given lower bound are allowed. In other words, the delay of a token is 'sampled' from the corresponding delay interval. Note that the term 'sampled' may be confusing, because the modeller does not specify a probability distribution, merely an upper and lower bound.

Moreover, it is possible that the modeller specifies a delay interval which is too wide, because of a lack of detailed information. In this case, the actual delays (in the real system) only range over a part of the delay interval.

The number of tokens produced by the firing of a transition may depend upon the values of the consumed tokens. Moreover, the values and delays of the produced tokens may also depend upon the values of the consumed tokens. The relation between the *multi-set* of consumed tokens and the *multi-set* of produced tokens is described by the **transition function**. Function $F(t_1)$ specifies transition t_1 in the net shown

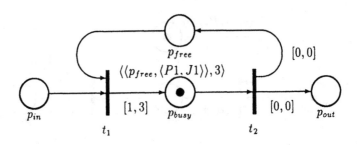

Figure 2: Transition t_1 has fired

in figure 1: $dom(F(t_1)) = \{\langle p_{in}, j \rangle + \langle p_{free}, p \rangle \mid j \in \mathcal{J} \text{ and } p \in \mathcal{P}\}$. For $j \in \mathcal{J}$ and $p \in \mathcal{P}$, we have:[1] $F(t_1)(\langle p_{in}, j \rangle + \langle p_{free}, p \rangle) = \langle \langle p_{busy}, \langle p, j \rangle \rangle, [1,3] \rangle$.

The domain of $F(t_1)$ describes the condition on which transition t_1 is enabled, i.e. t_1 is enabled if there is (at least) one token in place p_{in} and one token in p_{free}. This means that transition t_1 may occur if there is a job waiting and one of the processors is free. Note that, in this case, the enabling of the transition does not depend upon the values of the tokens consumed. The enabling time of transition t_1 depends upon the timestamps of the tokens to be consumed. If t_1 occurs, it consumes one token from place p_{in} and one token from p_{free} and it produces one token for place p_{busy}. The colour of the produced token is a pair $\langle p, j \rangle$, where p represents the processor and j represents the job. The delay of this token is an arbitrary value between 1 and 3, e.g. 2, 2.55 or 4/3. The situation shown in figure 2 is the result of firing t_1 in the state shown in figure 1. In this case the delay of the token produced for p_{busy} was equal to 2.

Transition t_2 is specified as follows: $dom(F(t_2)) = \{\langle p_{busy}, \langle p, j \rangle \rangle \mid j \in \mathcal{J} \text{ and } p \in \mathcal{P}\}$. For $j \in \mathcal{J}$ and $p \in \mathcal{P}$, we have: $F(t_2)(\langle p_{busy}, \langle p, j \rangle \rangle) = \langle \langle p_{free}, p \rangle, [0,0] \rangle + \langle \langle p_{out}, j \rangle, [0,0] \rangle$.

Transition t_2 represents the completion of a job. If t_2 occurs, it consumes one token from place p_{busy} and it produces two tokens (one for p_{free} and one for p_{out}) both with a delay equal to zero. If t_2 occurs in the state shown in figure 2, then the resulting state contains two tokens: $\langle \langle p_{out}, J1 \rangle, 3 \rangle$ and $\langle \langle p_{free}, P1 \rangle, 3 \rangle$.

3 The MTSRT Method

In the ITCPN model, a delay is described by an interval rather than a fixed value or some delay distribution. On the one hand, interval delays allow for the modelling of variable delays, on the other hand, it is not necessary to determine some artificial delay distribution (as opposed to stochastic delays). Instead, we have to specify bounds. These bounds are used to specify and to verify time constraints. This is very important when modelling time-critical systems, i.e. *real-time* systems with 'hard' deadlines. These deadlines have to be met for a safe operation of the system. An acceptable behaviour of the system depends not only on the logical correctness

---

[1]Note that $\langle p_{in}, j \rangle + \langle p_{free}, p \rangle$ and $\langle \langle p_{busy}, \langle p, j \rangle \rangle, [1,3] \rangle$ are multi-sets, see appendix A.1.

of the results, but also on the time at which the results are produced. Therefore, we are interested in techniques to verify these deadlines and to calculate upper and lower bounds for all sorts of performance criteria. This is the reason we developed the **Modified Transition System Reduction Technique** (MTSRT), which was presented in [2] and [3]. Before giving a short description of this analysis method, we provide a brief survey of existing techniques which can be used to analyse the dynamic behaviour of timed and coloured Petri nets. The techniques may be subdivided into three classes: simulation, reachability analysis and Markovian analysis.

Simulation is a technique to analyse a system by conducting controlled experiments. Because simulation does not require difficult mathematical techniques, it is easy to understand for people with a non-technical background. Simulation is also a very powerful analysis technique, since it does not set additional restraints. However, sometimes simulation is expensive in terms of the computer time necessary to obtain reliable results. Another drawback is the fact that (in general) it is not possible to use simulation to *prove* that the system has the desired set of properties.

Recent developments in computer technology stimulate the use of simulation for the analysis of timed coloured Petri nets. The increased processing power allows for the simulation of large nets. Modern graphical screens are fast and have a high resolution. Therefore, it is possible to visualize a simulation graphically (i.e. animation).

Reachability analysis is a technique which constructs a reachability graph, sometimes referred to as reachability tree or occurrence graph (cf. Jensen [11, 13]). Such a reachability graph contains a node for each possible state and an arc for each possible state change. Reachability analysis is a very powerful method in the sense that it can be used to prove all kinds of properties. Another advantage is the fact that it does not set additional restraints. Obviously, the reachability graph needed to prove these properties may, even for small nets, become very large (and often infinite). If we want to inspect the reachability graph by means of a computer, we have to solve this problem. This is the reason several authors developed reduction techniques (Hubner et al. [10] and Valmari [22]). Unfortunately, it is not known how to apply these techniques to timed coloured Petri nets.

For timed coloured Petri nets with certain types of stochastic delays it is possible to translate the net into a *continuous time Markov chain*. This Markov chain can be used to calculate performance measures like the average number of tokens in a place and the average firing rate of a transition. If all the delays are sampled from a negative exponential probability distribution, then it is easy to translate the timed coloured Petri net into a continuous time Markov chain. Several authors attempted to increase the modelling power by allowing other kinds of delays, for example mixed deterministic and negative exponential distributed delays, and phase-distributed delays (see Ajmone Marsan et al. [15]). Nearly all stochastic Petri net models (and related analysis techniques) do not allow for coloured tokens, because the increased modelling power is offset by computational difficulties. This is the reason stochastic high-level Petri nets are often used in a simulation context only.

Besides the aforementioned techniques to analyse the behaviour of timed coloured Petri nets, there are several analysis techniques for Petri nets without 'colour' or explicit 'time'. As an example, we mention the generation of place and transition

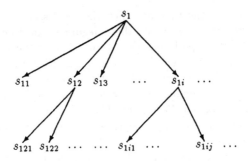

Figure 3: A reachability graph

invariants, which may be used to verify properties which are time independent. For more information about the calculation of invariants in a coloured Petri net, see Jensen [11, 13].

The **Modified Transition System Reduction Technique** is a technique which generates the **reduced reachability graph** to answer all kinds of questions.

If we try to construct the reachability graph of an ITCPN in a straightforward manner we get into problems. The basic idea of a reachability graph is to organize all reachable markings in a graph, where each node represents a state and each arc represents an event transforming one state into another state. Consider for example the reachability graph shown in figure 3. Suppose that s_1 is the initial state of the ITCPN we want to consider. This state is connected to a number of states $s_{11}, s_{12}, s_{13}, ..$ reachable from s_1 by the firing of some transition, i.e. $s_1 \longrightarrow s_{1i}$. These states are called the 'successors' (or children) of the s_1. Repeating this process produces the graphical representation of the reachability graph, see figure 3. Such a reachability graph contains all relevant information about the dynamic behaviour of the system. If we are able to generate this graph, we can answer many questions about the behaviour of the system. However, for an ITCPN this graph is generally *infinite*! This is mainly caused by the fact that we use interval timing. Consider an enabled transition. In general, there is an infinite number of allowed firing delays, all resulting in a different state. If a transition produces a token for a place with a delay x specified by the delay interval $[1, 3]$, then every delay x between 1 and 3 is allowed. Moreover, each x leads to a different state. Since one firing already results in a 'fan-out' of reachable states, the reachability graph cannot be used to analyse the system.

To avoid this fan-out problem, we propose a reduction which aggregates states into **state classes**. Informally speaking, state classes are defined as the union of similar states having the same token distribution (marking) but different timestamps (within certain bounds).

A state s of an ITCPN is a multi-set of tuples $\langle\langle p, v\rangle, x\rangle$. Each tuple $\langle\langle p, v\rangle, x\rangle$ corresponds to one token in the net; p is the location of the token (i.e. the place where it resides), v is the value (colour) of the token and x is the timestamp of the token (i.e. the time it becomes available).

A state class $\overline{s}$ is also a multi-set of tuples $\langle\langle p, v\rangle, [y, z]\rangle$. Each tuple also corresponds

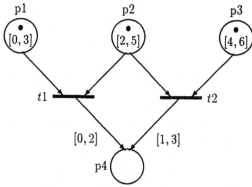

Figure 4: An example used to illustrate the modified firing rule

to one token in the net. However, instead of a timestamp each token has a **time-interval**. One state class corresponds to a set of states, state class $\overline{s}$ corresponds to state s if and only if there is a bijection between the tokens in $\overline{s}$ and s such that $\langle\langle p, v\rangle, [y, z]\rangle$ is mapped onto $\langle\langle p, v\rangle, x\rangle$ with $x \in [y, z]$ (i.e. a token residing in the same place, having the same value and a timestamp which is within the time-interval $[y, z]$). We can think of these state classes as some kind of equivalence classes.

Since tokens bear time-interval instead of timestamps, we have to modify the the firing rules given in section 2.

A transition is still enabled if there are enough tokens on each of its input places. However, the enabling time of a transition t is given by an interval! The lower bound of this interval is the **minimal enabling time** and upper bound of this interval is the **maximal enabling time** of t. These bounds are calculated by taking the maximum of the upper and lower bounds of the time-intervals of the tokens to be consumed respectively. We will use an example to clarify the modified firing rule.

Consider the net shown in figure 4. Initially, there is one token in place $p1$ with an interval of $[0, 3]$, there is one token in $p2$ with an interval of $[2, 5]$ and there is one token in $p3$ with an interval of $[4, 6]$. Note that this state class $\overline{s}$ corresponds to an infinite number of states in the original reachability graph, for instance the state with a token in $p1$ with timestamp 2.4 and a token in $p2$ with timestamp π and a token in $p3$ with timestamp 31/6. Both transitions are enabled. If transition $t1$ fires first, then the tokens in $p1$ and $p2$ are consumed, if transition $t2$ fires first, then the tokens in $p2$ and $p3$ are consumed. The enabling time of $t1$ is between 2 $(ET_{min}(t1))$ and 5 $(ET_{max}(t1))$, the enabling time of $t2$ is between 4 $(ET_{min}(t2))$ and 6 $(ET_{max}(t2))$. The transition with the smallest enabling time will fire first. Since the intervals associated to the enabling times of the transitions (i.e. $[2, 5]$ and $[4, 6]$) overlap it is not determined whether $t1$ or $t2$ fires first. However, the upper bound of the transition time $(MT_{max}(\overline{s}))$ is equal to 5, i.e. a transition will fire before or at time 5. If $t1$ fires, it will be between 2 $(ET_{min}(t1))$ and 5 $(MT_{max}(\overline{s}))$. If $t2$ fires, it will be between 4 $(ET_{min}(t2))$ and 5 $(MT_{max}(\overline{s}))$. In both cases a token is produced for place $p4$. There are two possible terminal states: one with a token in $p3$ and $p4$ and one with a token in $p1$ and $p4$. In the first case the time interval of the token in $p4$ is $[2, 7]$, because the delay interval of a token produced by $t1$ is $[0, 2]$. In the second case the time interval of the token in $p4$ is $[5, 8]$. Using intervals rather than timestamps prevented us from having to consider all possible delays in the intervals $[0, 2]$ and $[1, 3]$, i.e. it suffices to consider upper and lower bounds.

In Van der Aalst [2, 3] a formal definition of these alternative semantics are given. If we use these semantics to construct a reachability graph, we obtain the reduced reachability graph which is finite for any practical application (see [2]). The alternative semantics have been introduced for computational reasons only. However, calculating the reduced reachability graph only makes sense if the reduced reachability graph can be used to deduce properties of the original reachability graph which represents the behaviour of the ITCPN. Therefore, we have to prove that there exists some relationship between the original reachability graph and the reduced reachability graph. Fortunately, the alternative semantics are 'sound' which means that any state reachable in the original reachability graph is also reachable in the reduced reachability graph. A formal proof is given in [2] and [3]. In these references it is also shown that the opposite is not true, i.e. the alternative semantics are not 'complete'.

Despite the non-completeness, the soundness property allows us to answer various questions. We can *prove* that a system has a desired set of properties by proving it for the modified transition system. For example, we can often use the reduced reachability graph to prove boundedness, absence of traps and siphons (deadlocks), etc. The reduced reachability graph may also be used to analyse the *performance* of the system modelled by an ITCPN. With performance we mean characteristics, such as: response times, occupation rates, transfer rates, throughput times, failure rates, etc. The MTSRT method can be used to calculate *bounds* for these performance measures. Although these bounds are sound (i.e. safe) they do not have to be as tight as possible, because of possible dependencies between tokens (non-completeness). However, experimentation shows that the calculated bounds are often of great value and far from trivial. Moreover, we are able to answer questions which cannot be answered by simulation or the method proposed by Berthomieu et al. [6].

We have modelled and analysed many examples using the approach presented in this paper, see Van der Aalst [1, 2] and Odijk [19]. To facilitate the analysis of real-life systems we have developed an analysis tool, called *IAT* ([2]). This tool also supports more traditional kinds of analysis such as the generation of place and transition invariants. IAT is part of the software package *ExSpect* (see ASPT [5], Van Hee et al. [9] and Van der Aalst [2, 4]).

4   Case A: A Simple Computer System

In the remainder, we discuss two applications of the approach just presented. This section describes a model of a simple computer system. This system is composed of multiple CPU's and a number of disks. Each job which is processed by this computer system visits one CPU and one disk. Initially there are three parallel processors (CPU's) and three disks. Each processor has a service time between 50 and 60 milliseconds. The I/O service time of disk 1 is between 95 and 100 milliseconds. Disk 2 has a service time between 100 and 115 milliseconds and disk 3 has a service time between 95 and 110 milliseconds. For each job some memory has to be allocated. To swap a job into memory some time is required (between 20 and 25 ms.). Jobs that are 'swapped in' are called 'active'. Active jobs visit a CPU and a disk, then they are

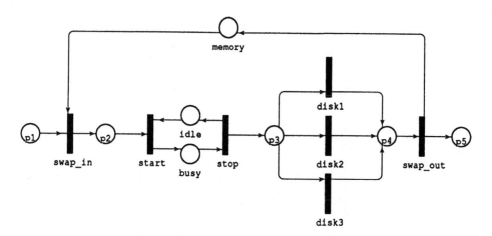

Figure 5: An ITCPN modelling a simple computer system

'swapped out' to allow other jobs to be swapped in. The time required to release the memory (swap-out time) is between 40 and 45 ms. The maximum number of active jobs is 10.

Figure 5 shows an ITCPN which models this system. Jobs enter the system via place p1 and leave the system via place p5. Transitions swap_in and swap_out take care of the memory management. Initially there are 10 tokens in place memory. (The maximum number of active jobs is 10.) The three disks are modelled by the transitions disk1, disk2 and disk3. Initially place idle contains 3 tokens, each representing one processor. Transition start represents the beginning of the execution of a job on a processor. Note that transition start is enabled if an active job is waiting in place p2 and one of the processors is free (i.e. there is a token in idle). Transition stop represents the termination of the execution of a job on a processor, i.e. stop releases the processor and sends the job to one of the three disks. The delay intervals are constructed in a straightforward manner on the basis of the information just given.

The MTSRT method allows us to calculate bounds for all kinds of performance measures, e.g. throughput times, waiting times, occupations rates, average number of active processes, etc. Figure 6 shows the relation between the load (i.e. the number of jobs arriving each second) and the average throughput times in milliseconds. (We assume batch arrivals.) The dashed line gives an upper bound for the average throughput time given a specific load. The solid line gives a lower bound for the average throughput time. If the load is 20 jobs per second, then the average throughput time is between 327.5 and 394.5 ms. Recall that all results are obtained by inspecting the reduced reachability graph. To calculate bounds for the average throughput time given a load of 20 jobs per second we need to construct a reduced reachability graph with 2304 states. This takes about 20 seconds on a SUN/SPARC-station (ELC). Further analysis shows that the three CPU's are the main bottle-neck.

To speed up this system we investigated two alternatives: (1) adding processors and memory and (2) replacing the processors by faster ones. In the first alternative we have added two processors (i.e. there are 5 identical processors) and the maximum number

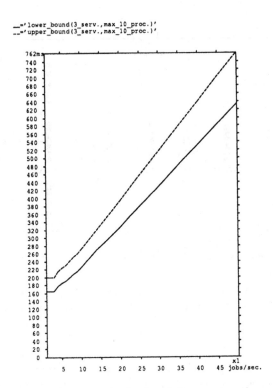

Figure 6: The relation between the load and the average throughput times

of active jobs is equal to 15 instead of 10. (Analysis shows that just adding extra processors is not sufficient, because in this case the memory management becomes a bottle-neck.) In the second alternative there are still three processors, however they are twice as fast. These processors have service times between 25 and 30 milliseconds.

It is easy to modify the ITCPN for each of the two alternatives. In figure 7 these two alternatives are compared with the original situation. The lines ...... and .... correspond to the bounds for the throughput times of alternative (1) and the lines ___ and _._ correspond to the bounds for the throughput times of alternative (2). If the load is 20 jobs per second, then the average throughput time is between 253.75 and 306.25 ms for alternative (1) and between 360.0 and 313.5 ms for alternative (2). Clearly, both alternatives give a significant speed-up. For low occupation rates (a load below +/- 15 jobs per sec.) alternative (2) is better. However, for higher loads alternative (1) is preferable.

It is also possible to verify all kinds of properties. The MTSRT can be used to prove boundedness, reachability, etc. This may be very useful. For example, we can decide to add a surplus of processors (e.g. 50) to the original system. In this case, place busy is still 10-bounded. This means that it makes no sense to have more than 10 CPU's without extending the memory.

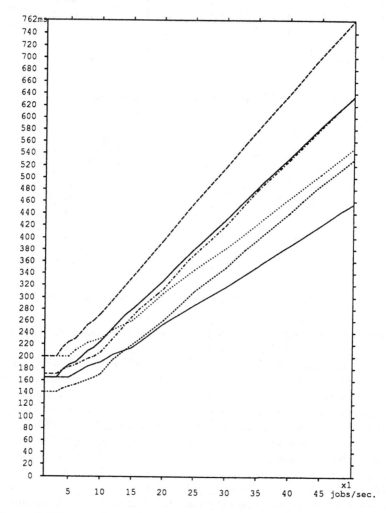

Figure 7: The two alternatives compared with the original situation

5   Case B: A Manufacturing System

The ITCPN model and related techniques have been used in many other applications areas, in particular logistics ([2]). Note that a just-in-time production system is also an example of a time-critical system. In this section we model and analyse a manufacturing system. This manufacturing system is divided into five units, see figure 8. The manufacturing system receives raw materials and transforms them into end-products. The raw materials are divided over two *production units*. Each production unit transforms raw materials into intermediate products. These intermediate

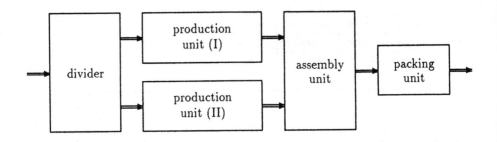

Figure 8: A manufacturing system

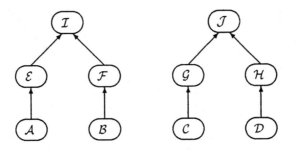

Figure 9: The bill of materials of end-products $\mathcal{I}$ and $\mathcal{J}$

products are assembled into end-products by the *assembly unit*. The *packing unit* prepares these products for shipment.

In this particular case, there are two kinds of end-products $\mathcal{I}$ and $\mathcal{J}$. To manufacture $\mathcal{I}$, we need two kinds of raw material: $\mathcal{A}$ and $\mathcal{B}$. $\mathcal{A}$ is transformed into $\mathcal{E}$, $\mathcal{B}$ is transformed into $\mathcal{F}$ and $\mathcal{E}$ and $\mathcal{F}$ are assembled into $\mathcal{I}$. $\mathcal{J}$ has a similar production process. The bill of materials of these two end-products is shown in figure 9.

We model this manufacturing process in terms of an ITCPN. This ITCPN has an 'input' place $p1$ to receive raw materials and an 'output' place $p18$ which contains end-products ready to be shipped. These two places are the only places having interactions with the environment of the manufacturing system, see figure 10.

Tokens in these places represent products (or materials) and have a value which describes, the kind of product it represents, the identification of the product and some status information. Therefore, the value of such a token is a three tuple $(pt, id, stat)$

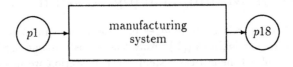

Figure 10: The interactions of the manufacturing system with the environment

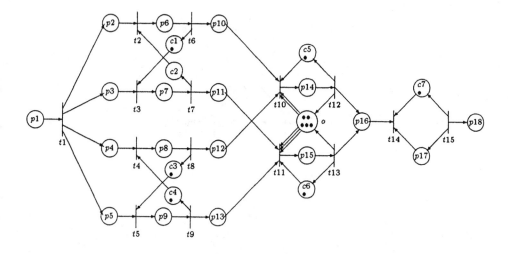

Figure 11: The ITCPN

where $pt \in \{\text{`}A\text{'}, \text{`}B\text{'}, \text{`}C\text{'}, \text{`}D\text{'}, \text{`}\mathcal{E}\text{'}, ..\}$, $id \in \mathbb{N}$ and $stat$ is a string value.

The ITCPN describing the entire manufacturing system is shown in figure 11. Note that this figure does not contain information on colour sets and transition functions. In this paper we will not a supply a detailed description of the net. Instead we will give a sketch of each of the subsystems. (A more detailed description of this case is given in [2]).

The *divider* works as follows: it takes raw materials from place p1 and distributes them over the two production units. Moreover, the divider differentiates between the four kinds of raw material. Figure 11 shows the divider which is modelled by a transition t1 dividing the raw material over four places p2, p3, p4 and p5. Transition t1 fires if there is some raw material available, material of kind A goes to place p2, material of kind B goes to place p4, material of kind C goes to place p3 and material of kind D goes to place p5. Note that we assume that distributing these goods takes no time.

The first *production unit* transforms products of type A into $\mathcal{E}$ and products of type C into $\mathcal{G}$. These transformations are performed by one machine alternately working on products of type A and C. This machine needs between 0.35 and 0.37 hours to transform A into $\mathcal{E}$ and between 0.78 and 0.81 hours to transform C into $\mathcal{G}$. Figure 12 shows this production unit in terms of an ITCPN. The machine has four states: (i) busy, transforming A into $\mathcal{E}$, (ii) busy, transforming C into $\mathcal{G}$, (iii) free, waiting for product A and (iv) free, waiting for product C.

Initially, the machine is in state (iv). In this example tokens in c1 and c2 are colourless and the tokens in the other places (p2, p3, p6, p7, p10 and p11) represent products. The delay of a token produced by t2 is between 0.35 and 0.37, the delay of a token produced by t3 is between 0.78 and 0.81.

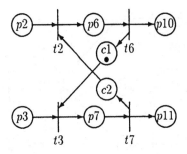

Figure 12: Production unit I

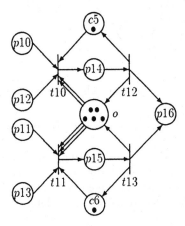

Figure 13: The assembly unit

The second *production unit* has a similar structure, instead of one machine there are are *two* identical machines. This is represented by the initial state, where there is one token in $c3$ and one token in $c4$. Both machines are capable of doing two kinds of transformations: B into F and D into H. Transforming B into F takes between 1.58 and 1.61 hours. Transforming D into H takes between 0.18 and 0.20 hours. Initially, one of the machines is ready to transform B into F, the other one is ready to transform D into H.

There is one *assembly unit*. This unit is capable of assembling E and F into I and G and H into J. Products are assembled in order of their arrival, i.e. the assembly unit uses a 'First Come First Served' discipline. The assembly unit consists of two dedicated assembly lines, one for end-product I and one for end-product J. Figure 13 shows these two assembly lines. The two assembly lines share a number of operators. Free operators are represented by tokens in the place o. Initially, there are five operators in the place o. To assemble E and F into I two operators are needed, this takes between 0.5 and 0.6 hours. Transition $t10$ consumes two tokens from place o and produces one token for place $p14$. Transition $t12$ produces two tokens for place

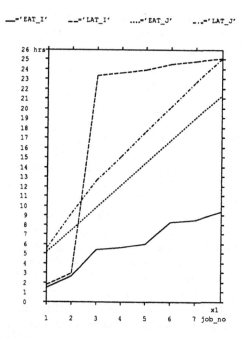

Figure 14: Upper and lower bounds for the completion time of products $\mathcal{I}$ and $\mathcal{J}$

$p14$ and one token for place $p16$. To assemble $\mathcal{G}$ and $\mathcal{H}$ into $\mathcal{J}$ three operators are needed, this takes between 1.3 and 1.4 hours. Transitions $t11$ and $t13$ represent the beginning and termination of this operation. Note that place $p16$ may contain two kinds of products: $\mathcal{I}$ and $\mathcal{J}$.

The subnet composed of the transitions $t14$ and $t15$ and the places $p16$, $p17$, $p18$ and $c7$ represents the *packing unit*. The packing unit is used to prepare end-products $\mathcal{I}$ and $\mathcal{J}$ for shipment. To prepare these products, they are packed in wooden crates. Moreover, end-products $\mathcal{J}$ have to be tuned. The time needed to prepare a product for shipment depends on the type of product: packing $\mathcal{I}$ takes between 0.2 and 0.3 hours, packing and tuning product $\mathcal{J}$ takes between 2.3 and 2.5 hours. The packing unit handles the products one by one.

Using the MTSRT method we can calculate several performance measures, for example upper and lower bounds for the arrival time of tokens in place $p18$. Assume that initially there are 32 pieces of raw material available (8 of each kind), i.e. in the initial state s there are 32 tokens in $p1$, eight with a value x such that $\pi_1(x) = $ '$\mathcal{A}$', .. etc. In figure 14 the upper and lower bounds for the completion times of both kinds of products are given. The eighth end-product of type $\mathcal{I}$ leaves the manufacturing system between 9.39 and 25.11 hours after the production started. The completion time of the eighth product of type $\mathcal{J}$ is between 21.28 and 25.11 hours. Note that the completion times of products of type $\mathcal{I}$ are quite uncertain. Nevertheless, we can *prove* that the entire production process finishes within 25.11 hours!

The reduced reachability graph needed to compute these results contains 21237 states. It takes about 12 minutes to produce these results on a SUN/SPARC-station. Experience shows that the memory required to store the reduced reachability graph is more restrictive than the processing time required to construct and inspect this graph.

6 Conclusion

In this paper, we have used ITCPNs to model and analyse two non-trivial cases, a computer system and a manufacturing system. The ITCPN model uses a new timing mechanism where time is associated with tokens and transitions determine a delay specified by an interval. Specifying each delay by an interval rather than a deterministic value or stochastic variable is promising, since it is possible to model uncertainty without having to bother about the delay distribution.

We have used the MTSRT method to analyse the two cases. This analysis method constructs a reduced reachability graph. In such a graph a node corresponds to a set of (similar) states, instead of a single state. The reduced reachability graph can be used to *prove* certain properties or to calculate accurate bounds for all kinds of performance measures (e.g. throughput times, waiting times, occupation rates).

The application of the ITCPN model and the MTSRT method to the two cases described in this paper shows that the results obtained by using this approach are quite meaningful. Moreover, the bounds calculated for performance measures just mentioned are always valid. Therefore, the proposed approach is extremely useful when evaluating the design of a time-critical system.

References

1. W.M.P. van der Aalst. Modelling and Analysis of Complex Logistic Systems. In H.J. Pels and J.C. Wortmann, editors, *Integration in Production Management Systems*, volume B-7 of *IFIP Transactions*, pages 277–292. Elsevier Science Publishers, Amsterdam, 1992.

2. W.M.P. van der Aalst. *Timed coloured Petri nets and their application to logistics.* PhD thesis, Eindhoven University of Technology, Eindhoven, 1992.

3. W.M.P. van der Aalst. Interval Timed Coloured Petri Nets and their Analysis. In M. Ajmone Marsan, editor, *Application and Theory of Petri Nets 1993*, volume 691 of *Lecture Notes in Computer Science*, pages 453–472. Springer-Verlag, New York, 1993.

4. W.M.P. van der Aalst and A.W. Waltmans. Modelling logistic systems with EXSPECT. In H.G. Sol and K.M. van Hee, editors, *Dynamic Modelling of Information Systems*, pages 269–288. Elsevier Science Publishers, Amsterdam, 1991.

5. ASPT. *ExSpect 4.0 User Manual.* Eindhoven University of Technology, 1993.

6. B. Berthomieu and M. Diaz. Modelling and verification of time dependent systems using Time Petri Nets. *IEEE Transactions on Software Engineering*, 17(3):259–273, March 1991.

7. G. Florin and S. Natkin. Evaluation based upon Stochastic Petri Nets of the Maximum Throughput of a Full Duplex Protocol. In C. Girault and W. Reisig,

editors, *Application and theory of Petri nets: selected papers from the first and the second European workshop*, volume 52 of *Informatik Fachberichte*, pages 280–288, Berlin, 1982. Springer-Verlag, New York.

8. K.M. van Hee. *Information System Engineering: a Formal Approach.* Cambridge University Press, (to appear) 1994.

9. K.M. van Hee, L.J. Somers, and M. Voorhoeve. Executable specifications for distributed information systems. In E.D. Falkenberg and P. Lindgreen, editors, *Proceedings of the IFIP TC 8 / WG 8.1 Working Conference on Information System Concepts: An In-depth Analysis*, pages 139–156, Namur, Belgium, 1989. Elsevier Science Publishers, Amsterdam.

10. P. Hubner, A.M. Jensen, L.O. Jepsen, and K. Jensen. Reachability trees for high level Petri nets. *Theoretical Computer Science*, 45:261–292, 1986.

11. K. Jensen. Coloured Petri Nets. In W. Brauer, W. Reisig, and G. Rozenberg, editors, *Advances in Petri Nets 1986 Part I: Petri Nets, central models and their properties*, volume 254 of *Lecture Notes in Computer Science*, pages 248–299. Springer-Verlag, New York, 1987.

12. K. Jensen. Coloured Petri Nets: A High Level Language for System Design and Analysis. In G. Rozenberg, editor, *Advances in Petri Nets 1990*, volume 483 of *Lecture Notes in Computer Science*, pages 342–416. Springer-Verlag, New York, 1990.

13. K. Jensen. *Coloured Petri Nets. Basic concepts, analysis methods and practical use.* EATCS monographs on Theoretical Computer Science. Springer-Verlag, New York, 1992.

14. K. Jensen and G. Rozenberg, editors. *High-level Petri Nets: Theory and Application.* Springer-Verlag, New York, 1991.

15. M. Ajmone Marsan, G. Balbo, A. Bobbio, G. Chiola, G. Conte, and A. Cumani. On Petri Nets with Stochastic Timing. In *Proceedings of the International Workshop on Timed Petri Nets*, pages 80–87, Torino, 1985. IEEE Computer Society Press.

16. M. Ajmone Marsan, G. Balbo, and G. Conte. A Class of Generalised Stochastic Petri Nets for the Performance Evaluation of Multiprocessor Systems. *ACM Transactions on Computer Systems*, 2(2):93–122, May 1984.

17. P. Merlin. *A Study of the Recoverability of Computer Systems.* PhD thesis, University of California, Irvine, California, 1974.

18. S. Morasca, M. Pezzè, and M. Trubian. Timed High-Level Nets. *The Journal of Real-Time Systems*, 3:165–189, 1991.

19. M.A. Odijk. ITPN analysis of ExSpect specifications with respect to production logistics. Master's thesis, Eindhoven University of Technology, Eindhoven, 1991.

20. C. Ramchandani. *Performance Evaluation of Asynchronous Concurrent Systems by Timed Petri Nets.* PhD thesis, Massachusetts Institute of Technology, Cambridge, 1973.

21. W. Reisig. *Petri nets: an introduction.* Prentice-Hall, Englewood Cliffs, 1985.

22. A. Valmari. Stubborn sets for reduced state space generation. In *Proceedings of the 10th International Conference on Applications and Theory of Petri Nets*, Bonn, June 1989.

A   Formal Definition

In this section we define interval timed coloured Petri nets in mathematical terms, such as functions, multi-sets and relations. This definition was presented in [3].

A.1   Multi-sets

A *multi-set*, like a set, is a collection of elements over the same subset of some universe. However, unlike a set, a multi-set allows multiple occurrences of the same element. Another word for multi-set is *bag*. Bag theory is a natural extension of set theory (Jensen [12]).

Definition 1 (multi-sets)
A multi-set b, over a set A, is a function from A to $\mathbb{N}$, i.e. $b \in A \to \mathbb{N}$.[2] If $a \in A$ then $b(a)$ is the number of occurrences of a in the multi-set b. A_{MS} is the set of all multi-sets over A. The empty multi-set is denoted by $\emptyset_A$ (or $\emptyset$). We often represent a multi-set $b \in A_{MS}$ by the formal sum:[3]

$$\sum_{a \in A} b(a)\, a$$

Consider for example the set $A = \{a, b, c, ..\}$, the multi-sets $3a$, $a + b + c + d$, $1a + 2b + 3c + 4d$ and $\emptyset_A$ are members of A_{MS}.

Definition 2
We now introduce some operations on multi-sets. Most of the set operators can be extended to multi-sets in a rather straightforward way. Suppose A a set, $b_1, b_2 \in A_{MS}$ and $q \in A$:

$$\begin{aligned}
&q \in b_1 \text{ iff } b_1(q) \geq 1 &&\text{(membership)} \\
&b_1 \leq b_2 \text{ iff } \forall_{a \in A}\, b_1(a) \leq b_2(a) &&\text{(inclusion)} \\
&b_1 = b_2 \text{ iff } b_1 \leq b_2 \text{ and } b_2 \leq b_1 &&\text{(equality)} \\
&b_1 + b_2 = \sum_{a \in A} (b_1(a) + b_2(a))\, a &&\text{(summation)} \\
&b_1 - b_2 = \sum_{a \in A} ((b_1(a) - b_2(a)) \max 0)\, a &&\text{(subtraction)} \\
&\#b_1 = \sum_{a \in A} b_1(a) &&\text{(cardinality of a finite multi-set)}
\end{aligned}$$

See Jensen [12, 13] for more details.

A.2   Definition of Interval Timed Coloured Petri Nets

The ITCPN model presented in this paper is analogous to the model described in [2]. However, in this paper we give a definition which is closer to the definition of Coloured Petri Nets (CPN), see Jensen [11, 12, 13].

---

[2] $\mathbb{N} = \{0, 1, 2, ..\}$
[3] This notation has been adopted from Jensen [12].

Nearly all timed Petri net models use a continuous time domain, so do we.

Definition 3

TS is the **time set**, $TS = \{x \in \mathbb{R} \mid x \geq 0\}$, i.e. the set of all non-negative reals.
$INT = \{[y, z] \in TS \times TS \mid y \leq z\}$, represents the set of all closed intervals.
If $x \in TS$ and $[y, z] \in INT$, then $x \in [y, z]$ iff $y \leq x \leq z$.

We define an interval timed coloured Petri nets as follows:

Definition 4 (ITCPN)

An **Interval Timed Coloured Petri Net** is a five tuple ITCPN $= (\Sigma, P, T, C, F)$
satisfying the following requirements:

(i) Σ is a finite set of types, called **colour sets**.

(ii) P is a finite set of **places**.

(iii) T is a finite set of **transitions**.

(iv) C is a **colour function**. It is defined from P into Σ, i.e. $C \in P \to \Sigma$.

(v) $CT = \{\langle p, v \rangle \mid p \in P \land v \in C(p)\}$ is the set of all possible **coloured tokens**.

(vi) F is the **transition function**. It is defined from T into functions. If $t \in T$, then:[4]

$$F(t) \in CT_{MS} \nrightarrow (CT \times INT)_{MS}$$

(i) Σ is a set of types. Each type is a set of colours which may be attached to one of the places.

(ii) and (iii) The places and transitions are described by two disjoint sets, i.e. $P \cap T = \emptyset$.

(iv) Each place $p \in P$ has a set of allowed colours attached to it and this means that a token residing in p must have a value v which is an element of this set, i.e. $v \in C(p)$.

(v) CT is the set of all coloured tokens, i.e. all pairs $\langle p, v \rangle$ where p is the position of the token and v is the value of the token.

(vi) The transition function specifies each transition in the ITCPN. For a transition t, $F(t)$ specifies the relation between the multi-set of consumed tokens and the multi-set of produced tokens. The domain of $F(t)$ describes the condition on which transition t is enabled. Note that the produced tokens have a delay specified by an interval. In this paper, we require that both the multi-set of consumed tokens and the multi-set of produced tokens contain finitely many elements.

Apart from the interval timing and a transition function instead of incidence functions, this definition resembles the definition of a CP-matrix (see Jensen [11, 13]).

The formal semantics (i.e. a mathematical definition of the dynamic behaviour) of the ITCPN model are given in [2] and [3].

---

[4] $A \nrightarrow B$ denotes the set of all partial functions from A to B.

Analyzing PICL Trace Data with MEDEA**

Alessandro P. Merlo

Patrick H. Worley

University of Pavia
Via Abbiategrasso, 209
I–27100 Pavia, Italy

Oak Ridge National Laboratory
P.O. Box 2008, Bldg. 6012
Oak Ridge, TN 37831-6367

Abstract. A detailed performance analysis of the behavior of a computer system under its real workload can be achieved by means of event–driven monitors, i.e., tools that capture the events generated by a program and store them into trace files. Execution traces and performance statistics can be collected for parallel applications on a variety of multiprocessor platforms by using the Portable Instrumented Communication Library (PICL). Starting from these measurement data, the construction of accurate workload models requires the application of different types of statistical and numerical techniques interacting together to fully characterize the behavior of the applications submitted to a system. The static and dynamic performance characteristics of performance data can be analyzed easily and effectively with the facilities provided within the MEasurements Description Evaluation and Analysis tool (MEDEA). This paper outlines a case study that uses PICL and MEDEA to characterize the performance of a parallel benchmark code executed on different hardware platforms and using different parallel algorithms and communication protocols.

1   Introduction

The demands for hardware and software resources of a computer system significantly influence its performance. Therefore, the quantitative description of resource consumption when running an application plays a fundamental role in every performance evaluation study [3]. The best way to obtain such a quantitative description for a system is to take measurements while the system is processing its real workload. However, the set of data collected by the monitoring tools represents a detailed "discrete" description of the behavior of the measured applications. While such a characterization is very useful when used

** This research was supported by the Applied Mathematical Sciences Research Program, Office of Energy Research, US Department of Energy, under contract DE–AC05–84OR21400 with Martin Marietta Energy System Inc., by the Italian Research Council (C.N.R.) under Grant 92.01571.PF69, and by the Italian MURST under the 40% and 60% Projects.

as input to visualization tools, it is inappropriate when applied to system modeling, where a compact and manageable representation of the workload processed by the real system is needed.

The process of deriving a compact representation of the workload, *workload characterization*, can be subdivided into several phases [10] [3]. The input to the process is the data collected by monitoring the execution of a given application over the system. Output includes both standard data analysis results, which provide useful insights into the behavior of the application, and workload models, which can be used as input to either simulation or analytic system models. How the data is analyzed and how the model is derived are functions of the type of questions being addressed about the performance of the computer system, the type of data collected, and the level of detail at which the analysis will be performed. For example, at some point in the process, the basic unit of work that is considered in a quantitative description of the workload, the *workload component*, must be specified.

While the type of analysis that is appropriate for a particular workload characterization will vary as different questions are asked or different computer systems evaluated, many mathematical techniques are common to multiple analyses. To support this commonality, researchers at the University of Pavia have developed the MEasurements Description Evaluation and Analysis tool (MEDEA) [13]. The basic aim of MEDEA is to define an integrated environment in which to perform workload modeling studies. The different operations required to fully examine the behavior of the applications submitted to a system have been logically subdivided into *modules*, each performing a specific manipulation over the performance data and the intermediate results produced at each step of the workload characterization process.

The collection of performance data is often a difficult task in itself, especially on systems without dependable operating system or hardware support for the collection of useful data. One portable option for the collection of performance data for message passing computer systems is to use the Portable Instrumented Communication Library (PICL), developed at Oak Ridge National Laboratory, when implementing application codes [6]. PICL implements a generic message–passing interface able to support interprocessor communications on a variety of different hardware platforms. Furthermore, PICL tracing routines allow the user to collect detailed information on the behavior and performance of parallel programs. The trace files generated by PICL can be used as input both to performance visualization tools, e.g. ParaGraph [14], for performance tuning and debugging, or to performance evaluation tools.

This paper is organized as follows. Sections 2 and 3 give a brief description of the main features provided within PICL and MEDEA, respectively. Section 4 deals with the identification of possible workload components and the specification of the corresponding performance parameters. Section 5 outlines an experimental application. A few conclusions are summarized in Section 6.

2 The Portable Instrumented Communication Library

A detailed performance analysis of the behavior of a computer system under its real workload can be achieved by means of event–driven monitors, i.e., tools that capture the events generated by a program and store them into trace files. However, the trace file formats adopted by different monitoring tools are, in general, quite different from one to another (see, for example, [6], [9], [11]), with each developer defining a specific record format able to address those events of interest for the particular system being evaluated. This lack of standardization makes it difficult to easily analyze trace files collected on different systems, but is a reflection of system differences that can not simply be eliminated by a standardization process. Recently, there has been a movement toward establishing a standard metaformat in which to specify trace file formats [2].

The machine independent layer of PICL has proven to be a sufficient framework to support portability between different platforms, and the trace file format used by PICL is flexible enough to collect data for performance evaluation. The basic structure of PICL [1] trace records is shown in Table 1.

Record type [int]	Event type [int]	Timestamp [double]	Processor ID [int]	Task ID [int]	Number of data fields [int]	Data descriptor [int or string]	Data

Table 1: Basic structure of PICL trace records.

Four different record types are currently supported by PICL:

- *user–defined* record types are used to specify the data associated with user–defined events (for example, single loop statements or even single instructions);
- *event* record types are used to collect detailed information, for both system and user–defined events, needed for a visualization tool like ParaGraph or for the analysis of user events by means of MEDEA, as will be explained in §4.1;
- *statistics* record types are used to collect profile data of system and user–defined events;
- *subset–definition* record types are used to define subsets, e.g., of processors or processes, for which cumulative statistics are to be collected.

The tracing facilities provided within PICL allow the user to specify the amount and the type of data to store into trace files: if detailed data are needed, then for each event generated by the application (e.g., send/receive commands, I/O operations, etc.) timestamped entry/exit records are stored for the processor and the process associated with the event; if only global information are needed (e.g.,

---

[1] For a detailed description of PICL trace file format, see [15].

when it is not important to know the exact timing of the single events but when we are interested in the corresponding cumulative times), then the statistics records can be used to characterize the general behavior of an application.

The event types currently supported by PICL cover most of the event data utilized in performance evaluation studies of message-passing parallel applications. The most important categories of events follow:

- *user-defined* events allow the user to specify that the execution of a subroutine or even arbitrary code segments be considered an event of a certain type, allowing the logical structure of the application to be represented during subsequent analysis;
- *interprocessor communication* events represent PICL commands for enabling, disabling, or invoking interprocess communications, including, for example, send and receive;
- *file I/O* events are used to collect performance data on (physical) I/O, which strongly influence the performance of most real parallel applications;
- *synchronization* events currently supported include "clock normalization" and "barrier";
- *resource allocation* events deal with the allocation/deallocation of processors;
- *tracing* events are recorded with the dual goals of allowing a correct interpretation of the trace files and of providing a measure of the overhead implied by the tracing activity.

3   The MEasurements Description Evaluation and Analysis tool

The construction of accurate workload models requires the application of different types of statistical and numerical techniques interacting together to fully characterize the behavior of the applications submitted to a system. During the design phase of MEDEA[2], the need for integration between these different underlying techniques and the need for portability across a variety of computer platforms led to the choice of a standard development environment. As a consequence, MEDEA is currently implemented on UNIX systems running X Window/Motif[3]. Figure 1 shows the main window of the graphical interface provided by MEDEA.

In the active elements of this window (i.e., buttons) the main steps used in workload characterization studies can be easily identified.
The *data manipulation* module performs a preliminary analysis of the trace data in order to correlate the events recorded during the execution of an application. Traditional performance indices, such as computation and communication

---

[2] For a detailed description of the tool, see [12].
[3] MEDEA requires at least X11R5 and Motif 1.1.4 to work properly.

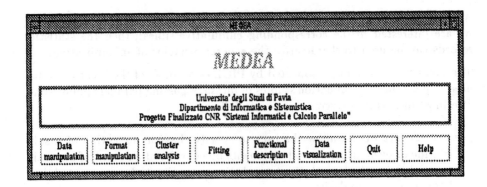

Fig. 1. Main window of the graphical interface of MEDEA.

times, and parallel metrics, such as speedup and efficiency, can then be derived by filtering the trace data.

Within MEDEA, a *format* is a subset of the performance parameters that can be associated with the specific workload component under study. The *format manipulation* module of MEDEA enables user-defined subsets of parameters to be stored in an internal library, allowing subsequent workload analyses to be performed with fewer interactions with the graphical interface.

The *cluster analysis* module is used to examine the statistical properties of the measured data set. Multidimensional cluster analysis techniques are used to identify groups of workload components having homogeneous characteristics with respect to some predefined parameters. The clustering algorithm implemented within MEDEA is the *k-means*, an iterative nonhierarchical method of partitioning data sets [7]. Each partition is derived by minimizing the distances between each workload component and the centroid of the cluster it belongs to. At the end of the analysis, the optimal partitions (if any) are derived according to the overall mean square ratios of the evaluated clusters.

Workload models must be compact and easily manageable. The *fitting* module provided within MEDEA allows the user to derive analytic descriptions of the dynamic behavior of the workload from the measured data. The analytic models are described in terms of one or more of the collected parameters, and are able to represent the variations of the workload parameters with respect to any independent variables, including time.

The process of workload characterization can be approached from two different viewpoints. The "physical" viewpoint describes the behavior of the system and the applications by means of indices related to resource consumptions, such as computation and communication times. This approach is the one realized by the data manipulation and the cluster analysis modules of MEDEA. The "functional" viewpoint gives a logical description of the workload. In this case, the classification of workload components is based, for example, on the membership of particular components in a specific cluster. The *functional description* module of MEDEA deals with this second viewpoint.

The graphical visualization of the parameter values derived from the trace data and of the results produced by the various analyses performed within MEDEA is often an important tool in understanding the characteristics and the behavior of the workload. The *data visualization* module of MEDEA provides this facility.

4 Workload Characterization of Parallel Applications

The selection of appropriate workload components and of the corresponding performance parameters is strongly dependent on the type of information collected into the trace files. Since PICL tracing routines allow the user to specify the level of detail and the amount of data to collect during the execution of an application, the information that can be derived in the workload characterization process may be different from trace file to trace file. If detailed trace files are used as input to MEDEA, then the tool looks for each single event entry/exit pair and, according to the event record type, correlates this new information to the previous ones in order to cumulate statistics that refer to the performance parameters used to characterize the workload components. If trace files are used that contain only statistics records, then MEDEA parses only those records that contain global information. In the following sections, specifications for the possible workload components and the corresponding parameters are given.

4.1 Workload Components

The workload submitted to a system may be analyzed at different levels of detail, according to the "granularity" of the components selected for the modeling activity. As mentioned in §1, a workload component is defined as the basic unit of work that is considered in a quantitative description of the workload. Three different approaches (or granularities) have been adopted in MEDEA for the analysis of PICL trace files: *program–oriented, processor-* or *task–oriented,* and *user–event–oriented.*

In the program–oriented approach, a trace file is analyzed from a global viewpoint and information about the behavior of the application considered as a whole can be derived. The basic workload component is the program itself. The second approach derives a more detailed analysis of a trace file, in which the tasks executed on each single processor are selected as representative workload components. Finally, in the user–event–oriented approach the facility provided within PICL of defining arbitrary code segments to represent distinct workload components allows MEDEA to use the "logical" or "user" view of the application when analyzing its behavior.

4.2   Parallel Metrics

Parallel profiles represent one of the best tools for analyzing the dynamic behavior of an application [4]. If detailed PICL trace files are used as input to MEDEA, then the number of processors in use as a function of the execution time can be evaluated with respect to the different types of operations performed by the processors. MEDEA allows the user to derive execution, computation, communication, receive, and transmit profiles.

When the same application is executed several times, varying the number of allocated processors, additional parallel metrics, such as speedup, efficiency, efficacy, and execution signature, can be use to characterize the behavior of the workload [5]. Note that one trace file has to refer to a single processor execution of the application itself, according to the definition of speedup, but that relative speed-up can be used when the application is too large (or too slow) to take measurements on a single processor.

4.3   Performance Parameters

The selection of meaningful parameters to be considered in the workload characterization phase represents one of the most critical steps of this process. Table 2 lists the parameters that are currently used to characterize the program–oriented and the processor–oriented approaches for PICL trace files.

Time parameters			
Execution time	(extime)	I/O time	(iotime)
Computation time	(cptime)	Enable/disable comm. time	(iptime)
Communication time	(cmtime)	Synchronization time	(cktime)
Receive time	(rctime)	Resource allocation time	(rstime)
Transmit time	(trtime)	System time	(sytime)
Volume parameters			
Volume of data exchanged	(ttdata)	Volume of transmitted data	(trdata)
Volume of received data	(rcdata)	Volume of I/O data	(iodata)
Occurrence parameters			
Number of receive requests	(rcnum)	Number of I/O requests	(ionum)
Number of transmit requests	(rcnum)	Number of processors	(prnum)

Table 2: Parameters for the program–oriented and the processor-oriented approaches.

Note that these parameters may be meaningful only with respect to a particular "granularity" of the workload components. For example, the number of processors allocated to an application (parameter *prnum*) makes no sense if applied to a processor-oriented approach, where we are interested in the behavior of each single task. Furthermore, according to the workload component selected, we can have different interpretations for the same parameter. As an example, consider the case of the computation time (parameter *cptime*). If the single task has been selected to be the workload component, then the value of this parameter can be calculated as the sum of all the time intervals during which a computation is performed. This time represents the difference between the total time of the application on the specific processor and the total time spent by this processor while executing noncomputational instructions (e.g., send/receive requests, synchronization commands, etc.). Alternatively, in the case of the program–oriented approach, this parameter can be defined only as a mean with respect to the number of processors.

Table 3 lists the parameters currently used within MEDEA to characterize the user–event–oriented approach.

Time parameters			
Total event time	(ctime)	User events time	(utime)
System events time	(stime)	Hidden system events time	(hstime)
Occurrence parameters			
Number of event occurrences	(cnum)	Number of hidden system events	(hsnum)
Number of system events	(snum)	Number of hidden user events	(hunum)
Number of user events	(unum)		

Table 3: Parameters for the user–event–oriented approach.

These parameters are completely different from those adopted for the other two approaches. In the following discussion, we use the trace records in Tab. 4 to explain meaning and usage of the parameters. Here, the first field in each record denotes an event entry (-3) or an event exit (-4), the second field denotes the event type id, and the third field denotes the timestamp for the record. The other fields can be ignored for the following discussion. System events have types less than -10, and user events have nonnegative types.

```
-3 0 0.000016 6 0 2 2 0 0          (timestamp a)
  -3 -52 0.000128 6 0 1 2 0        (timestamp b)
  -4 -52 0.000516 6 0 3 2 8 0 0    (timestamp c)
  -3 1 0.000711 6 0 2 2 0 0        (timestamp d)
    -3 -52 0.000818 6 0 1 2 1      (timestamp e)
    -4 -52 0.001643 6 0 3 2 8 1 5  (timestamp f)
    -3 -21 0.001665 6 0 3 2 8 1 7  (timestamp g)
    -4 -21 0.001711 6 0 0          (timestamp h)
    -3 2 0.001982 6 0 0            (timestamp i)
    -4 2 0.002005 6 0 0            (timestamp j)
  -4 1 0.002013 6 0 0              (timestamp k)
-4 0 0.002067 6 0 0                (timestamp l)
```

Table 4: Example trace records.

In PICL applications, user–defined events can correspond to any arbitrary code segment. As a consequence, the presence of nested user events is very common, especially if the user events are associated with the execution of program subroutines. With respect to the example trace records in Table 4, two nested events (of types 1 and 2) can be recognized within the "main" event of type 0.

When these PICL trace records are analyzed according to the user–event–oriented approach, the following meanings and values are assigned to the identified parameters for user events of type 0:

- *total event time* (ctime = 0.002051 secs) is the elapsed time between the entry record for a type 0 event (timestamp a) and the corresponding exit record (timestamp l) if the event type occurs once, or is the sum of the elapsed times if it occurs multiple times;
- *system events time* (stime = 0.000388 secs) is the sum of the execution times of any system events that are nested at the first level of type 0 events (a type −52 event starting at timestamp b);
- *user events time* (utime = 0.001302 secs) is the time spent executing user events nested at the first level of type 0 events (one type 1 event);
- *hidden system events time* (hstime = 0.000871 secs) is the time spent to execute system events that are detected in nested user events (type −52 and type −21 events nested in a type 1 event);
- *number of event occurrences* (cnum = 1) is the number of times type 0 events have been executed on a given processor;
- *number of system events* (snum = 1) is the number of system events that are nested at the first level of type 0 events (a type −52 event starting at timestamp b);
- *number of user events* (unum = 1) is the number of user events that are nested at the first level of type 0 user events (one type 1 event);
- *number of hidden system events* (hsnum = 2) is the number of system events occurring within nested user events (type −52 and type −21 events beginning at timestamps e and g, respectively, and nested within a type 1 event);
- *number of hidden user events* (hunum = 1) is the number of user events nested within user events at the first level (one type 2 event).

5   A Case Study

In the following, a brief sketch of a workload characterization study is given, to indicate how PICL trace data can be analyzed using MEDEA.

The application used for the study is PSTSWM, a message-passing benchmark code and parallel algorithm testbed that solves the nonlinear shallow water equations on the sphere [16]. This code models closely how CCM2, the Community Climate Model developed by the National Center for Atmospheric Research, handles the dynamical part of the primitive equations. PSTSWM was developed to compare parallel algorithms and to evaluate multiprocessor architectures for parallel implementations of CCM2.

PSTSWM uses the spectral transform method to solve the shallow water equations. During each timestep, the state variables of the problem are transformed between the physical domain, where the physical forces are calculated, and the spectral domain, where the terms of the differential equation are evaluated. The physical domain is a tensor product longitude-latitude grid. The spectral domain is the set of spectral coefficients in a spherical harmonic expansion of the state variables.

Transforming from physical coordinates to spectral coordinates involves performing a Fast Fourier transform (FFT) for each line of constant latitude, followed by integration over latitude using Gaussian quadrature, approximating the Legendre transform (LT), to obtain the spectral coefficients. The inverse transformation involves evaluating sums of spectral harmonics and inverse FFTs, algorithmically analogous to the forward transform.

Parallel algorithms are used to compute the FFTs and to compute the forward Legendre transforms. Processors are treated as a two dimensional grid, with the longitude dimension mapped onto row processors and the latitude dimension mapped onto column processors. Thus, the specified aspect ratio determines how many processors are allocated to computing the FFTs and the LTs. Many different parallel algorithms are embedded in the code, and the choice of algorithms is determined via input parameters at runtime.

In this study, variants of two parallel algorithms to compute the forward Legendre transforms are compared. Both parallel algorithms are based on (1) computing local contributions to the vector of spectral coefficients, (2) summing the "local" vectors element-wise over a logical ring of processors, and (3) broadcasting the result to the members of the ring. Both algorithms send $P - 1$ (equal-sized) messages per processor to compute the global sum and $P - 1$ messages to implement the broadcast, where P is the number of processors in a processor column. Each message in the summation is sent to the logical right neighbor, while each message in the broadcast is sent to the logical left neighbor. The algorithms differ in when the three stages are executed. The first algorithm, *ringsum*, first computes all local contributions, then computes the global sum, and finally broadcasts the results. The second algorithm, *ringpipe*, interleaves the calculation of the local contribution with the global summation in a pipeline fashion, and interleaves the broadcast with the computation that uses the result, also in a pipeline fashion. The ringpipe algorithm allows the communication and computation to be overlapped, and requires less memory than ringsum. The question that was posed in the study is whether attempting to overlap communication with computation is cost effective on a given architecture.

To address the question, PSTSWM was executed on four different platforms: the Intel iPSC/2, iPSC/860, Touchstone DELTA, and Paragon machines. The Intel iPSC/2 and iPSC/860 systems are distributed memory, hypercube–connected parallel architectures. The processor elements are the Intel i80386/387 with 4MB of local memory and the Intel i860 with 8MB of local memory, respectively. The communication hardware, based on bit–serial channels, is the same for both the systems. The Intel Touchstone DELTA and Paragon systems are distributed memory, wormhole–routed, mesh–connected parallel architectures. The processor elements are the Intel i860, the same microprocessors used by Intel iPSC/860 system, and the Intel i860SP, respectively, both with 16MB of local memory.

5.1 Measurements

On each architecture, PSTSWM was executed on a logical 1x16 mesh topology, thus calculating each FFT sequentially and each LT in parallel. Multiple runs were made using both ringsum and ringpipe algorithms, with varying implementations of the algorithms, underlying communication protocols, and number of communication buffers. The following naming convention will be used to identify a given experiment:

```
<application_name>.<algorithm_type>.<protocol_option>.<buffering_option>
```

A guideline for the interpretation of trace file names is as follows:

Algorithm type. Each stage of both algorithms is characterized by sending data to one neighbor, receiving from another, and using the data to update a running sum. The following options differ in the order of these operations.

- ringpipe:
  1) type 00: calculate local contribution (calc)/sum/send/receive
  2) type 01: calc/sum/send/receive or calc/sum/receive/send
  3) type 02: calc/receive/sum/send
- ringsum:
  1) type 10: send/receive/sum
  2) type 11: send/receive/sum or receive/send/sum
  3) type 12: same as 10, but posting receive requests early
  4) type 13: same as 11, but posting receive request early

Protocol option. On Intel multiprocessors, PICL supports both blocking and nonblocking communication requests and both regular and forcetype[4] communication protocols.

---

[4] The forcetype protocol assumes that a receive has been posted before a send request is made, thus allowing the elimination of some handshaking overhead, but requires that the user insure that this condition holds.

1) type 0: blocking send – blocking receive
2) type 1: nonblocking send – blocking receive
3) type 2: blocking send – nonblocking receive
4) type 3: nonblocking send – nonblocking receive
5) type 4: blocking send – nonblocking receive with forcetypes
6) type 5: nonblocking send – nonblocking receive with forcetypes
7) type 6: blocking synchronous send/receive (for algorithms of type 01, 11, and 13)

Buffering option. When nonblocking receives and/or sends are used and extra buffer space is available, some of the receive requests can be posted "early" and some sends completed "late", potentially eliminating system buffer copying overhead and allowing additional communication and computation to be overlapped.

1) type 0: use no extra communication buffers
2) type x: use the maximum number of extra communication buffers

As an example, the trace file pstswm.02.3.0 refers to the execution of PSTSWM using the ringpipe algorithm with computational paradigm "calc/[send/receive | receive/send]", assuming the "nonblocking send – nonblocking receive" communication protocol and no extra communication buffers.

5.2   Preliminary Analysis: Performance Parameters and Parallel Metrics

PSTSWM was executed on the parallel systems described in §5, varying the algorithm type and the protocol and buffer options. From each execution, a detailed trace file was collected by means of the tracing facilities provided by PICL. As outlined in §4.1, these trace files can be analyzed at different levels of detail, according to the granularity of the workload components selected, which, in turn, is a consequence of the objectives of the analysis. The usefulness of overlapping communications and computation can be evaluated by selecting the program–oriented approach: each trace file is analyzed by MEDEA from a global viewpoint and any subset of the performance parameters described in Tab. 2 can be used to characterize the behavior of the application.

In our study, total execution, computation, communication, receive and transmit times were selected as representative parameters. Their values were used as input to MEDEA, and the parallel metrics described in §4.2 were used to obtain a first insight into the dynamic behavior of the application runs.

As an example of the differences between the experimental runs, Fig. 2 and 3 show, respectively, the receive profiles derived from the execution on the Intel Paragon for pstswm.02.4.x and pstswm.12.4.0. The first is a ringpipe algorithm based on a "calc/receive/sum/send" execution paradigm.

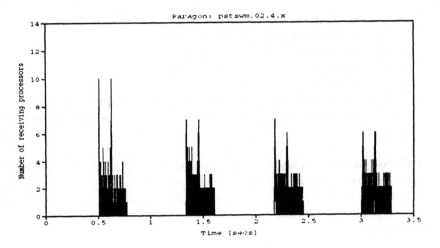

Fig. 2. Receive profile for the `pstswm.02.4.x` ringpipe algorithm.

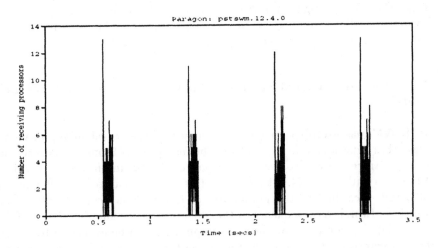

Fig. 3. Receive profile for the `pstswm.12.4.0` ringsum algorithm.

It uses the "blocking send – nonblocking receive with forcetypes" communication protocol options and the maximum number of extra communication buffers. This algorithm maximizes the opportunity for overlap of communication and computation phases. `pstswm.12.4.0` is a ringsum algorithm based on a "send/receive/sum" execution paradigm. It uses the same communication protocol as the ringpipe example, but without any extra communication buffers. The protocol option minimizes the overhead of interprocessor communication for any given send/receive pair, but the algorithmic options do not attempt to interleave the communication and computation or to eliminate all system buffer copying.

Figures 4 and 5 give a detailed view of the first communication phase shown in the receive profiles of these algorithms.

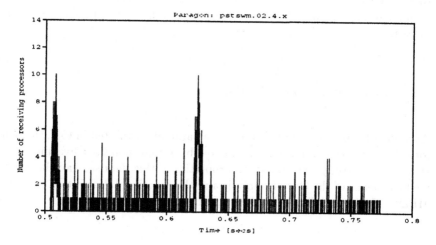

Fig. 4. Detailed view of the first communication phase for the `pstswm.02.4.x` ringpipe algorithm.

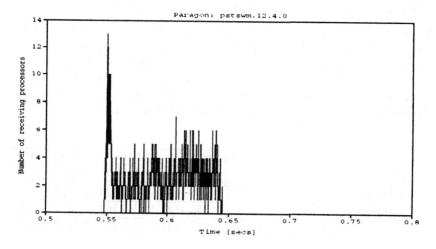

Fig. 5. Detailed view of the first communication phase for the `pstswm.12.4.0` ringsum algorithm.

Note that two subphases, summation and broadcast, can be easily identified for the ringpipe example (Fig. 4): each phase starts with a peak in the number of receiving processors, corresponding to the early posting of nonblocking receive requests by the single tasks, and then contains communications patterns involving a small number of processors at any one time as the explicit handshaking required when using forcetypes takes place. In the ringsum example (Fig. 5), there is only one peak when the summation/broadcast has been started, and then an almost continuous communication pattern can be identified as the messages move around the logical ring.

5.3 Workload Characterization

According to the number of performance parameters selected in §5.2, the behavior of each experimental run is represented by a single point in a five–dimensional space. In our study, the statistical properties of this data set have been examined by means of the cluster analysis and the functional description modules of MEDEA for each different hardware platform,

Workload Models Since the behavior of a real workload is very complex and difficult to reproduce, and since the amount of information collected into trace files is, in general, difficult to manage, a model is usually required. Even though the execution of a workload is usually a deterministic phenomenon, it is often modeled as a nondeterministic one and statistical techniques are applied. As outlined in §3, MEDEA classifies processes in preparation for construction of workload models by means of the *k–means* clustering algorithm [8].

In our study, 56 trace files (corresponding to the execution of PSTSWM for the different implementations of the ringsum and the ringpipe algorithms and varying the underlying communication protocol and the number of communication buffers) have been analyzed for each architecture. The optimal partitions of the workload components with respect to the overall mean square ratios of the evaluated clusters are summarized in the following tables.

Cluster	Percentage	Extime		Cptime		Cmtime	
		mean	std dev	mean	std dev	mean	std dev
Cluster 1	17.0%	139.577	0.561	137.659	0.225	1.712	0.797
Cluster 2	47.2%	144.483	0.350	141.450	0.169	3.013	0.334
Cluster 3	35.8%	142.571	0.247	137.376	0.167	5.514	0.303

Table 5: Workload model for the trace files collected on iPSC/2.

Cluster	Percentage	Extime		Cptime		Cmtime	
		mean	std dev	mean	std dev	mean	std dev
Cluster 1	1.8%	16.158	0.000	12.836	0.000	3.321	0.000
Cluster 2	28.6%	19.788	0.355	13.089	0.218	6.690	0.380
Cluster 3	19.6%	17.849	0.840	13.307	0.324	4.525	0.706
Cluster 4	3.6%	15.463	0.686	13.869	0.159	1.585	0.846
Cluster 5	46.4%	19.802	0.481	12.904	0.099	6.891	0.467

Table 6: Workload model for the trace files collected on iPSC/860.

Cluster	Percentage	Extime		Cptime		Cmtime	
		mean	std dev	mean	std dev	mean	std dev
Cluster 1	20.2%	5.227	0.147	3.947	0.069	1.278	0.139
Cluster 2	34.8%	5.849	0.169	4.185	0.140	1.663	0.136
Cluster 3	15.7%	5.835	0.126	4.756	0.181	1.074	0.191
Cluster 4	20.2%	5.679	0.139	4.205	0.152	1.471	0.140
Cluster 5	9.0%	5.152	0.159	3.947	0.042	1.204	0.129

Table 7: Workload model for the trace files collected on DELTA.

Cluster	Percentage	Extime		Cptime		Cmtime	
		mean	std dev	mean	std dev	mean	std dev
Cluster 1	49.1%	3.372	0.053	3.028	0.029	0.344	0.344
Cluster 2	33.3%	3.866	0.039	3.033	0.033	0.833	0.049
Cluster 3	8.8%	3.920	0.128	3.035	0.028	0.885	0.110
Cluster 4	8.8%	3.345	0.011	3.167	0.042	0.178	0.045

Table 8: Workload model for the trace files collected on Paragon.

The means of the execution, computation and communications times represent the values for the centroid of the corresponding cluster and they can be used, for example, together with the standard deviations, as input to either analytic or simulation system models to reproduce the behavior of real workload.

Note that the problem input for PSTSWM is not necessarily comparable between the different machines. These experiments are part of a larger exercise in determining optimal algorithm parameters for problems that will be used on the largest configurations of each machine. To capture the right granularity when running on only 16 processors, the problem sizes were scaled. Thus, there is some difference between the different sets of experiments.

Functional Description The composition of each cluster has been also investigated from a functional viewpoint.

A preliminary characterization of the behavior of the different program types was indicated by displaying the projections of the experimental runs on a subspace identified by two of the parameters selected for the workload modeling activity. Figure 6 shows the projections of the ringsum and the ringpipe algorithms within the *extime–cptime* subspace for experiments run on the Paragon (see Table 8). With respect to this figure, while the first and second cluster can be easily identified, the remaining partitions do not have well defined shapes. This simply indicates that the extime and cptime parameters are not able by themselves to characterize the workload generated by algorithms belonging to the third and to the fourth cluster. If we consider the projections within the *rctime–trtime* subspace (see Fig. 7), the third and fourth clusters are well shaped too.

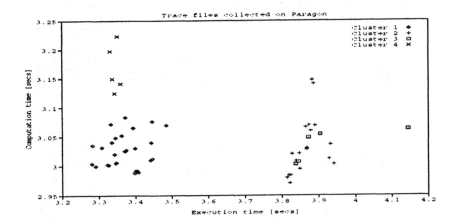

Fig. 6. Projections within the *extime–cptime* subspace.

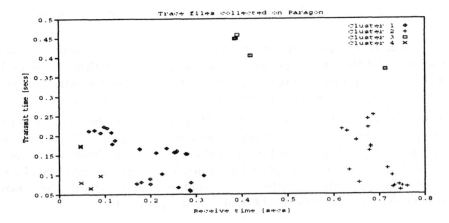

Fig. 7. Projections within the *rctime–trtime* subspace.

As a second step in our functional characterization process, the components belonging to a specific cluster can be listed in order to obtain better insights into the model of the workload being evaluated. As an example, Table 9 lists the applications grouped into the fourth cluster of the workload model for the Paragon.

Paragon: cluster 4	
pstswm.02.2.x	pstswm.02.5.0
pstswm.02.3.x	pstswm.02.5.x
pstswm.02.4.x	

Table 9: Composition of the fourth cluster of the workload model for the Paragon.

Note that all the components belonging to this cluster correspond to trace files derived from ringpipe algorithms based on the "send/calc/receive" execution paradigm. Furthermore, this cluster groups together those PSTSWM runs utilizing nonblocking receive communication protocols and extra communication buffers. The cluster also includes the experiment utilizing nonblocking send – nonblocking receive communications with forcetypes and no extra communication buffers (pstswm.02.5.0). These results imply that the forcetype protocol does not change the fundamental behavior of this algorithm when using extra communication buffers, but that extra buffers are unnecessary (on the Paragon, using this algorithm) when forcetypes are used with nonblocking sends and receives.

Results This case study has important implications on how these multiprocessors should be used. The preliminary results confirm that the utility of overlap varies across the platforms. Moreover, the techniques required to productively exploit overlapping communication with computation also vary between the architectures, even though their programming models are identical. For example, overlap is useful, and simple to characterize and exploit, on the iPSC/2. It is even more important for efficiency on the iPSC/860, but is more difficult to utilize effectively. Techniques maximizing the possibility of overlap have a marginal utility on the Touchstone DELTA, and it is doubtful whether overlap is the reason for the efficiency. The performance analysis on the Paragon currently changes with every operating system upgrade, but its performance characteristics, with regard to exploiting overlap, seem to lie between those of the Touchstone DELTA and the iPSC/860.

The purpose of introducing this case study is to demonstrate how MEDEA can be utilized to analyze PICL trace data, what types of analyses are possible, and, hopefully, how useful the insights available from the analysis are. Our experiments on the Paragon also point out the utility of having portable tools like PICL and MEDEA. While the Paragon will have a full suite of performance monitors and tools in future releases of the system software [5], they were not available for these experiments. This is typical in the analysis of early or experimental systems. But it is important to understand the performance of such systems quickly, and we were not hindered in our case study by the lack of vendor supplied tools.

6 Conclusions

In this paper, the integration of the Portable Instrumented Communication Library (PICL) trace file format into the MEasurements Description Evaluation and Analysis tool (MEDEA) has been presented, with the objective of defining a complete framework to be used in performance evaluation studies of parallel applications. This integration was motivated by the wide availability and utility of PICL trace files, and by the capabilities in MEDEA for easily analyzing the static and dynamic characteristics of parallel workloads from trace data. A workload characterization study was also presented, as a means to indicate exactly how PICL data can be analyzed using MEDEA, and to indicate what types of insight can be obtained from using these tools.

References

1. R. Anderson, W. Auld, D. Brezeal, K. Callaghan, E. Richards, and W. Smith. The Paragon Performance Monitoring Environment. In *Proc. of Supercomputing'93*, pages 850–859, Portland, November 1993.
2. R. A. Aydt. The Pablo Self–Defining Data Format. Technical Report, University of Illinois at Urbana–Champaign, Urbana, IL, March 1992.
3. M. Calzarossa and G. Serazzi. Workload Characterization: A Survey. *Proceedings of the IEEE*, 81(8):1136–1150, August 1993.
4. L. W. Dowdy, M. R. Leuze, and K. H. Park. Multiprogramming a distributed-memory multiprocessor. *Concurrency: Practice and Experience*, 1989.
5. D.L. Eager, J. Zahorjan, and E.D. Lazowska. Speedup Versus Efficiency in Parallel Systems. *IEEE Transactions on Computers*, 38(3):408–423, March 1989.
6. G. A. Geist, M. T. Heath, B. W. Peyton, and P. H. Worley. A User's Guide to PICL: A Portable Instrumented Communication Library. Technical Report ORNL/TM–11616, Oak Ridge National Laboratory, Oak Ridge, TN, August 1990.
7. J. A. Hartigan. *Clustering Algorithms*. John Wiley, 1975.
8. J. A. Hartigan and M. A. Wong. A K–Means Clustering Algorithm. *Applied Statistics*, 28:100–108, 1979.
9. V. Herrarte and E. Lusk. Studying parallel program behavior with upshot. Technical Report ANL/TM–91/15, Argonne National Laboratory, Argonne, IL, August 1991.
10. R. Jain. *The Art of Computer System Performance Analysis*. John Wiley & Sons, New York, 1991.

---

[5] A performance monitor for the Paragon has been recently presented in [1].

11. P. Lenzi and G. Serazzi. PARMON: PARallel MONitor – User's Guide Release 1.0. Technical Report R3/95, University of Milan, October 1992.
12. A. Merlo. MEDEA: MEasurements Description Evaluation and Analysis tool – User's Guide Release 1.0. Technical Report R3/117, Progetto Finalizzato C.N.R. "Sistemi Informatici e Calcolo Parallelo", Aprile 1993.
13. A. Merlo and P. Rossaro. MEDEA: Design Document. Technical Report R3/92, Progetto Finalizzato C.N.R. "Sistemi Informatici e Calcolo Parallelo", Settembre 1992.
14. Heath M. T. and J. A. Etheridge. Visualizing the performance of parallel programs. *IEEE Software*, (8), 1991.
15. P. H. Worley. A New PICL Trace File Format. Technical Report ORNL/TM–12125, Oak Ridge National Laboratory, Oak Ridge, TN, October 1992.
16. P. H. Worley and I. T. Foster. PSTSWM: A Parallel Algorithm Testbed and Benchmark Code for Spectral General Circulation Models. Technical Report ORNL/TM–12393, Oak Ridge National Laboratory, Oak Ridge, TN. (in preparation).

Springer-Verlag
and the Environment

We at Springer-Verlag firmly believe that an international science publisher has a special obligation to the environment, and our corporate policies consistently reflect this conviction.

We also expect our business partners – paper mills, printers, packaging manufacturers, etc. – to commit themselves to using environmentally friendly materials and production processes.

The paper in this book is made from low- or no-chlorine pulp and is acid free, in conformance with international standards for paper permanency.

Lecture Notes in Computer Science

For information about Vols. 1–724
please contact your bookseller or Springer-Verlag